Nonlinear Waves

Main Editors

A. Jeffrey, University of Newcastle-upon-Tyne
R. G. Douglas, State University of New York at Stony Brook

Editorial Board

F. F. Bonsall, University of Edinburgh
H. Brezis, Université de Paris
G. Fichera, Università di Roma
R. P. Gilbert, University of Delaware
K. Kirchgässner, Universität Stuttgart
R. E. Meyer, University of Wisconsin-Madison
J. Nitsche, Universität Freiburg
L. E. Payne, Cornell University
I. N. Stewart, University of Warwick
S. J. Taylor, University of Liverpool

Nonlinear Waves

Tosiya Taniuti
Nagoya University
and
Katsunobu Nishihara
Osaka University

Translation from Japanese prepared by
Tosiya Taniuti and
Alan Jeffrey
University of Newcastle-upon-Tyne

Pitman Advanced Publishing Program
Boston · London · Melbourne

PITMAN PUBLISHING LIMITED
128 Long Acre, London WC2E 9AN

PITMAN PUBLISHING INC.
1020 Plain Street, Marshfield, Massachusetts 02050

Associated Companies
Pitman Publishing Pty Ltd, Melbourne
Pitman Publishing New Zealand Ltd, Wellington
Copp Clark Pitman, Toronto

© Tosiya Taniuti and Katsunobu Nishihara 1977
English translation © Pitman Books Limited 1983

Translated by Tosiya Taniuti and Alan Jeffrey
from NONLINEAR WAVES by Tosiya Taniuti and Katsunobu Nishihara

First published in Japanese by IWANAMI SHOTEN, Publishers, Tokyo
First English edition published by Pitman Books Limited, 1983

AMS Subject Classifications: (main) 35-02, 35K55, 35L60
(subsidiary) 73D99, 76J99, 76W05

Library of Congress Cataloging in Publication Data

Taniuti, Tosiya, 1924–
  Nonlinear waves.
  (Monographs and studies in mathematics; 15)
  Translation of: Hisenkei hadō.
  Bibliography: p.
  Includes index.
  1. Wave-motion, Theory of. 2. Wave equation. 3. Nonlinear theories.
I. Nishihara, Katsunobu, 1945–. II. Title. III. series.
QA927.T3613   530.1'4   82–5197
ISBN 0-273-08466-6   AACR2

British Library Cataloguing in Publication Data

Taniuti, Tosiya
  Nonlinear waves.—(Monographs and studies in
  mathematics; 15)
  1. Wave motion, Theory of   2. Nonlinear
  theories
  I. Title   II. Nishihara, Katsunobu
  III. Series
  531'.1133   QA927
  ISBN 0-273-08466-6

All rights reserved. No part of this publication may be reproduced, stored in a retrieval system, or transmitted, in any form or by any means, electronic, mechanical, photocopying, recording and/or otherwise, without the prior written permission of the publishers. This book may not be lent, resold or hired out or otherwise disposed of by way of trade in any form of binding or cover other than that in which it is published, without the prior consent of the publishers.

Filmset and printed in Northern Ireland at The Universities Press (Belfast) Ltd., and bound at the Pitman Press, Bath, Avon.

Contents

Preface to the English edition   vii

Preface   ix

1 Considerations involving model equations   **1**

   1.1 Theory of overtaking waves   1
   1.2 Physical examples   8
   1.3 The limit of weak dissipation and shock waves   14

2 Wave equations   **23**

   2.1 Linear wave equations and the method of characteristics   23
   2.2 Hydrodynamics   32
   2.3 Nonlinear hyperbolic equations—multi-component systems and characteristic curves   41
   2.4 Shock waves in gas dynamics   55
   2.5 Extension to the multi-dimensional case   72

3 Asymptotic methods   **89**

   3.1 Far fields for hyperbolic equations   89
   3.2 Dissipative systems   95
   3.3 Weakly dispersive systems   99
   3.4 The reductive perturbation method for long waves and its extension   107
   3.5 A far field for linear wave modulation—the Schrödinger field   114
   3.6 Strongly dispersive systems   116
   3.7 Self-focusing of waves   130
   3.8 Three-wave interaction   133
   3.9 Self-induced transparency and the Sine–Gordon equation   151

4 The inverse scattering method for the initial value problem for a nonlinear evolution equation 159

- 4.1 The inverse scattering method for the Korteweg–de Vries equation 159
- 4.2 The conservation laws and the canonical form of the KdV equation 180
- 4.3 Periodic solutions and lattice dynamics 188
- 4.4 The theory of the Lax pair 197
- 4.5 Application of the inverse scattering method to the nonlinear Schrödinger equation 200
- 4.6 The Sine–Gordon equation and the Bäcklund transformation 211
- 4.7 Bäcklund transformation and the inverse scattering method 223
- 4.8 Hirota's method 231

Appendix Two-fluid model of a plasma (electron fluid and ion fluid) 242

Bibliography 245

Index 251

Preface to the English Edition

Since the original book in Japanese was published, a number of interesting papers have appeared concerning soliton theory. However, the Japanese text was written as concisely as possible; in order to preserve this conciseness in this English edition, supplements to the original work are strictly restricted to those subjects which can be introduced briefly or should be explained mathematically in some detail. The last section to appear in this edition, concerning Hirota's method, was omitted from the Japanese text on account of a severe restriction on the number of pages permitted by the Japanese publisher.

February 1983

Tosiya Taniuti
Alan Jeffrey

Preface

Problems involving the propagation of nonlinear waves have become of increasing interest in various branches of science and engineering. In general, waves of finite amplitude governed by a nonlinear evolution equation are called nonlinear waves, and the purpose of this book is to discuss the propagation of such waves. As is well known, the superposition of solutions is not valid in nonlinear equations, so that the methods of solution familiar to physicists and engineers, like the use of Fourier or Laplace transforms, are no longer directly applicable, with the result that the study of nonlinear waves has not yet become well established. However, in recent years, a number of interesting phenomena involving nonlinear waves have been found, and with the development of the digital computer remarkable progress has been made in research into nonlinear waves.

Nonlinear waves, as well as linear waves, may be considered in the two ideal cases comprising the dissipative and the dispersive—though, of course, both dissipation and dispersion arise in reality. In contrast to linear wave behaviour, however, in the case of nonlinear waves without dissipation and dispersion, the phase velocity of a wave frequently depends only on its amplitude. As a direct consequence of this the wave speed is greater for waves of larger amplitude, so that a wave of large amplitude to the rear catches up with a wave of smaller amplitude in front, leading to a steepening of the wave-form and, finally, to wave breaking. Hence, in a nonlinear evolution equation without dissipation and dispersion (a nonlinear hyperbolic equation) it is, in general, possible that initially smooth waves become discontinuous after a finite time, so that solutions (in the usual sense) do not exist for all time. This is one of the conspicuous properties of nonlinear wave equations whose general behaviour is markedly different from that of linear equations. Dissipation and dispersion each play an important role in balancing the steepening due to the nonlinearity, so that when these effects are present a steep but smooth wave may be formed which then propagates for all time.

A typical example of dissipative nonlinear waves is the shock wave produced by a supersonic object. In that case the gas across the shock front is heated by viscosity, and the dissipation so caused then balances the steepening of the wave produced by the nonlinearity in the governing equations. Shock waves propagate in liquids and solids, as well as in gases and, in particular, the shock wave produced by an explosion is of considerable importance in many applications. Also, in a plasma, which is a fully ionized gas of high temperature, the magnetohydrodynamic shock wave involving an interaction with the electromagnetic field must be considered. On the other hand, a typical example of a nonlinear wave propagating in a dispersive medium, in which the phase velocity varies with the wavelength, is the solitary wave. This is a pulse formed as a result of a balancing occurring between the steepening effect due to the nonlinearity and the smoothing effect of the dispersion.

The solitary wave phenomenon has actually been observed for many years in the form of a surface wave in shallow water, the first observation having been recorded by Scott Russell in 1844. It was, in fact, in the last century (1895) that the Korteweg–de Vries (KdV) equation, the nonlinear equation which approximately describes this solitary wave propagation, was discovered. However, it was only just over a decade ago that the solitary wave was found to behave like a particle, which led to this nonlinear wave-packet being called a 'soliton'. It was by means of numerical integration of the KdV equation that Zabusky and Kruskal first discovered that solitary waves propagating in the same direction retain their separate identities after repeated overtaking, and behave as if they are particles. Gardner *et al.* then succeeded in solving analytically the KdV equation by using the inverse scattering method developed for quantum mechanics, and found as a result that the soliton solution corresponds to the bound state of the Schrödinger equation. Consequently, the particle-like behaviour of solitons was proved analytically. Since then, it has been established that a wide class of nonlinear evolution equations of dispersive type can be reduced asymptotically to the KdV equation. Thus the concept of the soliton has been established as a universal one.

Another conspicuous property which was revealed by study of the KdV equation is the recurrence phenomenon which occurs when it is subjected to periodic boundary conditions, for then the initial condition is recovered after a finite time. This is inconsistent with the conjecture that, in nonlinear systems, when a mode of a wave is excited initially, ultimately the wave-energy will be distributed equally over all modes through the process of nonlinear interactions between different modes. This fact was found first in nonlinear lattice dynamics where it was known as the Fermi, Pasta and Ulam paradox. In fact they found this property by numerically integrating a classical dynamical system comprising a series of harmonic oscillators between pairs of which there occurred nonlinear interactions. This, in turn, led to the discovery by Toda of a completely integrable one-dimensional lattice system (the Toda lattice).

The recent progress in soliton research is remarkable, and the particle aspect of the soliton and the property of recurrence have been observed experimentally. In this decade the theory of nonlinear waves centred around the soliton has appeared in modern dress, and this has been extensively applied to numerous branches of science and engineering. It is now applied to particle physics and solid state physics as well as to the mechanics of a classical continuum, to optics, plasma physics, ecology and elsewhere. In connection with the recent attempts to consider the soliton as a model of an elementary particle, it should be noted that for the Sine–Gordon equation, which is a nonlinear generalization of the Klein–Gordon equation, Perring and Skyrme also found the particle-like behaviour of the solitary wave by numerical methods.

In this book, the nonlinear wave propagation appearing in so many fields is dealt with in a unified manner, and general methods of solution are presented in a way which renders them applicable to a variety of different problems. We believe that in this way the essential points underlying phenomena which are common to a variety of different fields are clarified. The book is written for research workers and students who are either working in, or intending to work on, theoretical problems concerning nonlinear wave propagation in science and engineering. It is also intended for readers who are interested in the intermediate region between physics and mathematics. However, it was never our intention to write a book as one would a table of formulae, summarizing various technical methods of solution and giving examples. Rather, we have attempted to show the characteristic features of the various approaches by means of typical examples, and then to develop the general theories. In doing this, however, we have not aimed to establish a mathematically rigorous approach. On the other hand, we do not accept conventional methods of solution which are used in practice by physicists and engineers without first carefully examining their validity. We hope that by adopting this approach we have built a bridge, in the form of a book, linking these two extreme points of view.

The book is divided into three parts. One of them discusses nonlinear hyperbolic equations in the limit of sufficiently weak dissipation. Contained in this first part, together with other matters, is the method of characteristics, the weak extension of a solution to permit the inclusion of discontinuous solutions and shock propagation in the context of weak solutions. In the second part, we discuss the asymptotic methods which make it possible to reduce general nonlinear evolution equations to some tractable nonlinear equations, while the third part describes the methods which permit the exact solution of these equations, such as the inverse scattering method.

In the first chapter, the steepening of a wave, which is such a characteristic property of nonlinear waves, is introduced by means of a simple model, and the dissipative and dispersive effects which smooth out such steepening are also explained by means of simple models. In particular, the Burgers equation is introduced as an exactly solvable model of a dissipative nonlinear equation, and its properties are investigated in some detail. The characteristic features of a shock wave may be deduced from the considerations given in this chapter.

The second chapter is divided into five sections; Sections 2.1 to 2.3 deal with the general properties of nonlinear hyperbolic equations and with the method of characteristics. In Section 2.4, using the gas shock as a model, the properties, structure and propagation of a shock wave are explained. In particular, the evolutionary condition is formulated as a general selection principle designed to choose from among the weak solutions (satisfying a conservation law) a physically relevant solution. This is an extension of the law of entropy increase in thermodynamics, and the relationship between the direction of time and the dissipation is discussed. In Section 2.5 the characteristic manifold for a hyperbolic equation in four dimensional space-time is studied in relation to the speed of the wavefront. Sections 2.4 and 2.5 may, if the reader wishes, be omitted without detriment to the understanding of the succeeding chapters.

In Chapter 3, by means of an asymptotic method called the reductive perturbation method, general systems of nonlinear evolution equations are reduced to some exactly solvable nonlinear equations, on the assumption that the amplitudes of the waves are finite but small. In this singular perturbation method, the space and time coordinates are stretched in terms of a small expansion parameter. To clarify this approach we begin with an explanation of the physical meaning of the stretching process, and then introduce the concept of a far field which is the basis of the asymptotic approach. Then it is shown that a general dissipative system of nonlinear equations may be reduced to the Burgers equation, and a general weakly dispersive system to the KdV equation.

The succeeding sections, from 3.5 to 3.7, deal with a strongly dispersive system, in which slow modulations of amplitude and phase of a plane wave are considered. In this case, the asymptotic time evolution of the modulations is governed by the nonlinear Schrödinger equation. This is the Schrödinger equation with a potential proportional to the amplitude of the wave function (cubic nonlinearity), and it also admits soliton solutions. In this case these comprise the propagation of a modulated high-frequency oscillation called an envelope soliton. Results obtained for propagation in one-dimensional space are then generalized to multi-dimensional systems. By way of example, the self-focusing of a wave, which is the trapping of the wave by a potential produced by itself, is explained in terms of the envelope soliton. In Section 3.8 we are concerned with the study of the parametric interaction of waves, and with the resonance interaction among three waves, as our main objective. Here we introduce the method of averaging, which enables us to obtain nonlinear equations for the slowly varying parts after the elimination of the rapidly oscillating parts by means of the averaging process. Employing nonlinear optics as an example, we derive a system of equations for the slowly varying amplitudes and phases of the three waves and, in this way, we demonstrate the mechanism of parametric excitation. Finally, we consider the phenomenon of self-induced transparency (SIT), which is a remarkable nonlinear effect which occurs in nonlinear optics. It arises when an ultra-short light pulse is incident upon a medium of two-level atoms, with its frequency nearly equal to the resonance frequency of the medium. The medium then transmits the pulse without attenuation, provided its power is above a certain threshold. It is shown that in an ideal case this phenomenon is described by the Sine–Gordon equation, and as a result the self-induced transparency effect can be explained in terms of the propagation of a soliton.

In Chapter 4 the simple dispersive nonlinear evolution equations derived in Chapter 3 are solved exactly and the results are used to show the characteristic properties of solitons. Firstly, the solution of the KdV equation by means of the inverse scattering method is explained, on the basis of which the existence of an infinite number of conservation laws is then established. In this context the KdV equation is written in canonical form and it is shown that the Hamiltonian is separable; that is, the system is completely integrable. Moreover, the recurrence of solutions under periodic boundary conditions is discussed in physical terms, and the relationship to lattice dynamics is demonstrated using the Toda lattice as a model. In Section 4.4 the theory due to Lax is introduced, which formulates the method of inverse scattering in a very general framework. In Section 4.5, as an example of the application of Lax's general theory, the nonlinear Schrödinger equation is solved by the inverse scattering method. In Section 4.6, the Bäcklund transformation is explained as an extension of the principle of superposition in a linear system, and the Sine–Gordon equation is used as a model equation for the purpose of illustration. There exist some properties common to equations which are capable of solution by the inverse scattering method, and in the final section these are discussed collectively by means of the formulation called the AKNS scheme, which originated in the work of Ablowitz, Kaup,

Newell, and Segur. This permits coverage of a wider class of completely integrable nonlinear evolution equations.

This book has been written from the viewpoint of graduate students in physics and applied mathematics, but a considerable amount of detailed explanation has been included so that even undergraduate students in these disciplines should be able to follow many parts of the book. However, in order to make the book concise, a number of interesting problems have been omitted. For example, many problems arising from propagation in multi-dimensional space have been omitted, along with nonlinear waves of random phase.

On account of the level at which the book has been written, no attempt has been made at completeness of the reference list. The selection has, in the main, been made on the basis of contributions which offer a clear account of topics of interest and which are readily available.

March 1977

Tosiya Taniuti
Katsunobu Nishihara

1

Considerations involving model equations

1.1 Theory of overtaking of waves

Let N people of different height be lined up on a straight line, on which they move in the same direction. Then the profile obtained by linking the heads to those people will change in time. In order to represent the evolution of this profile, we denote the initial position of the person with height h_i by x_i. If the N people all move with the same constant speed v, their paths are parallel straight lines in the (x, t)-space, and the profile translates with speed v without distortion, because h_i is constant along the straight line $x - vt = x_i$, for $i = 1, 2, \ldots, N$. This process may be considered as a continuum in the limit of large N.

Let $h(x, t)$ be a function of x and t, varying continuously and smoothly in space and time, representing a profile such as the one above translating with the speed v. Then h is constant along each straight line $x - vt =$ constant, namely

$$\frac{dh(x, t)}{dt} = 0, \tag{1.1.1}$$

where d/dt is the directional derivative along the straight line $x - vt =$ constant, defined by the limit

$$\lim_{\Delta t \to 0} [h(x + v\,\Delta t, t + \Delta t) - h(x, t)]/\Delta t.$$

Consequently, we have

$$\frac{\partial h}{\partial t} + v \frac{\partial h}{\partial x} = 0. \tag{1.1.2}$$

In graphical terms, the general solution of this partial differential equation is simply a profile translating with the speed v and it may be written

$$h = \Phi(x - vt), \tag{1.1.3}$$

where Φ is an arbitrary function of $x-vt$. This can be readily confirmed by substituting Eq. (1.1.3) into Eq. (1.1.2). The solution describes a wave propagating in the positive x direction with the phase velocity v, and the function $\Phi(x)$ is just the initial profile when $t=0$. In general, a wave propagating in one direction is called the progressive wave, and a progressive wave propagating without deformation, such as the one given by Eq. (1.1.3), is often called a permanent (progressive) wave.

More generally, we have the situation in which the speed v with which each part of the profile translates is not constant. If v varies in space and time, the profile $h(x, t)$ becomes distorted with the passage of time, but it is still true that h remains constant along the curves given by integrating

$$\frac{dx}{dt} = v(x, t) \tag{1.1.4}$$

though it is, in general, a different constant along each different curve. Of course, we assume here that v is single valued everywhere. Hence, if the family of curved paths is denoted by $\varphi(x, t) = \xi$, with ξ a parameter, then h becomes a function $h(\varphi)$ of φ only. Thus h is constant if φ remains constant. In fact, along each curve of the family

$$d\varphi = \frac{\partial \varphi}{\partial t} dt + \frac{\partial \varphi}{\partial x} dx = 0,$$

and so from Eq. (1.1.4) it follows that

$$\frac{\partial \varphi}{\partial t} + v(x, t) \frac{\partial \varphi}{\partial x} = 0. \tag{1.1.5}$$

Consequently, it is easily seen that $h(\varphi)$ satisfies Eq. (1.1.2), so that

$$\frac{\partial h}{\partial t} + v \frac{\partial h}{\partial x} = \frac{\partial \varphi}{\partial t} \frac{\partial h}{\partial \varphi} + v \frac{\partial \varphi}{\partial x} \frac{\partial h}{\partial \varphi} = \left(\frac{\partial \varphi}{\partial t} + v \frac{\partial \varphi}{\partial x} \right) \frac{\partial h}{\partial \varphi} = 0.$$

In this way we find that for Eq. (1.1.2) the family of curves $\varphi(x, t) = \xi$ in the (x, t)-plane, given by Eqns (1.1.4) or (1.1.5), play an important role. These curves will be called the characteristic curves of Eq. (1.1.2) or, more simply, the characteristics.

However, in general, the speed of walking differs from person to person. For example, we may assume that the speed v is proportional to the height h of a person. In this case, a person of greater height walking behind one of lesser height will catch up and, as a result, overtaking occurs. We then encounter an entirely different phenomenon, since the continuous deformation of the profile we have assumed so far ceases to be valid after a finite time. When, in this simple model, the overtaking process is disturbed for some reason, it gives rise to a chaotic state, a shock! This happens, for example, when an emergency occurs in a theatre and people rush to a narrow exit where a shock occurs. However, before the overtaking process occurs, the evolution of the profile h is described by Eq. (1.1.2).

For simplicity let us take $v = h$, then we have

$$\frac{\partial h}{\partial t} + h \frac{\partial h}{\partial x} = 0. \qquad (1.1.6)$$

In this case, along a characteristic (path) determined by integrating

$$\frac{dx}{dt} = h, \qquad (1.1.7)$$

the height h is constant. Consequently, this equation can be integrated to give the family of straight lines

$$x = h(\xi, 0)t + \xi, \qquad (1.1.8)$$

in which the gradient $h(\xi, 0)$ is the initial value of h at the point $t = 0$, $x = \xi$. Hence Eq. (1.1.8) represents a one-parameter family of straight lines with parameter ξ, each of which, in general, issues out of a point on the x axis with a different gradient. The members of this family of straight lines are, of course, not all parallel.

By way of illustration, consider the two characteristics $x_1 = h(\xi_1, 0)t + \xi_1$ and $x_2 = h(\xi_2, 0)t + \xi_2$. Then if $\xi_1 < \xi_2$, $h_1 > h_2$ ($h_i \equiv h(\xi_i, 0)$, $i = 1, 2$), these two characteristics intersect at a time $t > 0$ given by $t = (\xi_2 - \xi_1)/(h_1 - h_2)$. At that point h takes at least the two different values h_1 and h_2, since these values are transported as constants along their respective characteristics, and so h becomes multi-valued. Therefore, in general, the solution of the nonlinear equation (1.1.6) does not necessarily exist uniquely for all time. In point of fact the derivative $\partial h/\partial x$ ceases to be defined at this time so that then the partial differential equation (1.1.6) no longer has a well defined meaning.

For example, let us consider the initial condition at $t = 0$,

$$\text{(A)} \qquad h = \begin{cases} 1 & (x \leq -a) \\ -x/a & (-a \leq x \leq 0) \\ 0 & (0 \leq x), \end{cases}$$

where $a > 0$. Then the characteristic starting from the point $x = x_0$ at $t = 0$ belongs to one of the three groups

(A.1) $\qquad x = t + x_0 \qquad\qquad (x_0 \leq -a)$

(A.2) $\qquad x = -(x_0/a)t + x_0 \qquad (-a \leq x_0 \leq 0)$

(A.3) $\qquad x = x_0 \qquad\qquad (0 \leq x_0).$

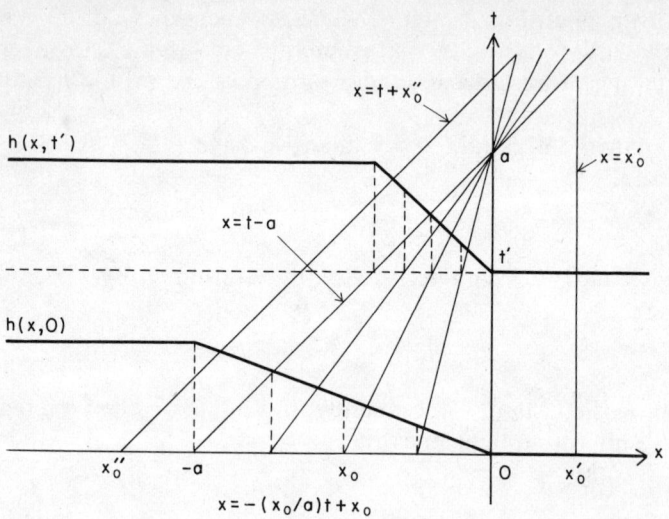

Fig. 1.1

The straight lines belonging to the second group (A.2) all pass through the point $x = 0$, $t = a$. That is all the paths from the domain $-a \leq x_0 \leq 0$ concentrate at this single point as a result of the overtaking process (Fig. 1.1). At that time, h becomes

$h = 1 \quad (x < 0),$
$h = 0 \quad (x > 0),$

and so is discontinuous at $x = 0$.

In general, a wave which becomes steeper on account of the overtaking process will be called a compression (converging) wave. In a compression wave, the waveform becomes discontinuous after a finite time because of the overtaking process, even when the initial waveform is arbitrarily smooth. Since the governing equation is reversible in time, by following time in the reverse direction, we conclude that an initial discontinuity can be smeared out instantaneously, and the waveform then becomes smoother as time evolves. Corresponding to a compression wave, there is a wave with divergent characteristics called an expansion (divergent) wave. In particular, an expansion wave with characteristics diverging from a single point is called a centred expansion (divergent) wave. If, in the above example, the time is reversed, it is readily seen that an initial discontinuity is instantaneously smoothed out and propagates in the form of a centred expansion wave.

1.1 THEORY OF OVERTAKING OF WAVES

The solutions of the initial value problem for Eq. (1.1.6) can be obtained easily by a geometrical method involving tracing the characteristics. In order to solve it analytically, we transform from the coordinates (x, t) to the new coordinates (φ, t'). Here we let $\varphi(x, t) = x_0$ denote the equation of the characteristic starting from $x = x_0$ at $t = 0$, and we set $t' = t$. Consequently, at $t = 0$, the φ axis is to be identified with the x axis, and φ is then determined by the equation

$$\frac{\partial \varphi}{\partial t} + h \frac{\partial \varphi}{\partial x} = 0, \qquad (1.1.9a)$$

and by the initial condition

$$\varphi(x, 0) = x. \qquad (1.1.9b)$$

Taking note of the relationship

$$\frac{\partial}{\partial t} = \frac{\partial \varphi}{\partial t} \cdot \frac{\partial}{\partial \varphi}\bigg|_{t} + \frac{\partial}{\partial t}\bigg|_{\varphi},$$

where $(\ )|_t$ and $(\ )|_\varphi$ signify, respectively, that t is held constant and φ is held constant, and using Eq. (1.1.9a), we see that Eq. (1.1.6) is equivalent to

$$\frac{\partial x}{\partial t}\bigg|_{\varphi} = h, \qquad (1.1.10a)$$

$$\frac{\partial h}{\partial t}\bigg|_{\varphi} = 0. \qquad (1.1.10b)$$

Integrating Eq. (1.1.10b) gives us at once the result $h = f(\varphi)$, where $f(\varphi)$ is an arbitrary function of φ. If the initial value of h is specified as $h_0(x)$, then by means of the initial condition Eq. (1.1.9b) we find that $f(\varphi(x, 0)) = f(x) = h_0(x)$, so that $f(\varphi) = h_0(\varphi)$. Introducing this into Eq. (1.1.10a), and again using Eq. (1.1.9b), we obtain

$$x = h_0(\varphi)t + \varphi. \qquad (1.1.11)$$

In general, this is an implicit equation for φ, but when φ is known as a function of x and t we arrive at the formal solution $h = h(\varphi(x, t))$. For example, if the initial condition is given by $h_0(x) = -(x/a)$ $(a > 0)$, then from Eq. (1.1.11) it follows that $\varphi = x/[1 - (t/a)]$, leading to the solution

$$h = -\frac{\varphi}{a} = \frac{x}{t - a}. \qquad (1.1.12)$$

This shows that all the characteristics pass through the point $x = 0$ at $t = a$. Also, noticing that $h = 1$ on the line $x = t - a$, and that $h = 0$ on the line $x = 0$, we are able to construct from this particular solution the solution for initial conditions (A). So, for $t<a$, we find that

in the region $x-t<-a$, $\qquad h=1$,

in the region $x-t>-a, x<0$, $\qquad h=\dfrac{x}{t-a}$,

in the region $x>0$, $\qquad h=0$.

If $a<0$, the solution (1.1.12) becomes a centred expansion wave in which all the characteristics diverge from the point $x=0$, $t=a\,(<0)$. Also, for the solution subject to the initial condition

(B) $\qquad h = \begin{cases} 0 & (x<0) \\ -x/a & (0<x<|a|) \\ 1 & (|a|<x), \end{cases}$

by taking the limit as $a \to 0$, we obtain the following solution for the discontinuous initial condition at $t=0$ comprising $h=0$ for $x<0$ and $h=1$ for $x>0$,

in the region $x<0$, $\qquad h=0$,

in the region $x>0, x-t<0$, $\qquad h=\dfrac{x}{t}$,

in the region $x-t>0$, $\qquad h=1$.

It should be noticed here that the centred expansion wave $h=x/t$ obtained above is, in fact, a similarity solution of Eq. (1.1.6), which is invariant under the affine transformation $t \to \varepsilon t$, $x \to \varepsilon x$.

As an extension of Eq. (1.1.6) we may consider the following equation in which v is a function of h,

$$\frac{\partial h}{\partial t} + v(h)\frac{\partial h}{\partial x} = 0. \qquad (1.1.13)$$

In this case, arguing as before, h is constant along each characteristic, the gradient of which is now given by $v(h)$, and so the characteristics are again straight lines. Consequently, the initial value problem for Eq. (1.1.13) can be solved by means of a similar geometrical method.

For a more general equation like

$$\frac{\partial h}{\partial t} + v(h)\frac{\partial h}{\partial x} = g(h), \qquad (1.1.14)$$

in which g is a function of h, the solution is no longer constant along a characteristic. However, as can be seen when compared with Eqns (1.1.9) and (1.1.10), this equation is equivalent to the set of ordinary differential

1.1 THEORY OF OVERTAKING OF WAVES

equations

$$\frac{dx}{dt} = v(h), \tag{1.1.15a}$$

$$\frac{dh}{dt} = g(h). \tag{1.1.15b}$$

Here d/dt is again used, as in Eq. (1.1.10), to denote the directional derivative along the characteristics $\varphi(x, t) = $ constant, given by

$$\frac{\partial \varphi}{\partial t} + v(h) \frac{\partial \varphi}{\partial x} = 0, \tag{1.1.16a}$$

$$\varphi(x, 0) = x. \tag{1.1.16b}$$

We may apply to Eqns (1.1.15) an analytical method of solution similar to that which may be applied to Eq. (1.1.6). Namely, integrating Eq. (1.1.15b) and using the initial condition, h becomes a function of φ and t, which may be substituted into Eq. (1.1.15a) to give

$$x = \int_0^t v(h(\varphi, \tau)) \, d\tau + \varphi.$$

When this implicit equation is solved for φ, h is given as a function of x and t.

We remark that Eqns (1.1.15) may be regarded as a set of differential equations for a family of spatial curves in the (x, t, h)-space. Usually these curves are called the characteristic curves of Eq. (1.1.14), and Eqns (1.1.15) are called the characteristic equations. The envelope of these special curves is the integral surface of the initial value problem for Eq. (1.1.14).

Finally, we notice that these ideas also have an important meaning in multi-dimensional cases. For example, for the equation

$$\frac{\partial h}{\partial t} + \sum_{i=1}^{N} v_i \frac{\partial h}{\partial x_i} = 0, \tag{1.1.17}$$

along the characteristics (paths) determined by integrating

$$\frac{dx_i}{dt} = v_i \quad (i = 1, 2, \ldots, N) \tag{1.1.18a}$$

h is constant, since

$$\frac{dh}{dt} = 0, \tag{1.1.18b}$$

because the change of h along the path is given by

$$\mathrm{d}h = \frac{\partial h}{\partial t}\mathrm{d}t + \sum_i^N \frac{\partial h}{\partial x_i}\mathrm{d}x_i = \left(\frac{\partial h}{\partial t} + \sum_i^N \frac{\partial h}{\partial x_i}v_i\right)\mathrm{d}t = 0.$$

If the v_i do not depend on h, so that Eq. (1.1.17) is the linear, the characteristics (paths) are the integral curves of the ordinary differential equation (1.1.18a) and, in general (for example, if there is no singular point), they do not cross. Even in the nonlinear case, they are straight lines if v_i $(i = 1, 2, \ldots, N)$ are functions of h only. However, their gradients are then different and hence, in general, they cross. As a simple example assume $N = 2$, $v_i = h$, then from Eq. (1.1.18a) it follows that $\mathrm{d}x_2/\mathrm{d}x_1 = 1$, provided h does not vanish. Consequently, the projections of the characteristics on to the (x_1, x_2)-plane (the characteristic base curve) become the parallel straight lines $x_2 - x_1 = \alpha$. In this case the characteristics for different α, and hence for different characteristic base lines, do not cross in the (x_1, x_2, t)-space. However, the characteristics starting (at $t = 0$) from the same characteristic base line can cross. In fact, taking $x_2 - x_1 = \varphi$, with x_1 and t as independent variables, the derivative with respect to φ vanishes in Eq. (1.1.17) and it reduces to Eq. (1.1.2).

1.2 Physical examples

As a physical example which can be described by the partial differential equation (1.1.6), we may consider a rarefied gas comprising particles moving in one dimension (a neutral beam). If the density is so low that the mutual collisions amongst the constituent particles may be neglected, the velocity distribution function of the gas, $f(x, v, t)$, is governed by the collisionless Boltzmann equation:

$$\frac{\partial f}{\partial t} + v\frac{\partial f}{\partial x} + F\frac{\partial f}{\partial v} = 0. \tag{1.2.1}$$

Here v is the velocity, F is an external force, and the characteristic Eqns (1.1.18) express the physical law that f is constant along each Newtonian particle path

$$\frac{\mathrm{d}x}{\mathrm{d}t} = v, \tag{1.2.2a}$$

$$\frac{\mathrm{d}v}{\mathrm{d}t} = F, \tag{1.2.2b}$$

where x_1 and x_2 in Eqns (1.1.18) ($N = 2$) correspond to x and v, and v_1 and v_2 correspond to v and F, respectively. The characteristic curves, that is the orbits in the phase space (x, v), are given by integrating Eqns (1.2.2) for respective initial conditions, when F is prescribed as a function of x, t, and v. We may assume, as is usual in the classical dynamics, that the characteristics do not cross. Hence the solution of the initial value problem of the linear equation (1.2.1) may be considered to be uniquely determined for all time.

For simplicity, let us assume that the external force is not applied, so that $F = 0$. Moreover, suppose that all the particles at every point move with the same macroscopic velocity $u(x, t)$. That is, in physical terms, we consider the motion of a cold beam for which the microscopic variance of the velocity is negligible, so that the temperature is zero. The distribution function of such a gas is given by

$$f(x, v, t) = n(x, t)\, \delta(v - u(x, t)), \tag{1.2.3}$$

where $n(x, t)$ is the density given by

$$n(x, t) \equiv \int f(x, v, t)\, dv.$$

Introducing Eq. (1.2.1) into Eq. (1.2.3), and integrating with respect to v, then yields the continuity equation

$$\frac{\partial n}{\partial t} + \frac{\partial}{\partial x}(nu) = 0. \tag{1.2.4}$$

Moreover, multiplying Eq. (1.2.1) by v, and then integrating with respect to v and using Eq. (1.2.4), we arrive at the equation for the conservation of momentum which takes the same form as Eq. (1.1.6), namely,

$$\frac{\partial u}{\partial t} + u\frac{\partial u}{\partial x} = 0. \tag{1.2.5}$$

As was explained in terms of the solution of Eq. (1.1.6), Eq. (1.2.5) allows overtaking, so that, in general, the solution becomes multi-valued after finite time. In the present example, this corresponds to the fact that the microscopic velocities are no longer identical where overtaking occurs, so that the velocity distribution Eq. (1.2.3) ceases to be valid. This means that a variance arises in the velocity, thereby causing the beam to be no longer cold. This thermalization without collision is often called phase-mixing. Therefore, the cold beam Eq. (1.2.3) as a solution of Eq. (1.2.1) is valid only for a finite time, and because of the overtaking process the system changes into one with a more complicated velocity distribution.

We conclude from this that the original equation is linear and that its solution exists for all time. However, if the solution is restricted to the one such as given by Eq. (1.2.3), then it has meaning only for a finite time until overtaking occurs. After that time we must solve directly the original equation (1.2.1). It should be noticed that Eqns (1.2.4) and (1.2.5) are nonlinear, while Eq. (1.2.1) is linear. This appearance of nonlinear equations from a linear equation is due to the nonlinear transformation involved in Eq. (1.2.3) or, more explicitly, to the relationship $nu = \int vf \, dv$.

Next, let us consider the Hamilton–Jacobi equation for a free particle. Defining the action of the particle by S, and assuming for simplicity that its mass is equal to unity ($m = 1$), we have

$$\frac{\partial S}{\partial t} + \frac{1}{2}\left(\frac{\partial S}{\partial x}\right)^2 = 0. \tag{1.2.6}$$

Differentiating this equation with respect to x, to give the equation for $u = \partial S/\partial x$, we find

$$\frac{\partial u}{\partial t} + u\frac{\partial u}{\partial x} = 0. \tag{1.2.7}$$

As has already been shown, the solution of this equation does not necessarily exist for all time for an arbitrarily prescribed initial condition. However, in Newtonian mechanics it is not physically relevant to consider the general initial value problem in the case of Eq. (1.2.6), and it is sufficient to consider the special solution, $u = $ constant, because $\partial S/\partial x$ is equal to the momentum of the particle. In other words, S is the generating function of the canonical transformation which transforms the Hamiltonian to zero. It is a perfect integral of the Hamilton–Jacobi equation (1.2.6) which, in the present case, should be given as a solution involving one arbitrary constant, except for an insignificant additive constant. Hence this is not the general solution which involves an arbitrary function. The problem is then to find the relationship which connects the action S, the energy E and the momentum p, where $S = -Et + px$, $E = p^2/2$ and p is an arbitrary constant. Consequently, this corresponds to the case that the gradient of the action S, $u = \partial S/\partial x$, is constant.

Here we shall discuss how the canonical equation is related to other equations which have been considered in this chapter, such as Eq. (1.2.7). Following the results of the previous section, we transform Eq. (1.2.7) to the corresponding system of ordinary differential equations along the characteristics, which take the form

$$\frac{dx}{dt} = u, \tag{1.2.8}$$

$$\frac{du}{dt} = 0. \tag{1.2.9}$$

1.2 PHYSICAL EXAMPLES

Since $u = \partial S/\partial x$ is the momentum, Eqns (1.2.8) and (1.2.9) are simply the canonical equations for a free particle. This corresponds to the well known fact that the characteristic equations of the Hamilton–Jacobi equation are the canonical equations themselves. For a general Hamiltonian $H(\partial S/\partial x, x, t)$, the Hamilton–Jacobi equation takes the form

$$\frac{\partial S}{\partial t} + H\left(\frac{\partial S}{\partial x}, x, t\right) = 0, \qquad \frac{\partial S}{\partial x} = p.$$

Differentiating this equation with respect to x, and denoting $\partial H/\partial p$ by u, we get

$$\frac{\partial p}{\partial t} + u\frac{\partial p}{\partial x} + \frac{\partial H}{\partial x} = 0.$$

This equation has the same form as that of Eq. (1.1.14), and hence Eqns (1.1.15) become

$$\frac{dx}{dt} = u = \frac{\partial H}{\partial p}, \tag{1.2.10}$$

$$\frac{dp}{dt} = -\frac{\partial H}{\partial x}. \tag{1.2.11}$$

The Eqns (1.2.10) and (1.2.11) are just canonical equations for the Hamiltonian H.

Next, we consider the behaviour of a free particle in quantum mechanics. In classical (Newtonian) dynamics, the special solution $u = $ constant is considered to be a physically relevant solution. However, in quantum mechanics the probability distribution of the particle is considered, and hence u is, in general, not constant. That is, the wave function for the free particle ψ is given by the Schrödinger equation:

$$-i\frac{\partial \psi}{\partial t} = \frac{\hbar}{2}\frac{\partial^2 \psi}{\partial x^2}. \tag{1.2.12}$$

If ψ is expressed in terms of the phase S, and the square root of the amplitude as $\psi = \sqrt{\rho}\, e^{iS/\hbar}$, and u is defined to be $\partial S/\partial x$ as before, then from the real and imaginary parts of Eq. (1.2.12) we obtain, respectively,

$$\frac{\partial \rho}{\partial t} + \frac{\partial}{\partial x}(\rho u) = 0, \tag{1.2.13}$$

$$\frac{\partial u}{\partial t} + u\frac{\partial u}{\partial x} = \frac{\hbar^2}{4}\frac{\partial}{\partial x}\left\{\frac{1}{\sqrt{\rho}}\frac{\partial}{\partial x}\left(\frac{1}{\sqrt{\rho}}\frac{\partial \rho}{\partial x}\right)\right\}. \tag{1.2.14}$$

The results of classical mechanics can be obtained by taking the limit

$\hbar \to 0$ in quantum mechanics. This confirms that Eq. (1.2.14) then reduces to Eq. (1.2.7). Physically, the right-hand side of Eq. (1.2.14) characterizes the quantum effect. It is easily seen that in Eqns (1.2.13) and (1.2.14) a solution exists for all time for an arbitrarily prescribed initial condition, because the original equation is the Schrödinger equation which is a linear equation for ψ. This may be examined further as follows. When the right-hand side of Eq. (1.2.14) is neglected, overtaking occurs as was already shown, so that the solution ceases to be smooth and does not exist for all time. However, in a region where overtaking occurs, u and ρ vary rapidly, and hence even if $\hbar$ is small, the third order derivative with respect to x on the right-hand side of Eq. (1.2.14) becomes large, by virtue of which the evolution of a smooth solution becomes possible. As can be seen from this result (which will be explained later in some detail), the higher order derivative with respect to x on the right-hand side of Eq. (1.2.14) plays a crucial role in various cases.

A similar situation may be found in connection with the diffusion equation

$$\frac{\partial \psi}{\partial t} = \mu \frac{\partial^2 \psi}{\partial x^2} \quad (\mu > 0). \tag{1.2.15}$$

Let ψ be given by

$$\psi = \exp\left(-\frac{\alpha}{2\mu} \int u \, dx\right), \tag{1.2.16}$$

where α is an arbitrary constant. Then, introducing this into Eq. (1.2.15), and differentiating with respect to x, brings us to the equation

$$\frac{\partial u}{\partial t} + \alpha u \frac{\partial u}{\partial x} - \mu \frac{\partial^2 u}{\partial x^2} = 0. \tag{1.2.17}$$

This equation is called the Burgers equation, and it is one of the important equations which we shall encounter in connection with nonlinear wave phenomena. The solution u of the Burgers equation is connected with the solution ψ of the diffusion equation by Eq. (1.2.16), which is the Cole–Hopf transformation,

$$u = -\frac{2\mu}{\alpha} \frac{1}{\psi} \frac{\partial \psi}{\partial x}. \tag{1.2.18}$$

In this sense, the nonlinear Burgers equation is exactly solvable in terms of the Cole–Hopf transformation and the linear diffusion equation.

Comparing the form of the Schrödinger equation (1.2.12) with that of the diffusion equation (1.2.15), we find that the only significant difference is that in the Schrödinger equation the time-derivative is multiplied by the imaginary unit i. However, the physical properties of these two equations are entirely different. In particular, the Schrödinger equation (1.2.12) is invariant with respect to time reversal. Putting $\psi \propto e^{i(kx-\omega t)}$, we arrive at the dispersion relation $\omega = (\hbar/2)k^2$, which corresponds to a wave propagating with the real frequency ω proportional to the square of the wave number k. The phase velocity ω/k increases with k. Hence the Schrödinger equation is an equation of dispersive type. On the other hand, for the diffusion equation, we have $\omega = -i\mu k^2$, which corresponds to a wave decaying with time at a rate proportional to k^2. Consequently, the diffusion equation (1.2.15) is of the dissipative type.

The difference between the Schrödinger equation and the diffusion equation can be seen in the equation for u derived from those equations; that is, in the Eqns (1.2.13), (1.2.14), and (1.2.17). The terms $\partial u/\partial t + u\,\partial u/\partial x$ appear in each case, but in Eq. (1.2.14) the remaining term is a third order derivative with respect to x, while in Eq. (1.2.17) it is only a second order derivative. It is obvious that the second and third order derivatives are responsible for the dissipation and the dispersion, respectively, which arise in each case. For initial conditions for u and n which are sufficiently smooth functions of x, and for suitably small times, the second and third order derivatives in Eqns (1.2.14) and (1.2.17) may be neglected, so that in each equation u can become steep. However, as u steepens (and consequently so also does n), the second and third order derivatives become important and could act to halt the steepening process. However, these two higher order derivatives play such a role in quite different ways, that the resultant states are entirely different. One is a shock wave and the other a quantum fluctuation.

Unfortunately, however, the set of Eqns (1.2.13) and (1.2.14) is too special as a model of a dispersive nonlinear process to merit further study, because it does not allow the solitary wave type of solution which is so conspicuous in dispersive nonlinear equations. Note here a comparison with Eqns (3.6.32) for the nonlinear Schrödinger equation. To obtain a typical model equation we need to replace the second derivative in the Burgers equation by the third derivative to give the Korteweg–de Vries (KdV) equation:

$$\frac{\partial u}{\partial t} + u\frac{\partial u}{\partial x} - \mu\frac{\partial^3 u}{\partial x^3} = 0. \qquad (1.2.19)$$

This equation was first derived and studied in connection with the propagation of a shallow water wave (Problem 3.4.1). However, it has been found during the last decade that this same equation can be used as a general nonlinear evolution equation of dispersive type when seeking to describe the propagation of weakly nonlinear dispersive waves (cf. Section 3.3). In the Burgers equation (1.2.17) a steepening wave tends to a shock wave, but on the other hand, in the KdV equation, the steepening process is balanced by dispersion to give rise to a solitary wave. These facts will be explained in detail in subsequent chapters. The crucial point is that the second and third order derivatives with respect to x in the Burgers and the KdV equations correspond to the dissipation and dispersion, respectively, which are entirely different physical processes. (See Sections 3.2 to 3.4.)

1.3 The limit of weak dissipation and shock waves

Let us consider the stationary solution of the Burgers equation (1.2.17), which is given by assuming that u is a function of $x' = x - \tilde{\lambda}t$ only, with $\tilde{\lambda}$ an arbitrary constant. Then Eq. (1.2.17) becomes an ordinary differential equation, which can be integrated easily. For simplicity, setting $\alpha = 1$, we obtain the solution

$$u' = u'_\infty \tanh\left(-\frac{u'_\infty x'}{2\mu}\right), \tag{1.3.1}$$

which as $|x| \to \infty$ approaches different limits in an asymptotic sense. Here u' is

$$u' = u - \tilde{\lambda}, \tag{1.3.2}$$

and u'_∞ is the value of u' at $x' = \infty$ ($x = \infty$), and we take note of the fact that

$$u'_\infty = u_\infty - \tilde{\lambda} < 0. \tag{1.3.3}$$

In a system moving with the wave at a constant speed $\tilde{\lambda}$, the solution (1.3.1) decreases monotonically as x' increases, and approaches the two different values $|u'_\infty|$ at the rear ($x' \to -\infty$) and u'_∞ at the front ($x' \to +\infty$), respectively (cf. Fig. 1.2). The exponential decay is characterized by the exponent $2\mu/|u'_\infty|$, and this may be regarded as the width of the wave. Since the amplitude is inversely proportional to the width, waves of larger amplitude change more rapidly, and also the width is narrower for a smaller dissipation parameter μ. From the analogy with the hydrodynamics, we call the solution given by Eq. (1.3.1) a shock wave, or the Burgers shock wave. In a hydrodynamic shock wave, the variation of u corresponds to that of the gas density.

We now consider the limit $\mu \to 0$ in the solution (1.3.1) of the Burgers equation (1.2.17). In this limit the solution becomes a step function with the value $|u'_\infty|$ for $x' < 0$ and the value u'_∞ for $x' > 0$. On the other hand, as $\mu \to 0$, Eq. (1.2.17) reduces its order by unity to become Eq. (1.1.6). (Hereafter we shall replace the h of Eq. (1.1.6) by u.) On account of this

1.3 LIMIT OF WEAK DISSIPATION AND SHOCK WAVES   15

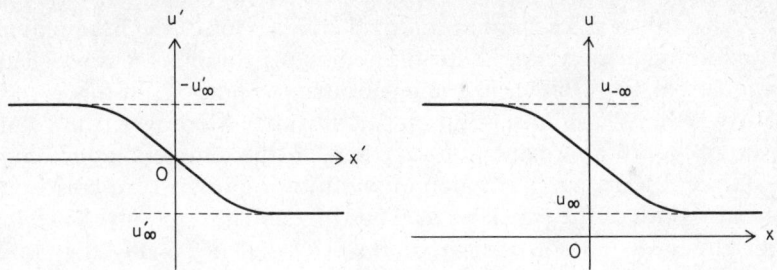

Fig. 1.2

it would be possible to define solutions of Eq. (1.1.6) as the limit as $\mu \to 0$ of (smooth) solutions of Eq. (1.2.17). When the limiting solution so obtained is also a smooth function, this leads to a smooth solution of Eq. (1.1.6) which is determined uniquely by its initial condition, and so there is no advantage in obtaining solutions of Eq. (1.1.6) by means of this approach. (Such a solution is called a strong solution.)

In the limit as $\mu \to 0$, however, the smooth solution (1.3.1) of Eq. (1.2.17) becomes discontinuous at $x' = 0$, and so the limit is not a smooth solution of Eq. (1.1.6). This result enables us to extend the set of smooth solutions of Eq. (1.1.6), so that they include discontinuous functions, by defining them as the limit as $\mu \to 0$ of smooth solutions of Eq. (1.2.17).

As was already shown in Section 1.1, in general, a solution of Eq. (1.1.6) becomes discontinuous because of the steepening process caused by the nonlinearity, so that a smooth solution can exist only for a finite time, even when evolving from an arbitrarily smooth initial condition. Accordingly, the solution describing the subsequent time evolution is not determined by Eq. (1.1.6). After overtaking has occurred, other effects like phase mixing may take place, as discussed in the example of a cold beam of particles in a rarefied gas. In that particular case we must return to Eq. (1.2.1) or a shock wave may be formed owing to the dissipation, for which Eq. (1.2.17) must then be used. So, the time evolution following overtaking cannot be determined from Eq. (1.1.6) alone, without taking into account further physical laws. Now, let us consider the limiting solutions of Eq. (1.2.17) as $\mu \to 0$ to be the solutions of Eq. (1.1.6) for all time. This approach corresponds to a physical law which causes the dissipation to work efficiently only in the very narrow region in which overtaking occurs. In consequence a very sharply varying but smooth shock wave is formed and then propagates.

Our aim is now to solve Eq. (1.1.6) in such a way that it admits discontinuous solutions, subject to the requirement that each solution is equivalent in the limit as $\mu \to 0$ of a solution of the Burgers equation. As will be shown later, for smooth solutions of Eq. (1.1.6) the requirement is satisfied automatically, whilst for discontinuous solutions it is satisfied by solving Eq. (1.1.6) subject to some subsidiary condition. In this way, Eq. (1.1.6) will have a unique solution for all time. Moreover, the solution will be a good approximation to that of the Burgers equation for $\mu \ll 1$, except for a narrow region of width of the order of μ corresponding to the width of the shock wave. Thus it is sufficient to solve Eq. (1.1.6) subject to the condition indicated above instead of solving the Burgers equation (1.2.17).

To present this condition in a concise form, let us investigate certain features of the shock wave solution (1.3.1). It follows from Eqns (1.3.2) and (1.3.3) that

(1) $\tilde{\lambda} > u_\infty (>0)$
(2) $u_{-\infty}$ is uniquely determined by $\tilde{\lambda}$ and u_∞ by the relationship $u_{-\infty} = 2\tilde{\lambda} - u_\infty$.

As $x \to \infty$, so the solution tends to a uniform state, and hence the third term of Eq. (1.2.17) may be taken to be zero under such conditions. That is, at infinity, u_∞ corresponds to the slope of the characteristics of Eq. (1.1.6), which we shall call the characteristic speed associated with Eq. (1.1.6). The above property (1) then shows that the speed $\tilde{\lambda}$ of the shock wave is always larger than the characteristic speed u_∞. It may also be seen from (1) and (2) that $u_{-\infty} > \tilde{\lambda}$, so that to the rear the characteristic speed exceeds the speed of the shock wave. This corresponds to the fact that a hydrodynamical shock wave has an associated flow which is supersonic at the front of a shock and subsonic at the rear, respectively (Section 2.4). Furthermore, as can be seen from (1) and (2), the value of u at the rear of the shock is larger than that at the front. This corresponds to the fact that a hydrodynamical shock wave is a compression wave, and it also implies the irreversibility of the system, giving rise to the increase of entropy across a shock wave. We shall say more about this matter later on. Here, the important point is that these properties of a shock wave also hold for the step function obtained as the limit as $\mu \to 0$ in Eq. (1.3.1). Thus, it suffices to pick out from the discontinuous solutions of Eq. (1.1.6) those solutions which satisfy the conditions (1) and (2). However, mathematically, we must first extend the class of solutions of the differential equation (1.1.6) so that it includes discontinuous solutions. This is done, as usual, by introducing a class of test functions, w, which are differentiable as many times as necessary, and which vanish identically outside a bounded space-time domain. So, in place of Eq. (1.1.6), we must now consider

$$\int_0^\infty \int_{-\infty}^\infty w \left(\frac{\partial u}{\partial t} + u \frac{\partial u}{\partial x} \right) dx\, dt = 0. \qquad (1.3.4)$$

1.3 LIMIT OF WEAK DISSIPATION AND SHOCK WAVES

Integrating Eq. (1.3.4) by parts, we can transfer the differentiation operation from u to w to obtain

$$-\int_{-\infty}^{\infty} w(x,0)u(x,0)\,dx - \int_0^{\infty}\int_{-\infty}^{\infty}\left\{\frac{\partial w}{\partial t}u + \frac{\partial w}{\partial x}\frac{u^2}{2}\right\}dx\,dt = 0. \quad (1.3.5)$$

Here, $u(x,0)$ is the initial condition at $t=0$.

Since Eq. (1.3.5) does not contain a derivative of u, it is also well defined when u is discontinuous. If we require that Eq. (1.3.5) holds for any test function w, subject to a given initial condition $u(x,0)$, then it may be regarded as an equation for u. The function u satisfying Eq. (1.3.5) for any w is called a weak solution of Eq. (1.1.6), whilst a smooth solution satisfying Eq. (1.1.6) is called a strong solution. A weak solution defined by Eq. (1.3.5) reduces to a strong solution, provided u is smooth. For, in such a case, we can derive Eq. (1.3.4) from Eq. (1.3.5) in the reverse order, and it turns out that Eq. (1.3.4) must hold for any w. We thus obtain Eq. (1.1.6). That is, a weak solution may be considered to be a kind of extension of a solution of Eq. (1.1.6) which includes a strong solution as a special case. This is often said to be an extension in the weak sense. Now, when we define a weak solution in this way, the step function considered above is, of course, to be regarded as a weak solution of Eq. (1.1.6).

However, the uniqueness of the solution is lost on account of the extension and, in consequence, physically meaningless solutions are included amongst such weak solutions. In order to show this by example, we consider a step function as the initial condition $u(x,0)$, for which $u = u_0$ for $x>0$ and $u = u_1$ for $x<0$. We assume $u = u_0$ for $x - \tilde{\lambda}t > 0$ and $u = u_1$ for $x - \tilde{\lambda}t < 0$ to be the solution for $t \geq 0$, and then substitute this into Eq. (1.3.5). The integration can be easily carried out and gives

$$-\int_C w[u]\,dx + \int_C w[\tfrac{1}{2}u^2]\,dt = 0.$$

Here, the contour C is taken along the line $x - \tilde{\lambda}t = 0$, and $[u]$ and $[u^2]$ denote the jumps, $u_1 - u_0$ and $u_1^2 - u_0^2$, across this line, respectively. Since $dx = \tilde{\lambda}\,dt$ along the contour C, the above equation becomes

$$\int_C w\left\{-\tilde{\lambda}[u] + \left[\frac{u^2}{2}\right]\right\}dt = 0.$$

Therefore, we have

$$-\tilde{\lambda}[u] + \left[\frac{u^2}{2}\right] = 0, \quad (1.3.6)$$

giving

$$\tilde{\lambda} = \tfrac{1}{2}(u_1 + u_0). \quad (1.3.7)$$

Thus a weak solution has been found which is discontinuous across $x - \tilde{\lambda}t = 0$. Eq. (1.3.7) is just the condition (2) for the shock wave.

Now, if $u_1 > u_0$ (>0), the other condition (1) is certainly satisfied, and $\tilde{\lambda} > u_0$. Therefore, the above weak solution agrees with the one obtained as the limit as $\mu \to 0$ in solutions of the Burgers equation. The problem arises, however, when $u_1 < u_0$. As is immediately seen from the discussion following Eq. (1.1.11), in this case, the following continuous solution (centred expansion wave) exists for $t > 0$ as well as the above step function:

$$u = u_0 \quad \text{for} \quad x - u_0 t > 0$$
$$u = u_1 \quad \text{for} \quad x - u_1 t < 0$$
$$u = x/t \quad \text{for} \quad u_1 t < x < u_0 t.$$

This is a strong solution and hence it also satisfies Eq. (1.3.5), as was already mentioned. Accordingly, in this case, there exist two different (weak) solutions satisfying the same initial condition for all t together with others which may be found, and so the uniqueness of the solution is lost. For the discontinuous solution given by the step function, we have from Eq. (1.3.7) that $\tilde{\lambda} < u_0$, and it follows that the condition (1) is not satisfied. Therefore this solution does not correspond to a shock wave which is obtained as the limit of solutions of the Burgers equation. In fact, the solution represents an expansion shock wave, since u decreases from u_0 to u_1 across the discontinuity. From the viewpoint that the Burgers equation is to be regarded as a physical model, such a mathematical solution may be called a non-physical solution.

As has been clarified by the above discussion, the class of weak solutions permitted by Eq. (1.3.5) is not necessarily determined uniquely by the initial conditions, and they contain amongst them non-physical solutions. Solutions with physical meaning must be the ones satisfying the conditions (1) and (2). As may be seen, however, from the process used when deriving Eq. (1.3.7) from Eq. (1.3.5), weak solutions always satisfy condition (2). Therefore, it is sufficient to require condition (1) as a selection principle for finding a solution with physical meaning from amongst the weak ones. In fact, the condition (2) is an equality, while the condition (1) is a relationship expressed by means of an inequality, and they are essentially different in their mathematical properties. (An inequality, such as the law of increase of entropy, cannot be derived from an equality like condition (2).)

Now it is known that the condition (2) is equivalent to condition (1.3.6) for the jump across a shock wave. As will be discussed in detail in the next chapter, this condition (1.3.6) can be derived from the conservation laws of physical quantities such as mass, momentum, and energy in the

1.3 LIMIT OF WEAK DISSIPATION AND SHOCK WAVES

case of a hydrodynamical shock wave, and it is usually referred to as the shock condition. Here, we shall simply call it a conservation law. This conservation law is a solution of Eq. (1.3.5) in the weak sense, though it must be noticed that it also admits a non-physical solution. Therefore, in order to select only solutions with physical meaning from amongst the class of weak solutions, we need some kind of condition expressed in terms of an inequality (such as condition (1)) in addition to Eq. (1.3.6). Such a condition is called an evolutionary condition and it will be discussed in detail later on. In gas dynamics the evolutionary condition is equivalent to the increase of entropy required by the second law of thermodynamics. We will show here that condition (1.3.6) corresponds to a conservation law, even for a general discontinuity surface $\varphi(x, t) = 0$.

First, we observe that Eq. (1.1.6), namely

$$\frac{\partial u}{\partial t} + u \frac{\partial u}{\partial x} = 0,$$

can be rewritten in the form

$$\frac{\partial u}{\partial t} + \frac{\partial}{\partial x}\left(\frac{u^2}{2}\right) = 0. \tag{1.3.8}$$

This indicates explicitly that the quantity $\int_{x_0}^{x_1} u \, dx$ is conserved with respect to time, provided that u has the same value at the limits x_0 and x_1. In general, an equation in the divergence form $(\partial u/\partial t) + \mathrm{div}\, f = 0$ is called a conservation law, so that we see Eq. (1.3.8) is itself expressed in conservation form.

Let us now introduce the two-dimensional vector $\boldsymbol{u}$ whose components are u and $u^2/2$, so that $\boldsymbol{u} = (u, u^2/2)$. Eq. (1.3.8) may then be written in the divergence form

$$\nabla \cdot \boldsymbol{u} = 0, \qquad \nabla = \left(\frac{\partial}{\partial t}, \frac{\partial}{\partial x}\right). \tag{1.3.9}$$

Integrating Eq. (1.3.8) over an arbitrarily narrow strip enclosing the discontinuity surface $\varphi(x, t) = 0$, we find from Gauss' theorem that

$$n_t[u] + n_x\left[\frac{u^2}{2}\right] = 0. \tag{1.3.10}$$

Here $\boldsymbol{n} = (n_t, n_x)$ is the unit normal to the discontinuity surface $\varphi(x, t) = 0$, and the components of $\boldsymbol{n}$ are given by

$$n_t = \frac{\partial \varphi/\partial t}{\sqrt{(\partial \varphi/\partial t)^2 + (\partial \varphi/\partial x)^2}}, \qquad n_x = \frac{\partial \varphi/\partial x}{\sqrt{(\partial \varphi/\partial t)^2 + (\partial \varphi/\partial x)^2}}.$$

Here again, $[u]$ denotes the jump of u across the discontinuity surface.

When we define the speed $\tilde{\lambda}$ of the surface by

$$\tilde{\lambda} = -\frac{\partial \varphi}{\partial t} \Big/ \frac{\partial \varphi}{\partial x}, \tag{1.3.11}$$

Eq. (1.3.10) reduces to

$$\tilde{\lambda}[u] = [\tfrac{1}{2}u^2]. \tag{1.3.6'}$$

It can be seen from this that the conservation law (1.3.6) holds locally, even when the speed $\tilde{\lambda}$ of the shock wave is not constant.

It has turned out from the above discussion that, in the limit as $\mu \to 0$ in the dissipative term, solutions of the Burgers equation (1.2.17) are obtained by requiring the evolutionary condition (1) to hold in conjunction with the simpler hyperbolic equation (1.1.6) (or in conjunction with the conservation law Eq. (1.3.8)). Mathematically rigorous theories for this have been developed by Olejnik (1954, 1955, 1956) and Ladyzhenskaya (1956). (Cf. Lax (1957, 1971); Dafermos (1974) for a general hyperbolic equation.)

As an example, we consider here the following initial value problem. At $t=0$, we suppose $u_0=0$ for $x>x_1$, $u_1>0$ for $0 \leq x \leq x_1$, and $u_2<0$ for $x<0$, as shown in Fig. 1.3. In such a case, the speeds of the characteristics starting from the regions $x>x_1$, $x_1>x>0$, and $x<0$ are given by $u_0=0$, u_1 and u_2, respectively. Since $u_1>u_0$, a shock wave is formed at the boundary between the constant states u_0 and u_1 and, for small t, the propagation speed of the shock is $\tilde{\lambda} = u_1/2$, as obtained from the relation (1.3.6). Also, for a characteristic starting from the origin $x=0$, the phase velocity $x/t = m$ is in the range $u_2 \leq m \leq u_1$, since $u_2 < u_1$ and u is constant along it. That is, a centred expansion wave is formed. However, $u_1 > \tilde{\lambda} = u_1/2$, so that at the time $t_c = 2x_1/u_1$ the centred expansion wave catches up with the shock. Since the value of u in the expansion wave (which is between u_1 and u_2) is larger than $u_0 = 0$, even after the elapsed time t_c, the shock keeps propagating, bounding the expansion wave behind and the constant state $u=0$ in front.

Let us find the propagation speed of the shock from the conservation law. In this case the propagation speed of the shock is not constant, since the value of u behind the shock varies with time. Thus, Eq. (1.3.11) becomes

$$\frac{dx}{dt} = -\frac{\partial \varphi}{\partial t} \Big/ \frac{\partial \varphi}{\partial x} = \frac{[u^2/2]}{[u]}. \tag{1.3.12}$$

Noticing that the value of u in front of the shock is the constant value $u_0 = 0$, whereas the value behind is in the range $u_2 < x/t < u_1$, we obtain

1.3 LIMIT OF WEAK DISSIPATION AND SHOCK WAVES

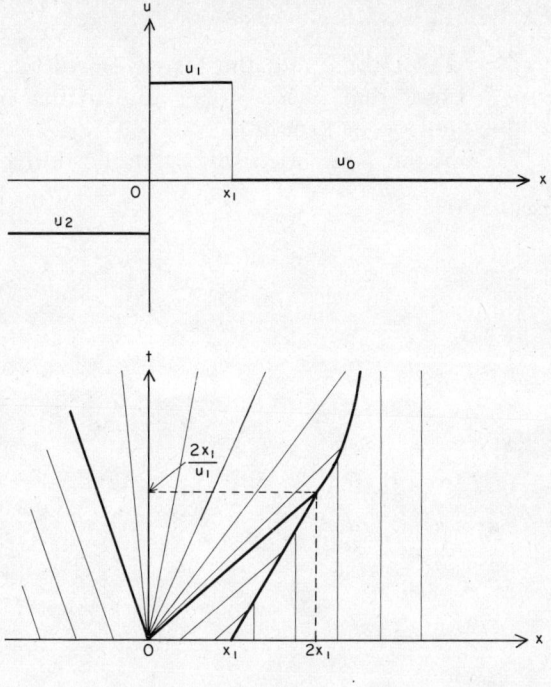

Fig. 1.3

the propagation speed of the shock from the relation (1.3.12) as

$$\frac{dx}{dt} = \frac{1}{2}\frac{x}{t}. \tag{1.3.13}$$

The locus of the shock in the (x, t)-plane is found by integrating Eq. (1.3.13). Using the boundary condition $(x = 2x_1)$ at $t_c = 2x_1/u_1$, this locus is found to be

$$x = (2x_1 u_1 t)^{1/2}. \tag{1.3.14}$$

The value of u behind the shock is given by x/t, and hence from the locus (1.3.14) we have as $t \to \infty$,

$$u = \frac{x}{t} = \frac{(2x_1 u_1 t)^{1/2}}{t} \propto t^{-1/2} \to 0 = u_0.$$

We have thus found that the shock attenuates with time.

Problem 1.3.1

For a plasma of cold electrons, find the restriction to be placed on the initial condition in order that phase-mixing due to the crossing of the electron-orbits does not occur (Taniuti (1963)).

Hint In Eq. (A.3) of the Appendix, set $T = 0$ and derive the result

$$d^2 u/dt^2 + \omega_p^2 u = 0$$

where

$$\frac{d}{dt} \equiv \frac{\partial}{\partial t} + u \frac{\partial}{\partial x}$$

and

$$\omega_p^2 = 4\pi e^2 n_0/m_e.$$

2
Wave equations

2.1 Linear wave equations and the method of characteristics

In the previous chapter we have taken as our model a wave propagating in one direction alone. However, in general, waves in one space dimension will propagate both to the right and left. In addition, there often coexist several waves with different velocities and in the case of an elastic material these might describe, for example, the transverse and longitudinal wave motion in that medium. In fact, in a plasma composed of ions and electrons, there are many modes of wave propagation, all of which interact with each other.

Before we proceed to an examination of such complex problems, let us start with the simple and familiar linear wave equation

$$\frac{\partial^2 \phi}{\partial t^2} - \frac{\partial^2 \phi}{\partial x^2} = 0. \tag{2.1.1}$$

As is well known, the general solution of Eq. (2.1.1) is given by the superposition of waves travelling in opposite directions. That is, by an expression of the form

$$\phi = f(x-t) + g(x+t),$$

where f and g are arbitrary twice differentiable functions. In this chapter, however, we shall apply to the wave equation the method of characteristics mentioned previously.

To begin with, let us factorize Eq. (2.1.1) as follows:

$$\left(\frac{\partial}{\partial t} + \frac{\partial}{\partial x}\right)\left(\frac{\partial}{\partial t} - \frac{\partial}{\partial x}\right)\phi = 0, \tag{2.1.2}$$

and introduce the new functions u and v by setting

$$u \equiv \frac{\partial \phi}{\partial t}, \qquad v \equiv \frac{\partial \phi}{\partial x}. \tag{2.1.3}$$

Combining Eqns (2.1.2) and (2.1.3) then gives Eq. (2.1.4a) below, while exchanging the order of the differentiation in Eq. (2.1.2), and again using Eq. (2.1.3) gives Eq. (2.1.4b), so that we have two first-order equations for u and v

$$\left(\frac{\partial}{\partial t}+\frac{\partial}{\partial x}\right)(v-u)=0, \tag{2.1.4a}$$

$$\left(\frac{\partial}{\partial t}-\frac{\partial}{\partial x}\right)(v+u)=0. \tag{2.1.4b}$$

These first-order equations are equivalent to the second-order wave equation and they show that the quantities $v \mp u$ remain constant along the two families of lines correponding to $dx/dt = \pm 1$. Hereafter, we shall call these the characteristic curves or, more simply, the characteristics of Eq. (2.1.1) or of the equivalent system (2.1.4).

We remark here that a function which is constant along a characteristic is called a Riemann invariant, so that $v-u$ and $v+u$ are the two Riemann invariants associated with Eqns (2.1.4a, b). So, along the family of straight line characteristics $C^{(+)}$ whose slope is $dx/dt = 1$, and which has the equation $x - t = \xi$, with ξ a parameter, $v - u$ is a function of ξ only. By setting $t = 0$, we see that the parameter ξ along a $C^{(+)}$ characteristic is merely the value of x at the point at which that characteristic intersects the x-axis. Similarly, $v+u$ is a function of η only along the family of straight line characteristics $C^{(-)}$, whose slope is $dx/dt = -1$, and which have the equation $x + t = \eta$, with η a parameter. Here again, by setting $t = 0$, we see that the parameter η along a $C^{(-)}$ characteristic is merely the value of x at the point at which that characteristic intersects the x-axis. We thus see that Eqns (2.1.4a,b) are equivalent to

$$v - u = r(\xi), \tag{2.1.5a}$$
$$v + u = s(\eta), \tag{2.1.5b}$$

where the arbitrary functions $r(\xi)$ and $s(\eta)$ are determined once the initial conditions, $u(x, 0) = u_0(x)$ and $v(x, 0) = v_0(x)$, for u and v at $t = 0$ have been specified. That is, we have

$$r(\xi) = v_0(\xi) - u_0(\xi), \tag{2.1.6a}$$
$$s(\eta) = v_0(\eta) + u_0(\eta). \tag{2.1.6b}$$

Expressed differently, the values of $v - u$ and $v + u$ on the corresponding $C^{(+)}$ characteristics $x - t = \xi$, and the $C^{(-)}$ characteristics $x + t = \eta$, are uniquely determined by the initial conditions for u and v. Thus, by solving Eqns (2.1.5) algebraically, the values of u and v at the intersection of a $C^{(+)}$ characteristic and a $C^{(-)}$ one can be found in the following

2.1 LINEAR WAVE EQUATIONS

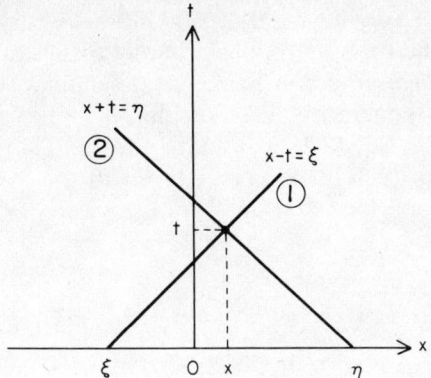

Fig. 2.1 Along ① $v - u = r(\xi) = v_0(\xi) - u_0(\xi)$; along ② $v + u = s(\eta) = v_0(\eta) + u_0(\eta)$.

form (see Fig. 2.1):

$$u = -\tfrac{1}{2}(r(\xi) - s(\eta)) = -\tfrac{1}{2}(v_0(\xi) - u_0(\xi) - v_0(\eta) - u_0(\eta)), \qquad (2.1.7a)$$

$$v = \tfrac{1}{2}(r(\xi) + s(\eta)) = \tfrac{1}{2}(v_0(\xi) - u_0(\xi) + v_0(\eta) + u_0(\eta)). \qquad (2.1.7b)$$

Taking into account the definitions Eq. (2.1.3) of u and v we see that ϕ is also a function of ξ and η, and that $\phi = f(\xi) + g(\eta)$. Since $\xi = x - t$ and $\eta = x + t$, we can easily differentiate ϕ partially with respect to t and x, and combine the results with Eqns (2.1.7) to find

$$f'(\xi) = \tfrac{1}{2}[v_0(\xi) - u_0(\xi)],$$
$$g'(\eta) = \tfrac{1}{2}[v_0(\eta) + u_0(\eta)].$$

Here a prime indicates differentiation of a function with respect to its argument. Denoting the initial condition $\phi(x, 0)$ of ϕ by $\phi_0(x)$ and using the result that $v_0(x) = d\phi_0(x)/dx$ or, equivalently, $v_0(\xi) = \phi_0'(\xi)$, we find

$$f(\xi) = \frac{1}{2}\left[\phi_0(\xi) - \int u_0(\xi)\,d\xi\right],$$

$$g(\eta) = \frac{1}{2}\left[\phi_0(\eta) + \int u_0(\eta)\,d\eta\right].$$

We thus obtain the following solution satisfying the initial conditions $\phi(x, 0) = \phi_0(x)$, $\partial \phi/\partial t|_{t=0} = u_0(x)$

$$\phi = \tfrac{1}{2}[\phi_0(x - t) + \phi_0(x + t)] + \frac{1}{2}\int_{x-t}^{x+t} u_0(\eta)\,d\eta. \qquad (2.1.7c)$$

This is known as D'Alembert's solution, and it shows the dependence of the solution ϕ at the point (x, t) on the initial conditions specified on the x-axis at $t = 0$ (Coulson and Jeffrey (1977)).

This method of solving the wave equation by using the characteristic curves $C^{(\pm)}$ is called the method of characteristics, and it will be applied to a nonlinear wave equation in the next section. With this objective in mind, let us now generalize these results.

We first rewrite Eq. (2.1.1) in matrix form. Taking Eqns (2.1.3) into consideration, Eq. (2.1.1) may be reduced to the following matrix equation for the vector

$$U = \begin{pmatrix} u \\ v \end{pmatrix}, \tag{2.1.8}$$

$$\frac{\partial U}{\partial t} + A \frac{\partial U}{\partial x} = 0, \tag{2.1.9}$$

where the matrix A is

$$A = \begin{pmatrix} 0 & -1 \\ -1 & 0 \end{pmatrix}. \tag{2.1.10}$$

Now we already know that the values of u and v at the intersection of the two characteristics, $x - t = \xi$ and $x + t = \eta$, can be found from the initial conditions they must satisfy. In addition, we have seen that it is the characteristic speeds, that is the phase velocities of the waves,

$$\frac{dx}{dt} = \pm 1, \tag{2.1.11}$$

which characterize these families of characteristics.

It should now be noticed that these characteristic speeds are just the eigenvalues of the matrix A. To show this, let λ and R be an eigenvalue and the corresponding right eigenvector of A, respectively. Then we have

$$AR = \lambda R. \tag{2.1.12}$$

From this equation we find

$$\lambda = \pm 1, \tag{2.1.13}$$

and the corresponding right eigenvectors may be given by

$$R_{\pm} = \begin{pmatrix} 1 \\ \mp 1 \end{pmatrix}. \tag{2.1.14}$$

In order to examine what the states characterized by $R_{\pm}$ represent, we write U in the form $U^{(\pm)} = \varphi R_{\pm}$, where φ is an arbitrary scalar function of x and t. Substituting this into Eq. (2.1.9), we can easily obtain the following equation for φ:

$$\frac{\partial \varphi}{\partial t} + \lambda \frac{\partial \varphi}{\partial x} = 0.$$

2.1 LINEAR WAVE EQUATIONS

Therefore solutions of Eq. (2.1.9), corresponding to the eigenvalues ± 1, become

$$U^{(+)} = \varphi_+(x-t)R_+ \quad \text{and} \quad U^{(-)} = \varphi_-(x+t)R_-. \tag{2.1.15}$$

which represent waves travelling to the right and left with the speeds ± 1, respectively. Thus, these are travelling waves.

A general solution to this linear system of equations may be expressed as a superposition of the two possible eigenmodes

$$U = \varphi_+(x-t)R_+ + \varphi_-(x+t)R_-, \tag{2.1.15'}$$

and when the initial condition $U = U_0(x)$ at $t = 0$ is given, Eq. (2.1.15') leads to the following equation connecting $\varphi_\pm$:

$$U_0(x) = \varphi_+(x)R_+ + \varphi_-(x)R_-.$$

Since the $R_\pm$ are linearly independent, the functions $\varphi_\pm$ can be uniquely determined by solving this equation algebraically. The solution Eq. (2.1.15') so obtained is, of course, the same as Eqns (2.1.7a, b). Nevertheless, it is worth emphasizing that the solution represents the superposition of two suitably combined eigenstates of matrix A.

In order to derive Eq. (2.1.5) involving the Riemann invariants of the system Eq. (2.1.9), it is convenient to introduce the left eigenvectors L satisfying

$$LA = \lambda L, \tag{2.1.16}$$

namely,

$$L_\pm = (1, \mp 1). \tag{2.1.17}$$

Then, multiplying Eq. (2.1.9) by $L_\pm$ from the left, we obtain

$$L_\pm \left(\frac{\partial}{\partial t} + \lambda_\pm \frac{\partial}{\partial x} \right) U_\pm = 0. \tag{2.1.18}$$

The differential operators in Eqns (2.1.18) represent the total derivative along the $C^{(+)}$ characteristics $x - t = \xi$ and the $C^{(-)}$ characteristics $x + t = \eta$, corresponding to the eigenvalues $\lambda_+ = 1$ and $\lambda_- = -1$, respectively. Hence we have for $\xi = \text{const.}$, or $\eta = \text{const.}$, that is along the characteristics $C^{(+)}$ or $C^{(-)}$, respectively,

$$L_\pm \, dU_\pm = 0. \tag{2.1.19}$$

Integrating Eq. (2.1.19) with the aid of expression (2.1.17) for $L_\pm$, and denoting the integration constants by $r(\xi), s(\eta)$, we obtain equations (2.1.5a) and (2.1.5b) for $\lambda_\pm = \pm 1$, respectively.

The above results for a two-element vector U can be extended at once to the case of a general vector U. Let us introduce an n element vector

$$U = \begin{pmatrix} u_1 \\ u_2 \\ \vdots \\ u_n \end{pmatrix}, \tag{2.1.20}$$

and consider the matrix partial differential equation for U

$$\frac{\partial U}{\partial t} + A \frac{\partial U}{\partial x} = 0. \tag{2.1.21}$$

When all of the n eigenvalues $\lambda_1, \lambda_2, \ldots, \lambda_n$ of the $(n \times n)$ matrix A are real and distinct, and the corresponding eigenvectors $R_1, R_2, \ldots, R_n$ are linearly independent, Eq. (2.1.21) is said to be totally hyperbolic. By analogy with the wave equation we also call the families of curves given by integrating

$$\frac{dx}{dt} = \lambda_i \qquad (i = 1, 2, \ldots, n) \tag{2.1.21'}$$

the families of characteristic curves of the hyperbolic system Eq. (2.1.21).

Here, the elements of A may also depend on the elements $u_1, u_2, \ldots, u_n$ of U, as well as on x and t. In a linear equation, in which A is independent of U, λ and R become specific functions of x and t, and there are n wave modes whose (local) phase velocities are given by the eigenvalues $\lambda_1, \lambda_2, \ldots, \lambda_n$. In particular, when A is a constant matrix, the general solution U can be written in terms of a set of arbitrary functions $\varphi_j(x - \lambda_j t)$ of $x - \lambda_j t$ as the sum

$$U = \sum_{j=1}^{n} \varphi_j(x - \lambda_j t) R_j. \tag{2.1.22}$$

By virtue of the linear independence of the R_j ($j = 1, 2, \ldots, n$), the φ_j can be uniquely determined from the initial condition $U(x, 0) = \Phi(x)$ imposed on the system Eq. (2.1.21).

Next, we shall discuss a special type of wave, usually referred to as a simple wave, which arises as a solution of Eq. (2.1.9). Though the Riemann invariants $r(\xi), s(\eta)$ remain constant along their respective characteristic curves, in general these values are different on different $C^{(+)}$ or $C^{(-)}$ characteristics. If $r(\xi)$ and $s(\eta)$ are absolute constants along all of their respective characteristics, so that

$$r(\xi) \equiv r_0, \qquad s(\eta) \equiv s_0, \tag{2.1.23}$$

it follows from Eqns (2.1.7a) and (2.1.7b) that the solutions u, v are constant over the whole of the (x, t)-space (constant state).

2.1 LINEAR WAVE EQUATIONS

Let us now consider the case in which either $r(\xi)$ or $s(\eta)$ is an absolute constant, say, $r(\xi) = r_0$. As may be seen from the relationship (2.1.5a), in this case, we clearly have

$$v(x, t) - u(x, t) = r_0 \tag{2.1.24}$$

along all of the $C^{(+)}$ characteristics, and hence in all of the (x, t)-space covered by the $C^{(+)}$ characteristics. If the characteristics $C^{(+)}$, on which $r(\xi) = r_0$, cover the whole of the (x, t)-space, then this relation is applicable for all space and time. The initial values $u_0(x), v_0(x)$ of u, v cannot then be chosen arbitrarily, since they must be given so that they satisfy the condition

$$v_0(x) - u_0(x) = r_0, \tag{2.1.25}$$

for all x. Since $r(\xi)$ takes the constant value r_0, from Eqns (2.1.7a, b) it follows that $u(x, t)$ and $v(x, t)$ are functions of η only. In other words, u and v are constant along the $C^{(-)}$ characteristics $x + t = \eta$, so that we have

$$\begin{aligned} u(\eta) &= u(x+t), \\ v(\eta) &= v(x+t), \end{aligned} \tag{2.1.26}$$

which represents a wave travelling to the left.

In the case that $s(\eta) \equiv s_0$, it turns out, similarly, that the solution represents a wave travelling in the opposite direction; that is to the right. When the solution of a wave equation corresponds to a wave propagating in one direction alone, either to the right or to the left, we call such a solution a simple wave. We also often refer to the physical state represented by such a wave as a simple wave solution, or simple wave state. For a simple wave, an initial condition like Eq. (2.1.25), or the analogous condition corresponding to Eq. (2.1.6a), must be satisfied, and so this might appear to be an extremely special case. It is easily shown, however, that a simple wave is a state which appears very frequently in the study of practical physical phenomena.

For example, suppose that the disturbance of certain physical quantities u, v is localized in the domain $|x| \leq x_0$ at $t = 0$. Then we must consider the initial condition given by

$$\begin{aligned} u(x, t=0) &= \tilde{u}(x) & (|x| \leq x_0), \\ v(x, t=0) &= \tilde{v}(x) & (|x| \leq x_0), \\ u(x, t=0) &= u_0(\text{const.}) & (|x| > x_0), \\ v(x, t=0) &= v_0(\text{const.}) & (|x| > x_0). \end{aligned} \tag{2.1.27}$$

In the domains of the (x, t)-space covered by the characteristics expressed in terms of $|x - t| = |\xi| > x_0$ and $|x + t| = |\eta| > x_0$ (the regions denoted by

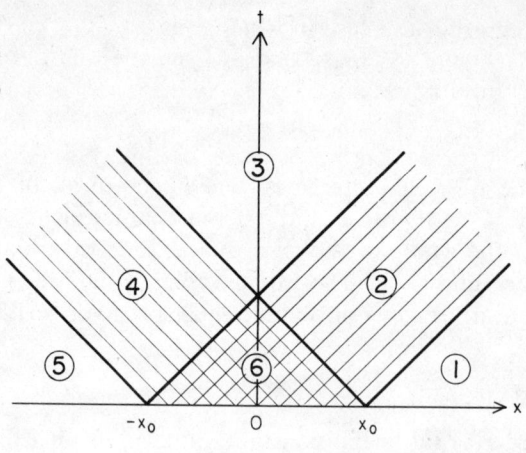

Fig. 2.2

the numbers 1, 3, and 5 in Fig. 2.2), $r(\xi)$ and $s(\eta)$ are constant, so that

$$r(\xi) = v_0 - u_0 \quad (|\xi| > x_0),$$
$$s(\eta) = v_0 + u_0 \quad (|\eta| > x_0),$$

and hence in those regions the solutions correspond to constant states.

In the regions covered by the characteristics $\{|x-t|=|\xi| \leq x_0, |x+t|=|\eta|>x_0\}$ or $\{|x-t|=|\xi|>x_0, |x+t|=|\eta| \leq x_0\}$ (the regions denoted by the numbers 2 and 4 in Fig. 2.2), we have $s(\eta) = v_0 + u_0$ or $r(\xi) = v_0 - u_0$, respectively. Consequently, the solutions in these regions are simple waves. That is, in each region, u and v are constant along $x - t = \xi$ or $x + t = \eta$, respectively, and represent a wave propagated either to the right or the left. In the remaining region u and v depend on both ξ and η, so this may be considered to be a general region. Thus, after a sufficient lapse of time $(t > x_0)$, the initial disturbance resolves itself into a simple wave travelling in either the positive or negative direction along the x-axis.

As may be seen from this example, the Riemann invariant r (or s) necessarily has a constant value r_0 (or s_0) in a region covered by the $C^{(+)}$ (or $C^{(-)}$) characteristics starting from a region of constant state. Therefore, in a region adjacent to a constant state, the solution is a simple wave. Thus the simple wave is an important class of wave which inevitably appears in a disturbance propagating into a constant state.

When all the right eigenvectors R_i are constant vectors in the general linear matrix hyperbolic equation (2.1.21) involving n elements in the vector U, we can define a simple wave in similar fashion as a travelling wave associated with a single mode whose (local) phase velocity is λ_i. To see this, let the solution U be a function of $\varphi(x, t)$ only, so that $U = U(\varphi)$, where φ is determined by the equation

$$\frac{\partial \varphi}{\partial t} + \lambda_i \frac{\partial \varphi}{\partial x} = 0. \tag{2.1.28}$$

Then Eq. (2.1.21) reduces to

$$\left(\frac{\partial \varphi}{\partial t} + A \frac{\partial \varphi}{\partial x}\right) \frac{dU}{d\varphi} = 0 \tag{2.1.29a}$$

and, using Eq. (2.1.28), we obtain

$$(A - \lambda_i I) \frac{dU}{d\varphi} = 0, \tag{2.1.29b}$$

where I is the unit matrix. Thus we have found the important relationship that, for the ith simple wave, $dU/d\varphi$ is proportional to the right eigenvector of A corresponding to λ_i, so that

$$\left(\frac{dU}{d\varphi}\right)_i \propto R_i. \tag{2.1.29c}$$

(We notice here that this is also valid for the case of a nonlinear equation, in which A depends on U, as will be shown in Section 2.2.) Since R_i is assumed to be a constant vector, we can choose any scalar function $f_i(\varphi)$ of φ as the ratio of $dU/d\varphi$ to R_i, and write $dU/d\varphi = f_i(\varphi)R_i$.

The integration with respect to φ can be easily carried out to give

$$U = \psi_i(\varphi) R_i + V_i, \tag{2.1.30a}$$

where $\psi_i = \int f_i(\varphi) \, d\varphi$, and V_i is a constant vector linearly independent of R_i. V_i may be expressed in terms of the normalized eigenvectors R_i as

$$V_i = \sum_{j=1}^{n}{}' \alpha_j R_j. \tag{2.1.30b}$$

Here, the coefficients α_j are arbitrary constants, and the prime in the summation notation signifies that the term with $j = i$ is omitted. In particular, if A is a constant matrix, from a comparison of Eq. (2.1.30) and the general solution Eq. (2.1.22), it is easily seen that if in the general solution all of the functions $\varphi_j(x - \lambda_j t)$ ($j = 1, 2, \ldots, n$) with the exception of $\varphi_i(x - \lambda_i t)$ take the constant values α_j, then the general solution reduces to a simple wave.

Multiplying the general solution Eq. (2.1.22) from the left by the normalized left eigenvector L_j, and using the orthogonality relationship $L_j R_i = \delta_{ji}$, we find that in this case, the Riemann invariants Eq. (2.1.5) become $L_j U = \varphi_j(x - \lambda_j t) = r_j(\xi_j)$. This shows that the Riemann invariant r_j

is constant along the family of characteristics $C^{(i)}$. Thus, we may say that the $n-1$ Riemann invariants r_j ($j \neq i$) have constant values α_j ($j \neq i$) in a simple wave of the ith mode. In this sense, a simple wave may be defined mathematically as a one-parameter family of solutions. That is, the ith such wave depends only on the single parameter ξ_i.

2.2 Hydrodynamics

A physical example of nonlinear hyperbolic equations is provided by the equations of hydrodynamics. As is well known, the fundamental equations of hydrodynamics consist of the conservation laws of mass, momentum, and energy:

$$\frac{\partial \rho}{\partial t} + \nabla \cdot (\rho \boldsymbol{u}) = 0, \tag{2.2.1a}$$

$$\frac{\partial (\rho \boldsymbol{u})}{\partial t} - \nabla : \boldsymbol{T} = 0, \tag{2.2.1b}$$

$$\frac{\partial W}{\partial t} + \nabla \cdot \boldsymbol{q} = 0. \tag{2.2.1c}$$

Here, ρ is the mass density (the mass per unit volume), $\boldsymbol{u}$ the flow velocity, and $\boldsymbol{T}$ the stress tensor which is expressed in the following form:

$$T_{ik} = -(p\delta_{ik} + \rho u_i u_k) + \Pi_{ik}, \tag{2.2.2a}$$

$$\Pi_{ik} = \zeta \left(\frac{\partial u_i}{\partial x_k} + \frac{\partial u_k}{\partial x_i} - \frac{2}{3} \delta_{ik} \nabla \cdot \boldsymbol{u} \right) + \zeta' \delta_{ik} \nabla \cdot \boldsymbol{u}. \tag{2.2.2b}$$

In the above equations, p is the pressure and the tensor $\boldsymbol{\Pi}$ represents the effect of viscosity, while ζ and ζ' are the coefficients of viscosity which generally depend on the density and temperature. W denotes the total energy density

$$W = \tfrac{1}{2} \rho u^2 + \rho e, \tag{2.2.2c}$$

where e is the internal energy per unit mass, which in a gas is usually proportional to the temperature, and q is the energy flux density given by

$$\boldsymbol{q} = \rho \boldsymbol{u} \left(\tfrac{1}{2} u^2 + e + \frac{p}{\rho} \right) - \boldsymbol{u} : \boldsymbol{\Pi} - \chi \nabla T. \tag{2.2.2d}$$

Here T and χ are the temperature and the thermal conductivity, respectively, and $\boldsymbol{u} : \boldsymbol{\Pi}$ is a vector whose ith component is given by $(\boldsymbol{u} : \boldsymbol{\Pi})_i = \sum_{k=1}^{3} u_k \Pi_{ki}$.

Let us compare these results with an adiabatic, reversible process in a perfect gas, so that all the dissipative effects can be neglected. The Eqns (2.2.1) reduce to

$$\frac{\partial \rho}{\partial t}+\nabla \cdot (\rho \boldsymbol{u})=0, \tag{2.2.3a}$$

$$\frac{\partial \boldsymbol{u}}{\partial t}+(\boldsymbol{u}\cdot\nabla)\boldsymbol{u}=-\frac{\nabla p}{\rho}, \tag{2.2.3b}$$

$$\frac{\partial S}{\partial t}+(\boldsymbol{u}\cdot\nabla)S=0, \tag{2.2.3c}$$

which may be called the dynamical equations of a perfect gas (cf. Section 2.4). Here S is the entropy per unit mass, and Eq. (2.2.2c) is the conservation law of entropy along each path followed by an infinitesimal fluid element as it moves in space. The pressure p is expressed by means of the adiabatic index γ (>1), as a function of ρ and S in the form

$$p=A(S)\rho^{\gamma}. \tag{2.2.3d}$$

The temperature T is determined as a function of ρ and S from the equation of state for a perfect gas $p=R\rho T$ (R is proportional to the gas constant).

One-dimensional unsteady gas dynamics

For simplicity we consider only one-dimensional fluid motion, and also suppose that the entropy S is everywhere constant. Eqns (2.2.3) then reduce to the following equations for ρ and u:

$$\frac{\partial \rho}{\partial t}+u\frac{\partial \rho}{\partial x}+\rho\frac{\partial u}{\partial x}=0, \tag{2.2.4a}$$

$$\frac{\partial u}{\partial t}+u\frac{\partial u}{\partial x}+\frac{a^2}{\rho}\frac{\partial \rho}{\partial x}=0. \tag{2.2.4b}$$

Here a is the velocity of sound, with $a^2=\partial p/\partial \rho$, and in the present case it takes the form

$$a^2=\gamma A(S)\rho^{\gamma-1}. \tag{2.2.5}$$

Let us solve the nonlinear equations (2.2.4) by means of the procedure we used in the case of a linear equation. To do so we write Eqns (2.2.4) in terms of the matrix vector

$$U=\begin{pmatrix}\rho\\u\end{pmatrix}. \tag{2.2.6}$$

We then have

$$\frac{\partial U}{\partial t} + A \frac{\partial U}{\partial x} = 0, \tag{2.2.7}$$

$$A = \begin{pmatrix} u & \rho \\ a^2/\rho & u \end{pmatrix}, \tag{2.2.8}$$

and the eigenvalues $\lambda_\pm$ of A are given by

$$\lambda_\pm = u \pm a. \tag{2.2.9}$$

The above equation shows that a sound wave is propagated both to the left and right with the sound speed a (in a fluid at rest), whereas in a fluid moving with speed u, the sound speed changes by an amount u because of the Doppler effect. For the eigenvalue $\lambda_+ = u + a$, the left eigenvector L_+ of the matrix A is (except for an arbitrary multiplicative constant)

$$L_+ = \left(\frac{a}{\rho}, 1\right). \tag{2.2.10a}$$

Similarly for the eigenvalue $\lambda_- = u - a$, the corresponding left eigenvector L_- is

$$L_- = \left(\frac{a}{\rho}, -1\right). \tag{2.2.10b}$$

It can be seen from Eqns (2.2.9) and (2.2.10) that the dynamical equations of a perfect gas (2.2.6) to (2.2.8) are hyperbolic. (We shall exclude the special case of a vacuum when $\rho = 0$.) Now we multiply Eq. (2.2.7) on the left by the eigenvector L, and then obtain the following equation which is now true in the nonlinear case as it was in the linear case (cf. Eq. (2.1.18)):

$$L\left(\frac{\partial}{\partial t} + \lambda \frac{\partial}{\partial x}\right) U = 0. \tag{2.2.11}$$

Though the differential operator in Eq. (2.2.11) corresponds to a directional derivative along a family of characteristics C corresponding to $dx/dt = \lambda$, in the present case the eigenvalue λ is a function of u and $a(\rho)$, and so is not constant. This is a crucial difference from the case of a linear equation. Nevertheless, Eq. (2.2.11) is formally expressed in the form of a differential along the family of curves $\varphi(x, t) = \xi$(constant) whose (local) slope is $dx/dt = \lambda$, and it may be written

$$L \frac{dU}{d\sigma} = 0$$

with

$$\frac{d}{d\sigma} \equiv \frac{\partial}{\partial t} + \lambda \frac{\partial}{\partial x}. \tag{2.2.12}$$

As in the linear case, we have

$$L \, dU = 0 \tag{2.2.13}$$

along $\varphi(x, t) = \xi$, so that

$$\int L \, dU = \text{constant}. \tag{2.2.14}$$

By using Eqns (2.2.5) and (2.2.10a), we may integrate Eq. (2.2.14) for a given eigenvalue, say $\lambda_+ = u + a$. Setting $m(\rho) = 2a/(\gamma - 1)$, the Riemann invariant is obtained in the form

$$m(\rho) + u = r(\xi). \tag{2.2.15a}$$

This shows that $m(\rho) + u$ remains constant along each of the $C^{(+)}$ characteristics $\varphi_+(x, t) = \xi$ expressed by integrating $dx/dt = u + a$. In the same way, for the eigenvalue $\lambda_- = u - a$ we find

$$m(\rho) - u = s(\eta) \tag{2.2.15b}$$

along each of the $C^{(-)}$ characteristics $\varphi_-(x, t) = \eta$, whose slope is $dx/dt = u - a$.

Results Eqns (2.2.15) correspond to the results Eqns (2.1.5) in the linear case, and the initial value problem is thus formally solved in the nonlinear case, provided the curves $\varphi_+(x, t) = \xi$ and $\varphi_-(x, t) = \eta$ are found. Thus we obtain the following expressions for ρ and u at the intersection of the two characteristics determined by specifying ξ and η, respectively,

$$m(\rho) = \tfrac{1}{2}(r(\xi) + s(\eta)), \tag{2.2.16a}$$

$$u = \tfrac{1}{2}(r(\xi) - s(\eta)). \tag{2.2.16b}$$

Since the eigenvalues $\lambda_\pm$ are functions of u and $a(\rho)$, the characteristics $\varphi_+(x, t) = \xi$ and $\varphi_-(x, t) = \eta$ cannot be found simply and, in general, the solution cannot be expressed in an explicit form.

Let us show, however, that the problem can be easily solved in the case of a solution involving a one-parameter family; that is, in the case of a simple wave, in which one of the Riemann invariants, r or s, is an absolute constant. Let us suppose, for example, that at $t = 0$ ρ and u are constant for $x > 0$ with $\rho = \rho_0$, $u = u_0 < a_0$ ($= a(\rho_0)$), and that these are functions of x for $x < 0$. In addition, we shall assume that $\rho(x)$ and $u(x)$ are smooth functions of their argument for all x. In such a case, it is evident that we have a constant state in the region of the (x, t)-plane determined by $x > (u_0 + a_0)t$; that is, for $r(\xi) = m(\rho_0) + u_0 = r_0$, $s(\eta) = m(\rho_0) - u_0 = s_0$.

In that region the $C^{(\pm)}$ characteristics become two families of straight lines $x - (u_0 + a_0)t = \xi$, $x - (u_0 - a_0)t = \eta$. Thus this constant state, I, corresponds to the region bounded by the $C^{(+)}$ characteristic $x - (u_0 + a_0)t = 0$, starting from the origin, and the positive x-axis. Now, since the Riemann invariant $s(\eta)$ remains constant along the $C^{(-)}$ characteristics passing through the constant state I, we have $s(\eta) \equiv s_0$ in the whole of the domain covered by these $C^{(-)}$ characteristics. That is, in a region S adjacent to the region I. Therefore, it follows that the region S corresponds to a simple wave region (cf. Fig. 2.3). Region S is bounded by the $C^{(\pm)}$ characteristics starting from the origin. In this case, the $C^{(+)}$ characteristic is a straight line, while the $C^{(-)}$ characteristic is a curve.

In the simple wave region S, Eqns (2.2.16) become

$$m(\rho) = \tfrac{1}{2}(r(\xi) + s_0), \tag{2.2.17a}$$

$$u = \tfrac{1}{2}(r(\xi) - s_0), \tag{2.2.17b}$$

and u and ρ are constant when ξ is constant (i.e. along the $C^{(+)}$ characteristics). Since the slope of the $C^{(+)}$ characteristic $\varphi_+(x, t) = \xi$ is $u + a(\rho)$, it is also constant on each $C^{(+)}$ characteristic. Thus the characteristic $\varphi_+(x, t) = \xi$ is a straight line whose slope is given by

$$\frac{dx}{dt} = -\frac{\varphi_{,t}}{\varphi_{,x}} = u(\xi) + a(\rho(\xi)) = \lambda_+(\xi), \tag{2.2.18}$$

where $\varphi_{,t} \equiv \partial\varphi/\partial t$, $\varphi_{,x} \equiv \partial\varphi/\partial x$. Here, and in what follows, we have used the usual notation of continuum mechanics, in which it is customary to indicate partial differentiation of the function ψ with respect to α by $\psi_{,\alpha}$. In purely mathematical discussions the comma is omitted and the same derivative is denoted by ψ_α.

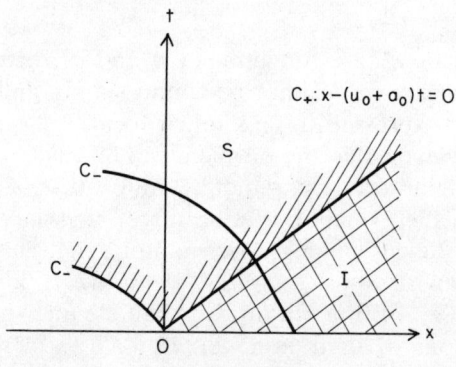

Fig. 2.3

In a simple wave region, therefore, the solution of Eq. (2.2.11) reduces to the problem of solving Eq. (1.1.16). In particular, let us consider $u(x, 0) = u_0 = 0$ as the initial condition, and seek the slope given by Eq. (2.2.18). The Riemann invariant $r(\xi)$ is determined from $\rho(x)$, $u(x) = 0$ at $t = 0$ as $r(\xi) = m(\rho(\xi))$. Since we are assuming $u_0 = 0$, the Riemann invariant s_0 is $m(\rho_0)$. Thus u and $m(\rho)$ may be found from Eqns (2.2.17) and, using $m(\rho) = 2a/(\gamma - 1)$, it follows that $C^{(+)}$ characteristics in the simple wave region are straight lines whose slope is given by

$$\left.\frac{dx}{dt}\right|_\xi = u + a = \frac{\gamma+1}{4} m(\rho(\xi)) + \frac{\gamma-3}{4} m(\rho_0). \tag{2.2.19}$$

If the initial value $\rho(x)$ is a monotonic decreasing function of x and φ is determined by the initial condition $\varphi(x, 0) = x$, then $m(\rho(\xi))$ decreases as ξ increases, and so the wave speed determined by Eq. (2.2.19) decreases as ξ increases. Consequently, the characteristics corresponding to smaller ξ overtake those for larger ξ, and in this case a shock is formed, as was shown in the previous chapter. The same form of reasoning shows that if $\rho(x)$ is a monotonic increasing function of x, then an expansion wave is formed (Coulson and Jeffrey (1977)).

The method of solution applied so far to the isentropic hydrodynamic equations can also be applied to a general system involving a two-element vector

$$U = \begin{pmatrix} u_1 \\ u_2 \end{pmatrix}, \tag{2.2.20}$$

provided the nonlinear equation

$$\frac{\partial U}{\partial t} + A(U) \frac{\partial U}{\partial x} = 0 \tag{2.2.21}$$

is hyperbolic. That is, provided the eigenvalues λ_1 and λ_2 of the matrix A are real and distinct so that $\lambda_1 \neq \lambda_2$.

Let us now show how this may be accomplished. Denote the left eigenvectors of the matrix A corresponding to the eigenvalues λ_1 and λ_2 by L_1 and L_2, respectively, and multiply the quasilinear matrix differential equation (2.2.21) by these from the left. We then find

$$L_i \left(\frac{\partial}{\partial t} + \lambda_i \frac{\partial}{\partial x} \right) U = 0 \quad (i = 1, 2). \tag{2.2.22}$$

Thus we have

$$L_i \, dU = 0 \quad (i = 1, 2) \tag{2.2.23}$$

along the characteristics $C^{(i)}$ whose slope is given by $dx/dt = \lambda_i$. So we arrive at the result

$$\frac{du_2}{du_1} = -\frac{l_{i1}(u_1, u_2)}{l_{i2}(u_1, u_2)}, \qquad (2.2.24)$$

where l_{i1} and l_{i2} are the components of L_i. By integrating this ordinary differential equation, we find the Riemann invariants

$$J_1(u_1, u_2) = r(\xi), \qquad (2.2.25a)$$

$$J_2(u_1, u_2) = s(\eta). \qquad (2.2.25b)$$

These results show that $J_1(u_1, u_2)$ and $J_2(u_1, u_2)$ remain constant along each member of the family of characteristics $\varphi_1(x, t) = \xi$ satisfying $\varphi_{1,t} + \lambda_1 \varphi_{1,x} = 0$, or along each member of the family of characteristics $\varphi_2(x, t) = \eta$ satisfying $\varphi_{2,t} + \lambda_2 \varphi_{2,x} = 0$ whose slopes correspond to λ_1 and λ_2, respectively.

As in the linear case, it is easily seen, moreover, that the solution represents a simple wave in the region covered by the characteristics, $C^{(1)}$ or $C^{(2)}$, issuing out from a constant state. Since one of the Riemann invariants, $r(\xi)$ or $s(\eta)$, is constant for a simple wave, the vector U is a function of η or ξ only. In other words, U is a function of φ_2 or φ_1 only.

Let us now suppose that U is a function of φ_i alone, and substitute it into Eq. (2.2.21). We then have $(\varphi_{i,t}I + A\varphi_{i,x}) dU/d\varphi_i = 0$, as in the linear case, where I is the unit matrix. By introducing the right eigenvector R_i of the matrix A, corresponding to the eigenvalue λ_i, we find

$$\frac{dU}{d\varphi_i} \propto R_i(U), \qquad \text{for } i = 1, 2. \qquad (2.2.26)$$

When $u_1(\varphi_i)$ is a monotonic function of φ_i (either monotonically increasing or decreasing with respect to φ_i), normalizing the vector R_i by choosing the first component of R_i to be unity, we obtain from the first component of Eq. (2.2.26)

$$\varphi_i = u_1, \qquad (2.2.27a)$$

and from the second component

$$\frac{du_2}{du_1} = r_2^{(i)}(u_1, u_2), \qquad (2.2.27b)$$

where $r_2^{(i)}$ is the second component of R_i. Therefore, by integrating Eq. (2.2.27b), we obtain the result $u_2 = u_2(u_1, C)$, where C is an integration constant. Thus, by means of Eqns (2.2.27a, b) the equation for the characteristics $\varphi_i = \text{const.}$, namely

$$\frac{\partial \varphi_i}{\partial t} + \lambda_i(u_1, u_2) \frac{\partial \varphi_i}{\partial x} = 0,$$

becomes the nonlinear equation for u_1

$$\frac{\partial u_1}{\partial t} + \lambda_i(u_1)\frac{\partial u_1}{\partial x} = 0. \tag{2.2.27c}$$

Therefore, for a simple wave, the matrix equation (2.2.21) can be reduced to a scalar equation of the type (1.1.13).

Example 2.2.1

Let us consider the second-order equation for the scalar ϕ

$$a\phi_{,tt} + 2b\phi_{,xt} + c\phi_{,xx} = 0. \tag{2.2.28}$$

If the coefficients a, b, and c are real functions of $\phi_{,x}$ and $\phi_{,t}$, this equation can be reduced to the system of the first-order equations (2.2.21) by introducing the variables u, v defined in Eq. (2.1.3) and the vector U of Eq. (2.1.8). Here

$$A = \begin{pmatrix} 2b/a & c/a \\ -1 & 0 \end{pmatrix}, \tag{2.2.29}$$

and the eigenvalues λ are the roots of the quadratic equation

$$a\lambda^2 - 2b\lambda + c = 0. \tag{2.2.30}$$

Equation (2.2.28) is called hyperbolic, elliptic, or parabolic, according as the discriminant $b^2 - ac$ is positive, negative, or zero. Noticing that c/a is equal to the product of the two roots, we can easily find

$$L_{1,2} = (1, \lambda_{2,1}), \tag{2.2.31}$$

where the indices 1 and 2 in L correspond to the indices 2 and 1 in λ, respectively. We therefore have

$$du + \lambda_{2,1}\, dv = 0 \tag{2.2.32}$$

along the characteristics $C^{(1,2)}$, respectively. From this we obtain

$$a(du)^2 + 2b\, du\, dv + c(dv)^2 = 0, \tag{2.2.33}$$

which corresponds to the characteristics $(dx/dt = \lambda)$

$$a(dx)^2 - 2b\, dx\, dt + c(dt)^2 = 0. \tag{2.2.34}$$

Problem 2.2.1 Vibration of a nonlinear elastic string

Denote the displacement from the equilibrium position of an infinitely long stretched string by y. Then the Lagrangian is $\mathscr{L} = (\tfrac{1}{2})y_{,t}^2 - (1+y_{,x}^2)^{1/2}$, and the equation of motion is $y_{,tt} - (1+y_{,x}^2)^{-3/2}y_{,xx} = 0$. Discuss the wave propagation which results when a disturbance occurs at the centre of the string (Taniuti (1959)).

Problem 2.2.2 Relativistic invariant fields

In a relativistically invariant scalar field, the Lagrangian $\mathscr{L}$ is a function of $Q \equiv (\tfrac{1}{2})(\phi_{,x}^2 - \phi_{,t}^2)$.

(a) Find the Riemann invariants of the Born–Infeld field $\mathscr{L} = (1-2Q)^{1/2}$. Prove also that the characteristics do not intersect (Taniuti (1957, 1958, 1959)).

Hint Show that λ_2 is constant along the characteristics $C^{(1)}$ (cf. Section 2.3).

(b) Show that for a simple wave propagating into a constant state in which $\phi = $ const., the characteristics do not usually intersect provided $\mathscr{L}'(Q)|_{Q=0} \neq 0$ (Taniuti (1959)).

Example 2.2.2 Shallow-water waves

We consider the propagation of a surface wave whose wavelength is large compared with the depth of water h_0, as shown in Fig. 2.4. In this case the motion of the water sustaining the wave occurs mainly in the horizontal direction, and the acceleration of the vertical water motion is negligibly small compared to the gravitational acceleration. This means that we have the result

$$0 \simeq \frac{d^2 y}{dt^2} = -g - \frac{1}{\rho}\frac{\partial p}{\partial y},$$

where g is the gravitational acceleration. Since we are considering an incompressible fluid, we have $\rho = $ const., and hence $p = g(\bar{y} - y)$, where $\bar{y}(x, t)$ denotes the height of the free-surface of the water. The boundary conditions are as follows. At the bottom of the water $y = 0$, the y-component of the velocity $\mathbf{v}(u, v)$ vanishes, so that $v = 0$. We also assume that all particles on the free-surface of the water remain there for all time,

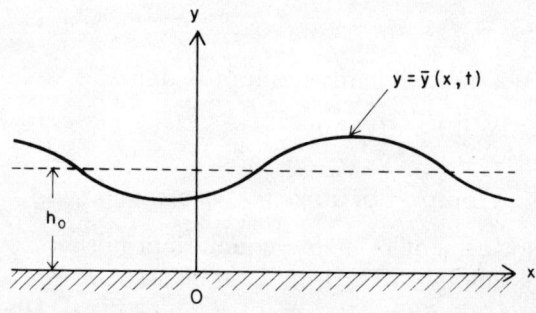

Fig. 2.4

so that

$$\left(\frac{\partial}{\partial t}+\boldsymbol{v}\cdot\nabla\right)(y-\bar{y})=0, \quad \text{or} \quad \frac{\partial \bar{y}}{\partial t}+u\frac{\partial \bar{y}}{\partial x}-v=0.$$

Since $\partial p/\partial x$ is independent of y, u does not depend on y. Consequently, the x-component of the equation of motion (2.2.3b) becomes

$$\frac{\partial u}{\partial t}+u\frac{\partial u}{\partial x}=-g\frac{\partial \bar{y}}{\partial x}. \qquad (2.2.35a)$$

From the incompressibility condition, we have $\nabla \cdot \boldsymbol{v}=0$. So, integrating this with respect to y, we find

$$[v]_0^{\bar{y}}=-\bar{y}\frac{\partial u}{\partial x},$$

and using the above boundary conditions, we finally obtain

$$\frac{\partial \bar{y}}{\partial t}+\frac{\partial \bar{y}u}{\partial x}=0. \qquad (2.2.35b)$$

The equations (2.2.35a, b) for $\bar{y}$ and u are equivalent to the dynamical equations of a perfect gas with $\gamma=2$. As the steepening of the wave occurs due to nonlinearity, the term of the next higher order in depth/wavelength becomes important and provides dispersion, thereby modifying the behaviour (cf. Problem 3.4.1).

2.3 Nonlinear hyperbolic equations—multi-component systems and characteristic curves

The results Eqns (2.2.26) and (2.2.27) derived for a simple wave in a quasilinear hyperbolic system with two components also hold for systems with vectors U having arbitrarily many components, provided only that the system is of the form Eq. (2.2.21). Hence the propagation of a simple wave (the ith simple wave) corresponding to a real eigenvalue λ_i ($i=1, 2, \ldots, n$) in a system with n components is described by equation (2.2.27c), and the ith family of characteristic curves becomes a family of straight lines. That is to say, let the quasilinear hyperbolic equation for the column vector U with n components $u_1, u_2, \ldots, u_n$ be given by

$$\frac{\partial U}{\partial t}+A(U)\frac{\partial U}{\partial x}=0, \qquad (2.2.21')$$

where A is a $n \times n$ matrix, the elements of which are functions of $u_1, u_2, \ldots, u_n$, the n eigenvalues of A denoted by $\lambda_1, \ldots, \lambda_n$ are all real and distinct, and the corresponding right eigenvectors $R_1, \ldots, R_n$ are all linearly independent.

Then we have

$$AR_i = \lambda_i R_i \qquad (i = 1, 2, \ldots, n),$$

and the ith family of characteristic curves is introduced through the equation

$$\frac{\partial \varphi_i}{\partial t} + \lambda_i \frac{\partial \varphi_i}{\partial x} = 0 \qquad (i = 1, 2, \ldots, n).$$

Let the ith simple wave be defined as a solution dependent on φ_i only, so that $U = U(\varphi_i)$. Then we know that we have the relationship

$$\frac{dU}{d\varphi_i} \propto R_i \qquad (i = 1, 2, \ldots, n). \tag{2.2.26'}$$

By the way of illustration, let us assume that R_i with elements $(r_1, r_2, \ldots, r_n)$ is normalized by requiring that $r_1 = 1$, so that

$$\varphi_i = u_1. \tag{2.2.27a'}$$

Then we have

$$\frac{dU}{du_1} = R_i \qquad (i = 1, 2, \ldots, n). \tag{2.2.27b'}$$

Since R_i is a given function of the elements $u_1, u_2, \ldots, u_n$ of U, this equation may be regarded as a system of ordinary differential equations for the $n-1$ dependent variables $u_2, \ldots, u_n$ in terms of the independent variable u_1. We may thus assume that a solution of the form $u_k = u_k(u_1)$ for $k = 2, \ldots, n$ will be obtained. Introducing this expression into the characteristic equation gives the scalar nonlinear equation for u_1,

$$\frac{\partial u_1}{\partial t} + \lambda(u_1) \frac{\partial u_1}{\partial x} = 0. \tag{2.2.27c'}$$

So, once u_1 has been determined from Eq. (2.2.27c') together with some suitable initial condition, all the elements $u_k = u_k(u_1)$ ($k = 2, \ldots, n$) of U are automatically determined.

This proof should be considered a purely formal one. For a mathematically rigorous account of simple waves we refer to Lax (1957), to the explanation and details provided by Jeffrey and Taniuti (1964) and to the account given by Jeffrey (1976).

We also see from Eq. (2.2.26) that the infinitesimal variation dU across each characteristic whose slope is λ_i is determined by

$$(-\lambda_i I + A) dU = 0. \tag{2.3.1}$$

2.3 NONLINEAR HYPERBOLIC EQUATIONS

Therefore, when we regard λ_i as a (local) phase velocity, it may be considered that the system is locally the same as a linear one unless the wave breaks. The (finite) variation of U results from an accumulation of infinitesimal ones. An important point which should be noticed, however, is that relations (2.2.25) showing the Riemann invariants are constant along their respective characteristics is no longer generally true. This is due to the fact that Eq. (2.2.23) is not generally integrable for a vector U of three or more components.

For example, for the Eqns (2.3.2) of a perfect gas, we have

$$U = \begin{pmatrix} \rho \\ u \\ S \end{pmatrix}, \quad A = \begin{pmatrix} u & \rho & 0 \\ p_{,\rho}/\rho & u & p_{,S}/\rho \\ 0 & 0 & u \end{pmatrix},$$

and the eigenvalues and the eigenvectors of A are, respectively,

$$\lambda_1 = u, \quad \lambda_2 = u + a, \quad \lambda_3 = u - a, \tag{2.3.2a}$$

$$R_1 = \begin{pmatrix} 1 \\ 0 \\ -p_{,\rho}/p_{,S} \end{pmatrix}, \quad R_2 = \begin{pmatrix} 1 \\ a/\rho \\ 0 \end{pmatrix}, \quad R_3 = \begin{pmatrix} 1 \\ -a/\rho \\ 0 \end{pmatrix}, \tag{2.3.2b}$$

$$L_1 = (0, 0, 1), \quad L_2 = (a/\rho, 1, p_{,S}/a\rho), \quad L_3 = (-a/\rho, 1, -p_{,S}/a\rho),$$

so that we have $\int L_{2,3} \, dU = \pm\int (a/\rho) \, d\rho + \int du \pm \int (p_{,S}/a\rho) \, dS$. Since p is not a function of S alone, these are not necessarily integrable (perfect differentials). As a consequence, Riemann invariants analogous to Eq. (2.2.25) do not exist. In this case we find only the relation $S = \text{const.}$ along the characteristics corresponding to $\lambda_1 = u$.

A simple wave is easily found, however. Since $dU = R_i \, d\rho$, for $i = 2$, say, we obtain

$$du - \frac{a}{\rho} d\rho = 0, \quad dS = 0, \tag{2.3.3a}$$

so that $u - m(\rho) = \text{const.}$ and $S = \text{const.}$ From this, Eq. (2.2.27c′) reduces to $\rho_{,t} + (\int (a/\rho) \, d\rho + a(\rho))\rho_{,x} = 0$, so that we obtain the results of the previous chapter. This corresponds to the fact that S becomes constant in the region covered by streamlines issuing out from a constant state, and in that region we found the simple wave solution of Eqns (2.2.4).

For $i = 1$, we also have

$$du = 0, \quad dS + \frac{p_{,\rho}}{p_{,S}} d\rho = 0, \tag{2.3.3b}$$

showing that $u = \text{const.}$ and $p = \text{const.}$, while Eq. (2.2.27c′) becomes $\rho_{,t} + u\rho_{,x} = 0$. The characteristics of this simple wave are parallel straight lines and, of course, they are streamlines along which the density and entropy are transported as invariant quantities. As a result, the wave steepening process does not occur in this case. In addition, the pressure is constant everywhere. A wave of this type is usually called an entropy wave and it is entirely different from a sound wave in its properties. It is evident from this example that the Riemann invariants of the type defined previously do not exist in general.

There do, however, exist Riemann invariants in a generalized sense. For the ith simple wave, let us consider a surface $J_i(u_1, \ldots, u_n) = \text{const.}$ in the n-dimensional U-space, when we have $dJ_i = \nabla_u J_i \cdot dU = 0$, where $\nabla_u \equiv (\partial/\partial u_1, \ldots, \partial/\partial u_n)$ denotes the gradient operator with respect to the dependent variables $u_1, u_2, \ldots, u_n$. Now, since dU is proportional to R_i for a simple wave, we get

$$\nabla_u J_i \cdot R_i = 0, \tag{2.3.4}$$

so that $\nabla_u J_i$ is orthogonal to R_i.

In this n-dimensional space there are $n-1$ linearly independent vectors which are orthogonal to R_i, and hence there exist $n-1$ independent relations $J_i^{(k)}(u_1, \ldots, u_n) = s_i^{(k)}$ ($k = 1, 2, \ldots, n-1$), where $s_i^{(k)}$ is a constant. By solving the above equations, we can express $u_2, \ldots, u_n$ in terms of u_1, and it is also seen from this that a simple wave corresponds to a solution of a one-parameter family. Naturally, if there are the n relations given by Eq. (2.3.4), the solution becomes a constant state. Following Lax (1957), we call $J_i^{(k)}$ the generalized Riemann invariants for the ith simple wave. Conversely, the solution becomes a simple wave in a region in which the $n-1$ generalized Riemann invariants remain constant.

Again taking the case of Eqns (2.2.3), it follows from Eq. (2.3.4) and the expressions for R_i that the generalized Riemann invariants are u and p for the entropy wave ($i = 1$) and $u \mp m(\rho)$ and S for the sound wave ($i = 2, 3$).

The theorem which asserts that the state adjacent to a constant state is a simple wave is also true for a multi-component system. However, before we proceed to the proof of this important theorem due to Friedrichs (1955), we show first that a constant state is bounded by characteristics. In the vicinity of a smooth boundary line C, let us introduce the curvilinear coordinates

$$\varphi(x, t) = \xi, \qquad \psi(x, t) = \eta, \tag{2.3.5a}$$

in which the boundary line C is embedded. We assume the necesssary smoothness for the functions $\varphi(x, t)$ and $\psi(x, t)$ together with a one-to-one correspondence with the Cartesian coordinates x and t.

Now choose $\varphi = \xi$ to be a family of curves parallel to C and $\psi = \eta$ to be a family of curves which is orthogonal to C. Moreover, we parameterize the new coordinates so that $\xi = 0$ corresponds to the curve C itself, while $\xi > 0$ corresponds to a region in which there is the state U_1, and $\xi < 0$ to a region in which there is a constant state U_0. Since we shall not consider solutions such as shocks, U is continuous across the curve C, and we have $U_1(C) = U_0$. However, there must necessarily exist a jump in some of the higher-order derivatives of U, because U_0 is constant. For simplicity, we shall assume that there is a finite jump in the first-order derivative of U, and that $0 < |\nabla U_1(C)| < \infty$. Since U_1 is constant on C, we have

$$\left.\frac{\partial^m U_1}{\partial \psi^m}\right|_{=0} = 0, \tag{2.3.5b}$$

where m is an arbitrary positive integer.

We mention here that a derivative with respect to ψ on $\varphi = 0$ is due to the variation of U along C, and for this reason we call it an interior derivative with respect to C. On the other hand, a derivative with respect to φ results from a variation in a direction normal to C, so that we call it an exterior derivative with respect to C.

Now let us rewrite derivatives with respect to x and t in terms of φ and ψ. Then, in the limit $\xi \to 0+$, the hyperbolic equation (2.2.21) with n components reduces to

$$(-\lambda I + A_0) U_{,\varphi}|_{\varphi = 0} = 0, \tag{2.3.6}$$

where λ is the slope of C and $A_0 \equiv A(U_0)$.

Since by assumption $U_{,\varphi}|_{\varphi=0} \neq 0$, we have $\det|-\lambda I + A_0| = 0$, showing that C is a characteristic curve which is, in fact, a straight line in this case (cf. Eq. (2.1.21)). At the same time, we also have $U_{,\varphi}|_{\varphi=0} \propto R|_{\varphi=0}$, and we see that the condition for a simple wave is satisfied on C. If C is not a characteristic, we find $U_{,\varphi}|_{\varphi=0} = 0$ from Eq. (2.3.6). Differentiating the Eq. (2.2.21) with respect to φ and repeating the same procedure, it follows that all the higher-order derivatives become zero on C, and so we can continuously extend the constant state U_0 beyond C to obtain the result that $U_1 = U_0$ in the vicinity of the C.

We will now prove that U_1 is a kth simple wave when C is a member of the kth family of characteristic curves ($\lambda = \lambda_k$). Multiplying Eq. (2.2.21′) by the ith eigenvector L_i from the left, we find

$$L_i U_{,i} = 0, \tag{2.3.7}$$

where $U_{,i} \equiv U_{,t} + \lambda_i U_{,x}$. Now we have the result $L_i R_k = 0$ ($i \neq k$) from the general property of the eigenvectors of a matrix A, and we also have the result Eq. (2.3.4). Hence, both L_i ($i \neq k$) and $\nabla_u J_k^{(j)}$ ($j = 1, 2, \ldots, n-1$) are vectors in the complementary space orthogonal to R_k, so that the vectors L_i ($i \neq k$) may be expressed as linear combinations of the $n-1$ vectors $\nabla_u J_k^{(j)}$. That is, for some constants b_{ij}, we have $L_i = \sum_{j=1}^{n-1} b_{ij} \nabla_u J_k^{(j)}$ ($i \neq k$)

with $\det|b_{ij}|\neq 0$. Inserting this into Eq. (2.3.7), we obtain

$$\sum_{j=1}^{n-1} b_{ij} J_{k,t}^{(j)} = 0 \qquad (i \neq k), \tag{2.3.8}$$

which are simultaneous partial differential equations for the $n-1$ dependent variables $J_k^{(j)}$.

The corresponding characteristic $\varphi' = \text{const.}$ is determined by the equation $\lambda = -\varphi'_{,t}/\varphi'_{,x}$, where λ satisfies $\det|b_{ij}(\lambda_i - \lambda)| = 0$. Thus it follows that Eq. (2.3.8) is a hyperbolic equation with $n-1$ distinct real characteristic speeds given by $\lambda = \lambda_i$ $(i \neq k)$. Now the $J_k^{(j)}$ are constant in the domain Ω_0 corresponding to U_0. If the $J_k^{(j)}$ are not constant in a domain Ω_1 adjacent to Ω_0 where $U = U_1$ is not constant, it turns out that the constant state of the hyperbolic equation (2.3.8) is bounded by a member of the kth family of characteristics of Eq. (2.2.21') for U. In addition, the characteristic speeds of Eq. (2.3.8) for the $J_k^{(j)}$ are λ_i $(i \neq k)$, and hence C is not a characteristic of Eq. (2.3.8). From this contradiction we conclude that the adjacent domain must be a domain of constant state for Eq. (2.3.8). So, in that domain, we have

$$J_k^{(j)} = s_j = \text{const.} \qquad (j = 1, 2, \ldots, n-1).$$

This shows that U_1 corresponds to a kth simple wave.

When the initial condition $U(x, 0) \equiv \Phi(x)$ at $t = 0$ has a constant value U_0 for $x \geq x_0$, it follows from this theorem that the leading edge of a wave (a wavefront) starting from x_0 is propagated into the constant state with one of the speeds $\{\lambda_i(U_0); i = 1, \ldots, n\}$ (in general, with the maximum value of $|\lambda_i|$) and there inevitably appears a simple wave behind the wavefront. Needless to say, in general, this is applicable only for a finite range of t, because after a finite time the solution can evolve into a discontinuous function such as a shock, even if $\Phi(x)$ is an arbitrarily smooth function of x.

Let us now add a few remarks about the analyticity required of $\Phi(x)$ for the above results to be valid. They are, in fact, true provided $\Phi(x)$ is smooth enough. However, if $\Phi(x)$ is discontinuous at x_0 we must consider a shock-like solution emanating from this point from the beginning. The speed of the wavefront is not then necessarily equal to the characteristic speed $\lambda_i(U_0)$, as can be inferred from the results of the previous chapter. We will return to this point again in the next chapter. Thus, in general, $\Phi(x)$ needs to be continuous, but this is not sufficient. According to the rigorous mathematical theory, it can be proved that $\Phi(x)$ must, in fact, be Lipschitz continuous. We remark here that a function $f(x)$ is said to be Lipschitz continuous in the interval $[a, b]$ if a positive constant M exists such that $|f(x_1) - f(x_2)| \leq M|x_1 - x_2|$ for all the points x_1, x_2 in the interval $[a, b]$. This definition of Lipschitz continuity implies that the first-order derivative of the function $f(x)$ may be discontinuous, but the magnitude of its jump <u>must</u> be finite. For example, if $f(x) = U_0$ for $x \geq x_0$ and $f(x) = U_0 + \sqrt{x_0 - x}$ for $x \leq x_0$, f is continuous but not Lipschitz continuous at $x = x_0$ since $(f_{,x})_{x \to x_0-} = \infty$.

The result that for a Lipschitz continuous solution the speed of the wavefront is equal to a characteristic speed is also true for the more general hyperbolic system

$$U_{,t} + A U_{,x} + B = 0. \tag{2.3.9}$$

Here U and B are column vectors with n components, A is a $(n \times n)$ matrix and, in general, A and B are functions of U, x, and t. We shall assume the necessary smoothness of A and B with respect to U, x, and t. When all the eigenvalues of A are real and the corresponding eigenvectors are linearly independent, the system (2.3.9) is said to be hyperbolic. In what follows we shall restrict ourselves to the case in which the eigenvalues of A are all distinct. This shows that the definition of a system (2.3.9) of hyperbolic type depends only on the properties of A above and does not involve B. Unlike system (2.2.21), however, the system (2.3.9) does not admit simple wave solutions because of the existence of the inhomogeneous term B. It can be proved, nevertheless, that the speed of a wavefront propagating into a constant state (constant in time) is still equal to a characteristic speed (one of the eigenvalues of A).

Assume that A and B do not explicitly depend on t, then for the constant state $U_0(x)$ we have

$$A_0 U_{0,x} + B_0 = 0, \tag{2.3.10}$$

where $A_0 \equiv A(U_0(x))$ and $B_0 \equiv B(U_0(x))$. When we denote the wavefront propagating into this state by $\varphi(x, t) = 0$, and introduce the curvilinear coordinates φ, ψ given by Eq. (2.3.5a) in the vicinity of the wavefront, system (2.3.9) reduces to

$$(\varphi_{,t} I + \varphi_{,x} A) U_{,\varphi} + (\psi_{,t} I + \psi_{,x} A) U_{,\psi} + B = 0. \tag{2.3.11}$$

Since U is continuous, we have $U(0-, \psi) = U(0+, \psi)$. We also consider that U varies smoothly along the wavefront, so that $U_{,\psi}$ is also continuous leading to $U_{,\psi}|_{\to 0-} = U_{,\psi}|_{\to 0+}$. Writing the above equation at two points located on either side of, but sufficiently close to, the wavefront and then subtracting one from the other, and then letting the two points approach a point on the wavefront, we find $(-\lambda I + A_0)[U_{,\varphi}] = 0$. Here $[U_{,\varphi}]$ is the jump of $U_{,\varphi}$ across the wavefront, and λ is the speed of the wavefront. Now U depends on t behind the wavefront, whereas it does not depend on t in the constant state. Hence, we may assume that there is a jump in the first-order derivative with respect to φ across the wavefront, so that $[U_{,\varphi}] \neq 0$. Thus λ must be equal to one of the eigenvalues of A_0, and we also have $(U_{,\varphi}) \propto R_0$.

We should notice that the speed of wavefront, that is the characteristic speed, is determined by the terms involving the derivatives (matrix A), and is independent of the vector B. For linear equations this corresponds to neither the phase nor the group velocity. This follows, because when we regard U as an infinitesimal variation about a constant state U_0, where $B(U_0) = 0$, and linearize system (2.3.9), both the phase and the group velocities are determined by a dispersion relation $\det|-i\omega I + ikA_0 + \nabla_u B_0| = 0$ for the wave number k and frequency ω. The phase and group velocities then necessarily depend on B, whereas the wavefront speed does not. For a Lipschitz continuous solution, we may consider that a linear approximation is valid in the neighbourhood of the wavefront, because $U(\delta\varphi, \psi) \simeq U(0, \psi) + U_{,\varphi}|_{\varphi=0} \delta\varphi$, and noting that $U_{,\varphi}|_{\varphi=0}$ is finite, we have $U(\delta\varphi, \psi) \simeq U(0, \psi) + \delta U$. Thus a characteristic speed may be regarded as the speed with which an infinitesimal jump is propagated.

We now mention some general results concerning the initial value problem for Eq. (2.3.9) (Lax (1954)). Suppose we specify initial data for $U(x, t)$ at $t = 0$, and seek the subsequent evolution of U according to Eq. (2.3.9). Then it is usually required on physical grounds that the solution of the initial value problem should depend continuously on the initial data. That is, in some suitable norm, when the initial data is changed by a small amount the solution also varies by a correspondingly small amount. This requirement will be discussed again in Chapter 3 in connection with the evolutionary condition we shall use to select a physical shock solution from amongst the mathematically possible weak solutions.

Let the initial data be denoted by $U(x, 0) = \Phi(x)$ $(-\infty < x < \infty)$, and all the eigenvalues of A, $\{\lambda_i(x, 0)\}$ be finite.

(i) Then for a finite time interval there exists a unique solution $U(x, t)$ satisfying Eq. (2.3.9) and depending continuously on the initial condition $\Phi(x)$, provided $\Phi(x)$ is a smooth function of x. At a certain time, however, the first-order derivative of $U(x, t)$ with respect to x can become infinite (as was seen for the case $B = 0$), and thereafter it is possible that the uniqueness of the solution is lost. Hence, in general, no classical solution exists for all time, even if $\Phi(x)$ is an analytic function.

(ii) If $\Phi(x)$ is Lipschitz continuous, we may approximate $\Phi(x)$ by a sequence of smooth functions $\{\Phi_n(x)\}$, and this enables us to define a Lipschitz continuous solution $U_l(x, t)$ corresponding to $\Phi(x)$ as the limit of a sequence of functions $\{U_n(x, t)\}$, each of which is a smooth solution for the corresponding $\Phi_n(x)$.

The Lipschitz continuous solution thus obtained exists uniquely for a finite time. If $d\Phi/dx$ at $t = 0$ is continuous almost everywhere, so that

2.3 NONLINEAR HYPERBOLIC EQUATIONS

there are finitely many jumps in $d\Phi/dx$ at some finite number of points $\{x_i\}$, these Lipschitz discontinuities are propagated along the characteristics radiating from each point x_i. That is to say, a discontinuity of the first-order derivative of the initial data is propagated with a characteristic speed. Thus the Lipschitz continuous solution U_l obtained as the limit of a sequence of smooth solutions depends continuously on the initial data, and there exists a smooth solution which is arbitrarily close to U_l. Consequently, we may consider Lipschitz continuous initial data in order to illustrate a smooth wavefront propagating into a stationary state.

This result is not necessarily true for initial data which is not Lipschitz continuous, however. For instance, a shock is not obtained as a limit of a sequence of smooth solutions. It is obvious that in the first-order differential equations considered so far we assume the existence of the first-order derivatives, and consequently the existence of a smooth solution. Therefore, when we consider solutions which are not smooth, the entire concept of a solution must be extended in some way. As was mentioned above in connection with system (2.3.9) of hyperbolic type, it is relatively easy to extend solutions to include Lipschitz continuous ones. However, when we try to extend solutions further so that they include discontinuous behaviour (shocks), we must use the concept of weak solutions discussed in Section 1.3, and then the uniqueness of solutions is lost.

Now suppose that $d\Phi/dx$ has finite jumps only at the points x_1, x_2 ($x_1 > x_2$) and for $x \geq x_1$ and $x \leq x_2$, Φ equals a function satisfying Eq. (2.3.10). Then the solution $U(x, t)$ will be time-dependent only in the region bounded by the extreme left and right characteristics issuing, respectively, from the points x_1, x_2 at $t = 0$. When we consider the case $x_2 \simeq x_1$, the region can be regarded as the triangular-shaped one bounded by the two extreme characteristics radiating out from the point x_1. In general, we call this region influenced by the initial value at a point x_1 the range of influence of x_1, and we denote it by $\mathcal{R}(x_1)$. Conversely, the interval $\mathcal{D}(P)$ cut out from the initial line $t = 0$ by tracing backwards the two extreme characteristics passing through a general point $P(x, t)$ is called the domain of dependence of P. The value of U at the point P depends only on the initial data contained within the domain of dependence of point P.

In the case of a two-component system, the domain of dependence and the range of influence are uniquely determined by the two distinct characteristic speeds. In a multi-component system, they are generally determined by the eigenvalues λ_1 and λ_n, where the eigenvalues have been ordered so that $\lambda_1 < \lambda_2 < \cdots < \lambda_n$ in the region of the U-space under consideration. Though the existence of the range of influence and the domain of dependence is intuitively evident from the fact that a disturbance is propagated with a finite (characteristic) speed (microscopic causality), an outline explanation may be given as follows.

We suppose that a smooth solution $U_P(x, t)$ exists in the region $\bar{\mathcal{D}}$ bounded by the two characteristics $C^{(1)}$, $C^{(n)}$ leaving the point P and the initial line $t = 0$. That is, the region supported by the domain of dependence $\mathcal{D}(P)$. We then try to extend this solution to the outside of $\bar{\mathcal{D}}$. In order to extend it outwards across $C^{(1)}$, with the equation $\varphi(x, t) = 0$ for example, let us again introduce curvilinear coordinates φ, ψ in the neighbourhood of $C^{(1)}$ and transform Eq. (2.3.9) into Eq. (2.3.11). Then, on $C^{(1)}$, the solution may be assumed to be a smooth function of ψ. Hence, if Eq. (2.3.11) is considered on curve $C^{(1)}$, it may be assumed that the second and third terms of this equation are known functions of ψ, and consequently Eq. (2.3.11) reduces to an algebraic equation for $U_{,\varphi}$. Since $C^{(1)}$ is the characteristic on which the determinant of the coefficients of $U_{,\varphi}$ vanishes, it should be noticed that the compatibility condition

$$L((\psi_{,t}I + \psi_{,x}A)U_{,\psi} + B) = 0 \tag{2.3.11'}$$

is satisfied, where L is the left eigenvector of A corresponding to $C^{(1)}$. In this case, $U_{,\varphi}$ is not determined uniquely, so that there are arbitrarily many ways of extending the solution U_P defined on $\bar{\mathcal{D}}$. It is the value in the region outside $\mathcal{D}$ that enables us to select a unique method of extension from amongst all the possibilities. In other words, U_P is not influenced by changing the initial conditions outside $\mathcal{D}$. If $C^{(1)}$ is not a characteristic, $U_{,\varphi}$ is uniquely determined on $C^{(1)}$ and the solution can be extended outside $\bar{\mathcal{D}}$ in a unique fashion. Consequently the initial conditions outside $\mathcal{D}(P)$ must be given in connection with U_P, that is, in this case, U_P and the initial conditions outside $\mathcal{D}(P)$ are interdependent. It follows from the above that U at the point P depends only upon the initial conditions on $\mathcal{D}(P)$ provided $C^{(1)}$ and $C^{(n)}$ are characteristics.

The above discussion on the existence of a domain of dependence applies directly to the case when initial data is specified on a general initial curve. Let this be the curve C with equation $\varphi(x, t) = 0$ and suppose that initial data $U(C)$ is specified on C. Introducing the curvilinear coordinates φ, ψ in the neighbourhood of C, we can express $U(C)$ as a function $\Phi(\psi)$ of ψ. If Φ is smooth on the C, $U_{,\varphi}$ is determined uniquely as a function of ψ provided C is nowhere tangent to a characteristic (for all $\varphi_{,t}/\varphi_{,x} \neq -\lambda_i(\psi)(i = 1, 2, \ldots, n)$). Thus the solution $U(\varphi + \delta\varphi, \psi)$ for $\varphi + \delta\varphi$ may be found uniquely. It can be seen from the above results that we may continuously transform the initial straight line $t = 0$ into a curve C which is nowhere tangent to a characteristic as long as the characteristic speeds remain finite. In particular, if C is a characteristic, U cannot be specified arbitrarily on C, but must be given so that it satisfies the condition (2.3.11'). In this case the solution is not determined uniquely.

In general, in order to obtain uniqueness of the solution we need knowledge of the solution on a curve different from C. In particular, in a two-component system, we can seek a solution satisfying initial conditions specified on two characteristics $C^{(+)}$ and $C^{(-)}$ (the so-called characteristic initial value problem). In Eqns (2.1.9) and (2.1.10), for example, let us specify u_1 and u_2 by setting $u_1 = u_2 = f(\eta)$ and $u_1 = -u_2 = g(\xi)$ on the characteristics $x - t = 0$ and $x + t = 0$, respectively. Then the solution in the region defined by $x - t \leq 0$ and $x + t \geq 0$ is given uniquely by $u_1 = f(\eta) + g(\xi)$, $u_2 = f(\eta) - g(\xi)$.

When the second-order differential equation (2.1.1) is expressed in the form (2.1.9) as a pair of first-order simultaneous equations, the Lipschitz continuity of u_1 and u_2 requires the continuity of the first-order derivatives of ϕ. That is, ϕ is required to be smooth and a discontinuity is only permitted in a second or higher-order derivative of ϕ. The characteristics of the second-order equation for ϕ should be defined as those curves across which there may occur a jump in the second or higher-order derivative of ϕ. Thus, letting a, b, c and f be functions of x, t, ϕ, $\phi_{,x}$ and $\phi_{,t}$, the characteristics for the equation

$$a\phi_{,tt} + 2b\phi_{,xt} + c\phi_{,xx} + f = 0 \tag{2.3.12}$$

are determined, independently of f, by Eq. (2.2.30), with $\lambda \equiv -\varphi_{,t}/\varphi_{,x}$, and the equation is then classified as being of hyperbolic, elliptic, or parabolic type in the same way as Eq. (2.2.28). We see from this that both the diffusion equation and Burgers' equation are of parabolic type, and the characteristics become $\varphi_{,x} = 0$, leading to $t = \text{const}$.

We may, if we wish, discuss Eq. (2.3.12) in the form of a pair of first-order simultaneous equations. However, in general, the transformation of a second-order equation into a first-order system is not unique, and we need to take note of the fact that the non-uniqueness can give rise to redundant eigenvalues (characteristic speeds). For instance, in the Klein–Gordon type of equation

$$\phi_{,tt} - c^2 \phi_{,xx} + m^2 \phi = 0, \tag{2.3.13}$$

let u_1, u_2 and u_3 be defined as $u_1 \equiv \phi_{,t}$, $u_2 \equiv \phi_{,x}$, and $u_3 = \phi$. Then, in place of Eq. (2.3.13), we have the system

$$u_{1,t} - c^2 u_{2,x} + m^2 u_3 = 0, \tag{2.3.14a}$$
$$u_{2,t} - u_{1,x} \quad\quad = 0, \tag{2.3.14b}$$
$$u_{3,t} - u_1 \quad\quad = 0, \tag{2.3.14c}$$
$$u_{3,x} - u_2 \quad\quad = 0. \tag{2.3.14d}$$

Now, since the last equation (2.3.14d) does not contain a derivative with respect to t, we should regard it as a subsidiary condition. In fact, if this condition is satisfied initially by the initial condition, then by means of Eqns (2.3.14b, c) it can easily be proved that this remains true for all time t. Now, for Eqns (2.3.14a to c), the eigenvalues λ of the matrix A become $\pm c, 0$. However, the eigenvalue 0 is incompatible with the subsidiary condition Eq. (2.3.14d). This is so because, from Eq. (2.3.14d), u_2 becomes discontinuous if u_3 is Lipschitz continuous. (This is obvious because u_3 is merely ϕ itself.) Therefore, the case that $u_{3,\varphi}$ is discontinuous must be excluded on account of the subsidiary condition. As a result, the characteristics, $\varphi = \text{const.}$, must be defined as lines across which $u_{1,\varphi}$, $u_{2,\varphi}$, and $u_{3,\varphi\varphi}$ may be discontinuous. The case that $u_{3,\varphi\varphi}$ is discontinuous and $u_{2,\varphi}$ is continuous (consequently, $u_{1,\varphi}$ is also continuous) must also be excluded because of the subsidiary condition. From this, we see that the eigenvalue $\lambda = 0$ must be excluded. Incidentally, for the Klein–Gordon equation (2.3.13), the relation, $|\omega/k| > \text{characteristic speed} = |c| > |d\omega/dk|$, is always true provided m is real.

In a dispersive system, the group velocity generally corresponds to the velocity with which the centre of a wave packet moves, so that the above relationship indicates that the centre of the wave packet moves slower than the wavefront itself.

It will be obvious from the above results that, in general, the characteristic curves depend only on the coefficients of the highest-order derivatives of the equation. In this sense, we call the terms comprising the highest-order derivatives the principal part of the partial differential equation. If the principal part is linear but the remaining terms are nonlinear (for example, a, b, and c in Eq. (2.3.12) are independent of ϕ, $\phi_{,x}$ and $\phi_{,t}$, or A in Eq. (2.3.9) does not depend on U), we say the resulting nonlinear equation is semilinear. In this case the characteristics are determined *a priori* and independently of the dependent variables (the field variables), so that the characteristics do not then intersect because of the overtaking process caused by the nonlinearity of an equation. As already remarked, we call nonlinear equations such as Eq. (2.3.9) and Eq. (2.3.12), which are linear with respect to their highest order derivatives, quasilinear equations. Even for a semilinear equation, however, the solution of an initial-value problem does not necessarily exist for all time. For instance, in the equation $u_{,t} + u_{,x} - u^2 = 0$, u can become infinite after a finite time interval along the characteristics $x - t = \text{const.}$ (the so-called blow-up type of instability).

We should draw attention to the fact that there are exceptional cases in which the characteristics are parallel and do not intersect due to the overtaking process, even when the principal part is nonlinear, so that the equation is not semilinear. For example, for the simple wave (the entropy wave) corresponding to the eigenvalue u of Eqns (2.3.2) we have $u = \text{const.}$, and the steepening of the wave due to overtaking does not occur. In the Eq. (2.2.27c′) for a simple wave, in general, the condition under which λ_i is constant, and hence does not depend on u_1 becomes, by means of the result $d\lambda_i = \sum_{k=1}^{n} (\partial \lambda_i / \partial u_k) du_k = \nabla_u \lambda_i \cdot R_i$,

$$\nabla_u \lambda_i \cdot R_i = 0. \tag{2.3.15}$$

Therefore, even if λ_i and R_i depend on U, the characteristics of the ith simple wave become parallel provided $\nabla_u \lambda_i$ and R_i are orthogonal to each other. Eq. (2.3.15) may be interpreted geometrically. Differentiate the characteristic equation $x_{,t} = \lambda_i$ with respect to the characteristic coordinate $\varphi = \xi$ to obtain

$$x_{,\varphi t} = \nabla_u \lambda_i \cdot U_{,\varphi}, \tag{2.3.16}$$

and notice that $x_{,\varphi}$ corresponds to the interval between two adjacent characteristics. (This relationship enables us to extend the notion of the exceptional condition obtained for simple waves to the more general hyperbolic system of equations (2.3.9) which does not admit simple wave solutions.) (Taniuti (1959); Jeffrey and Taniuti (1964); Boillat (1965).)

When Eq. (2.3.15) is satisfied for a certain λ_i, the system (2.3.9) is said to be (nonlinearly) exceptional with respect to the ith characteristic field, and when Eq. (2.3.15) is satisfied for all λ_i, the system is said to be completely exceptional. Also, when Eq. (2.3.15) is not satisfied for any λ_i then, following Lax (1954), we say that Eq. (2.3.9) is genuinely nonlinear.

In physics, we often encounter exceptional cases. As well-known examples, we can refer to the entropy wave mentioned above and to the Alfvén wave (Problem 2.3.1) in magnetohydrodynamics (Jeffrey and Taniuti (1964)). In Eq. (2.2.3), the expression (2.3.2) for R_1 corresponding to $\lambda_1 = u$ shows immediately that condition (2.3.15) is satisfied. As an example of the completely exceptional case, we may take a system of two components in which condition (2.3.15) is satisfied for each of the eigenvalues λ_1 and λ_2. We then have $\nabla_u \lambda_1 \propto L_2$ and $\nabla_u \lambda_2 \propto L_1$, from which we find $\nabla_u \lambda_2 \cdot dU = 0$ and $\nabla_u \lambda_1 \cdot dU = 0$. That is, $\lambda_2 = $ const. and $\lambda_1 = $ const., along the families of characteristics $C^{(1)}$ and $C^{(2)}$, respectively, so that the eigenvalues themselves then become the Riemann invariants of the system. Accordingly, $C^{(1)}$ and $C^{(2)}$ become two families of parallel curves (which are not generally straight lines). Hence the simple wave propagates without distorting its form and moreover, simple-wave pulses persist through mutual collisions. Therefore, in the completely exceptional case, the behaviour of a nonlinear field is closely similar to those of the linear field (Taniuti (1959)). One of the simplest examples of this type is a scalar field of the Born–Infeld type given in Problem 2.2.2. In addition the nonlinear electromagnetic field proposed by Born and Infeld has the above property (cf. Section 2.5).

Problem 2.3.1 Magnetohydrodynamics

Magnetohydrodynamics is a system of equations dealing with the macroscopic interaction between the motion of a conducting fluid and an electromagnetic field. It is obtained by combining the Maxwell equations (without the displacement current) and the hydrodynamic equations. The basic equations may be written in the following conservation form:

$$\frac{\partial \rho}{\partial t} + \nabla \cdot (\rho \boldsymbol{u}) = 0, \qquad (2.3.17a)$$

$$\frac{\partial (\rho \boldsymbol{u})}{\partial t} - \nabla : \boldsymbol{T} = 0, \qquad (2.3.17b)$$

$$\frac{\partial W}{\partial t} + \nabla \cdot \boldsymbol{q} = 0, \qquad (2.3.17c)$$

$$\frac{\partial \boldsymbol{H}}{\partial t} + \nabla : \boldsymbol{S} = 0, \qquad (2.3.17d)$$

$$\nabla \cdot \boldsymbol{H} = 0, \qquad (2.3.17e)$$

where $\boldsymbol{H}$ is the strength of the magnetic field and, neglecting the effects of dissipation, we have

$$T_{ik} = \frac{\mu}{4\pi}\left(H_i H_k - \frac{H^2}{2}\delta_{ik}\right) - (p\delta_{ik} + \rho u_i u_k),$$

$$W = \frac{\rho u^2}{2} + \rho e + \frac{\mu}{8\pi} H^2,$$

$$\boldsymbol{q} = \rho \boldsymbol{u}\left(\frac{u^2}{2} + e + \frac{p}{\rho}\right) + \frac{\mu}{4\pi}\boldsymbol{H}\times[\boldsymbol{u}\times\boldsymbol{H}],$$

$$S_{ik} = H_i u_k - H_k u_i.$$

For an adiabatic gas law these equations reduce to

$$\frac{\partial \rho}{\partial t} + \nabla \cdot (\rho \boldsymbol{u}) = 0, \qquad (2.3.18a)$$

$$\frac{\partial \boldsymbol{u}}{\partial t} + (\boldsymbol{u}\cdot\nabla)\boldsymbol{u} = \frac{\mu}{4\pi\rho}[\nabla\times\boldsymbol{H}]\times\boldsymbol{H} - \nabla p/\rho, \qquad (2.3.18b)$$

$$\frac{\partial S}{\partial t} + \boldsymbol{u}\cdot\nabla S = 0, \qquad (2.3.18c)$$

$$\frac{\partial \boldsymbol{H}}{\partial t} = \nabla\times[\boldsymbol{u}\times\boldsymbol{H}], \qquad (2.3.18d)$$

$$\nabla \cdot \boldsymbol{H} = 0. \qquad (2.3.18e)$$

2.4 SHOCK WAVES IN GAS DYNAMICS

For waves propagating in the x-direction, the characteristic roots λ of Eqns (2.3.18) are real and take the forms $\lambda = u_x \pm c_f$, $\lambda = u_x \pm c_s$, $\lambda = u_x \pm b_x$ and $\lambda = u_x$. The corresponding waves are called the fast magnetoacoustic wave, the slow magnetoacoustic wave, the Alfvén wave, and the entropy wave, respectively. Show the explicit forms of c_f, c_s and b_x, prove $c_f \geq b_x \geq c_s$, and obtain the simple waves. Show also that the Alfvén wave is a transverse wave and that it corresponds to the exceptional case (Jeffrey and Taniuti (1964)). $b_x = (\mu H_x^2/4\pi\rho)^{1/2}$.

Problem 2.3.2

Show that if $a \propto 1/\rho$ the system of equations (2.2.4) is completely exceptional.

2.4 Shock waves in gas dynamics

In the dynamical equations (2.2.4) of a perfect gas, we know that the wave profile of a sound wave is distorted by the nonlinearity and that an initially smooth solution evolves into a discontinuous one. From the properties of the Burgers equation exhibited in the first chapter it is readily deduced that as the distortion of the wave profile steepens, so terms responsible for irreversibility, such as the viscosity, become effective and a shock (shock wave) with a smooth but rapid transition is formed. In fact, rewriting the momentum equation (2.2.1b) of hydrodynamics in the one-dimensional case, we find

$$u_{,t} + uu_{,x} + \frac{p_{,x}}{\rho} - \frac{\mathcal{R}^{-1} u_{,xx}}{\rho} = 0, \tag{2.4.1}$$

which is similar to the Burgers equation. Here, all quantities are dimensionless, being normalized in terms of a typical length L, speed V and density ρ_0, while $\mathcal{R}$ is the Reynolds number $\mathcal{R} = LV/\mu$, where $\mu = (4/3)\zeta/\rho_0$, and for simplicity we assume $\zeta = $ const. and $\zeta' = 0$. The Reynolds number $\mathcal{R}$ may be regarded as a measure of the ratio of the inertial term to the viscous term (or, the nonlinear term to the dissipation term). Thus, when the viscosity coefficient is small as in a gas, $\mathcal{R}$ is usually large. However, it also depends on the scale of the characteristic length and speed, which may be taken as a characteristic wavelength and the speed of a sound wave. That is, $\mathcal{R}$ is extremely large for a sound wave with a very long wavelength (or for very large-scale phenomena such as occur in astrophysics). Then, in such cases, Eqns (2.2.4) represent a good approximation.

When waves of short wavelength are excited, however, L becomes small and $\mathcal{R}$ is not large, even if μ is small. That is, when we consider the structure of a shock wave, we must deal with a solution which varies rapidly in a narrow region, so that the approximation $\mathcal{R} \to \infty$ is not applicable and the viscosity term plays an important role.

In the neighbourhood of a shock, the entropy S increases owing to the viscosity, and hence p can no longer be expressed as a function of ρ and S by means of the adiabatic law Eq. (2.2.3d). Thus we must make use of the conservation law of energy Eq. (2.2.1c) instead of that for entropy Eq. (2.2.3c), so that we must solve Eqns (2.2.1a to c). However, we can apply the same procedure we used for Burgers' equation in Section 1.3. That is, we disregard the dissipation terms (the viscosity and the thermal conductivity terms) in the conservation laws of momentum and energy Eqns (2.2.1b) and (2.2.1c).

Instead, we regard a shock as a discontinuous function obtained by means of a limit as the transport coefficients tend to zero, and seek to determine the relationships which exist amongst the jumps in the physical quantities across the shock. Hence, we seek the shock conditions which are, of course, the consequence of the conservation laws of mass, momentum, and energy, and correspond to the condition (2) in Section 1.3. The shock solution obtained in this way is a (weak) solution of Eqns (2.2.4), extended so as to admit discontinuous functions. Consequently it is not unique, and we need (evolutionary) conditions (which correspond to the condition (1) in Section 1.3) which enable us to select a physically significant solution from amongst the mathematically possible solutions. As will be seen later, in gas dynamics this condition is equivalent to the second law of thermodynamics which requires that entropy should not decrease.

Corresponding to Eq. (1.3.8), let us first write Eqns (2.2.1) in the conservation form (by neglecting the dissipation terms)

$$V_{,t} + F_{,x} = 0. \tag{2.4.2a}$$

Since we are assuming one-dimensional motion, V and F are given by

$$V = \begin{pmatrix} \rho \\ \rho u \\ \rho u^2/2 + \rho e \end{pmatrix}, \quad F = \begin{pmatrix} \rho u \\ \rho u^2 + p \\ u(\rho u^2/2 + \rho e + p) \end{pmatrix}, \tag{2.4.2b}$$

$$e = C_v T. \tag{2.4.2c}$$

Here C_v is the specific heat at constant volume. For simplicity we shall assume it to be constant, though it generally depends on the temperature T. Using the equation of state $p = R\rho T$, Eqns (2.4.2) are to be regarded as equations for ρ, u, and p (or T), and the first and the second components

2.4 SHOCK WAVES IN GAS DYNAMICS

provide the dynamical equations for a perfect gas, Eqns (2.2.3a, b). Also, with the aid of those equations, the third component reduces to

$$\frac{de}{dt} + p\frac{d(1/\rho)}{dt} = 0, \qquad \frac{d}{dt} \equiv \frac{\partial}{\partial t} + u\frac{\partial}{\partial x}, \qquad (2.4.3a)$$

which turns out to be equivalent to the conservation law of entropy Eq. (2.2.3c) because of the thermodynamic relation

$$T\,dS = de + p\,d\left(\frac{1}{\rho}\right). \qquad (2.4.3b)$$

Furthermore, integrating Eq. (2.4.3b) by means of the results $\gamma = C_p/C_v$, $R = C_p - C_v$ (where C_p is the specific heat at constant pressure which is assumed constant), and Eq. (2.4.2c), we find the adiabatic law Eq. (2.3.3d). Thus, if we restrict ourselves to smooth solutions in the sense discussed in the previous chapter, and Lipschitz continuous solutions defined as the limit of them, Eqns (2.4.2) and (2.2.3) are equivalent. They are, however, not equivalent for weak solutions. At first sight this result seems somewhat strange, but it can be easily understood by considering the following example:

$$(\tfrac{1}{2}u^2)_{,t} + (\tfrac{1}{3}u^3)_{,x} = 0.$$

The equation is the same as Eq. (1.3.8) for smooth solutions, but it gives the entirely different shock condition $-\tilde{\lambda}[u^2] + \tfrac{2}{3}[u^3] = 0$ from that given in Eq. (1.3.6') for weak solutions. Accordingly, the important point to be recognized is that we must formulate physical laws which establish what sort of quantities must be conserved.

If we now apply the rules of Eqns (1.3.9) and (1.3.10) to Eq. (2.4.2), then corresponding to Eq. (1.3.6') we obtain the shock condition

$$\tilde{\lambda}[V] = [F]. \qquad (2.4.4)$$

We have denoted the shock speed by $\tilde{\lambda}$ in order to distinguish it from a characteristic speed (eigenvalue) λ. Rewriting Eq. (2.4.4) in component form, we obtain, for the dynamical quantities ρ and u,

$$-\tilde{\lambda}[\rho] + [\rho u] = 0, \qquad (2.4.5a)$$
$$-\tilde{\lambda}[\rho u] + [\rho u^2 + p] = 0, \qquad (2.4.5b)$$

respectively. If we let [ ] correspond to an infinitesimal jump δ, and consider the case that p is expressed in terms of the adiabatic law Eq. (2.2.3d), then Eqns (2.4.5) are merely Eqns (2.3.1) to (2.3.3a) for an infinitesimal jump in a simple wave. In this case, of course, we have $\tilde{\lambda} = \lambda$. The shock conditions given by Eqns (2.4.5) are called the mechanical conditions, to distinguish them from the thermodynamic conditions which will be mentioned below.

Let us now introduce a coordinate system moving with the shock speed $\tilde{\lambda}$, and denote the flow velocity in this system by $\tilde{u} = u - \tilde{\lambda}$. Then the conservation law of mass Eq. (2.4.5a) reduces to

$$\rho\tilde{u} = \tilde{m}(=\text{const.}). \tag{2.4.6}$$

If we now set $\tau = 1/\rho$, we obtain the identity

$$\tilde{m}[\tau] - [u] = 0. \tag{2.4.7a}$$

Using $\tilde{m}$, Eq. (2.4.5b) becomes

$$\tilde{m}[u] + [p] = 0. \tag{2.4.7b}$$

As mentioned previously, the results of Eqns (2.4.7) are also valid for the infinitesimal variations in a simple wave. For instance, supposing $\tilde{m} \neq 0$, and replacing $\tilde{u}$ by $u - \lambda = \mp a$, Eq. (2.4.7a) becomes the result for the jump across characteristics, $\mp (a/\rho)\,\delta\rho + \delta u = 0$. A finite change across a simple wave occurs as the result of an accumulation of these jumps and leads to Eq. (2.2.23).

Also, eliminating $[u]$ from Eq. (2.4.7), we get

$$\tilde{m}^2 = -\frac{[p]}{[\tau]}. \tag{2.4.8}$$

We remark that for an infinitesimal variation, Eq. (2.4.8) reduces to

$$(u - \tilde{\lambda})^2 = \frac{\delta p}{\delta\rho}, \tag{2.4.9}$$

which is simply the definition of the sound speed. Moreover, if $\tilde{m} = 0$, we find from Eqns (2.4.6) and (2.4.7) that

$$[u] = [p] = 0, \qquad \tilde{\lambda} = u. \tag{2.4.10}$$

These results are identical to the results given in Eq. (2.3.3b) for an entropy wave. The fact that for the entropy wave both finite and infinitesimal jumps satisfy the same jump relations can be regarded as a special feature of the exceptional case defined in Eq. (2.3.15), in which the steepening of waves does not occur. As may be clearly seen from Eq. (2.4.10), there are no jumps in either u or p across the discontinuity surface. However, there is a jump in ρ, so that there is also a jump in the temperature $T\,(= p/\rho R)$, and the entropy S. In the case $\tilde{m} = 0$, there is no mass flow, so that infinitesimal elements of fluid do not cross the discontinuity surface. The two states with different densities and temperatures are then merely in contact with each other. We call this a contact discontinuity, and the dividing surface itself a contact surface. As time passes, this surface will, of course, disappear on account of thermal conduction.

Now let us return to the problem of a shock, and seek the thermodynamic shock condition from the conservation law for energy; that is, from the third component of Eq. (2.4.4). In general, we have the relationship $[XY] = \langle X \rangle [Y] + [X] \langle Y \rangle$, where when we denote the values of X on both sides of the shock by X_0 and X_1, respectively, $\langle X \rangle$ represents the mean value $(X_0 + X_1)/2$. So, with the aid of Eqns (2.4.6) and (2.4.7), the shock condition becomes

$$[e] + \langle p \rangle [\tau] = 0. \tag{2.4.11}$$

When we reduce [ ] by an infinitesimal jump δ, this shock condition reduces to $dS = 0$ because of the thermodynamic condition (2.4.3b). The adiabatic law Eq. (2.2.3a) then follows directly. Expressing the internal energy e in terms of p and τ, Eq. (2.4.11) takes the form

$$\frac{[p\tau]}{\gamma - 1} + \langle p \rangle [\tau] = 0. \tag{2.4.12}$$

Thus the shock conditions are given by Eqns (2.4.7a, b) and (2.4.12), together with the condition $\tilde{m} \neq 0$.

To pursue the analysis further, it is convenient to define the front and the rear sides of a shock as follows. For an observer moving with a shock, that is, in a frame in which the shock is at rest (the wave frame), we call the side through which the fluid enters the front of the shock, while the side through which the fluid leaves is called the back of the shock. We will distinguish physical quantities on the front and the back of the shock by the subscripts 0 and 1, respectively (hereafter we will define $[X] = X_1 - X_0$).

From the shock condition Eq. (2.4.12), the relationship between p and τ at the front and the back of the shock is found to be

$$(\tau_1 - \nu^2 \tau_0)p_1 - (\tau_0 - \nu^2 \tau_1)p_0 = 0, \tag{2.4.13}$$

where $\nu^2 = (\gamma - 1)/(\gamma + 1)$. We call this relation the Rankine–Hugoniot (R–H) relation. For instance, when we specify τ_0 and p_0 on the front of a shock, Eq. (2.4.13) provides a relationship between τ_1 and p_1 (cf. Fig. 2.5). As $p_1 \to \infty$, $\tau_{\min}$ becomes $\nu^2 \tau_0$, and it follows that a fluid cannot be compressed by means of a shock to a density above the value $(\nu^2 \tau_0)^{-1}$.

Using the shock conditions mentioned above, it follows that a shock is uniquely determined when ρ_0, p_0 and u_0 are specified on the front of a shock and one physical quantity is specified on the back. For example, when we specify ρ_1, the remaining physical quantities p_1 and u_1 on the back are determined together with the shock speed $\tilde{\lambda}$. That is, p_1 is found from the Rankine–Hugoniot relation, and u_1 and $\tilde{\lambda}$ are obtained from Eqns (2.4.7) and (2.4.8). If u_1 is specified, $\tilde{m}$ may be eliminated from Eq. (2.4.7) and the resulting equation $[p][\tau] = -[u]^2$ may then be plotted in

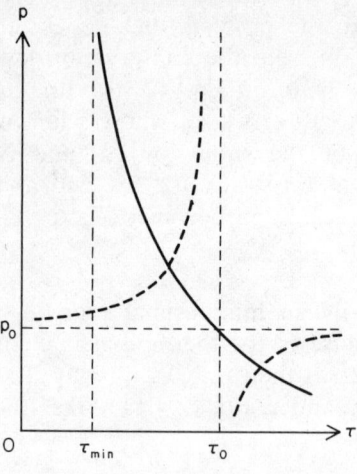

Fig. 2.5

the (p, τ)-space. The curve obtained is a hyperbola with the asymptotes $p = p_0$, $\tau = \tau_0$, and the other quantities may be found from the points of intersection of the hyperbola and the curve representing the Rankine–Hugoniot relation (cf. Fig. 2.5). Now, as is readily seen from Fig. 2.5, the Rankine–Hugoniot relation allows both the case $\tau_1 > \tau_0$ and the case $\tau_1 < \tau_0$. In other words, it is not possible on this basis to exclude expansion shocks merely by appeal to the Rankine–Hugoniot relations. However, if we require that entropy increases, so that $[S] \geq 0$ for a fluid element which crosses a shock, then it can be shown that we must have $[\tau] < 0$, and hence expansion shocks may be excluded.

To prove this, let us first integrate the thermodynamic relationship Eq. (2.4.3b) by means of Eq. (2.4.2c) and the equation of state, when the law of increase of entropy takes the form

$$\frac{p_1}{p_0} \geq \left(\frac{\rho_1}{\rho_0}\right)^\tau. \qquad (2.4.14)$$

Here the equality sign corresponds to $[S] = 0$, which is the adiabatic law Eq. (2.2.3d). The relationship between p and ρ (or τ) given by the equality is represented by the dotted line in Fig. 2.6. It follows that at the point (τ_0, p_0), this curve touches the solid line corresponding to the Rankine–Hugoniot relation. The dotted line lies below the Rankine–Hugoniot curve for $\tau < \tau_0$, and above it for $\tau > \tau_0$, respectively. The relationship Eq. (2.4.14) requires the increase in the pressure-ratio due to a shock be larger than that given by the adiabatic gas law, and hence it follows from Fig. 2.6 that an expansion shock must be excluded. Conversely, if we require compressibility, it is easily seen that the law of

2.4 SHOCK WAVES IN GAS DYNAMICS

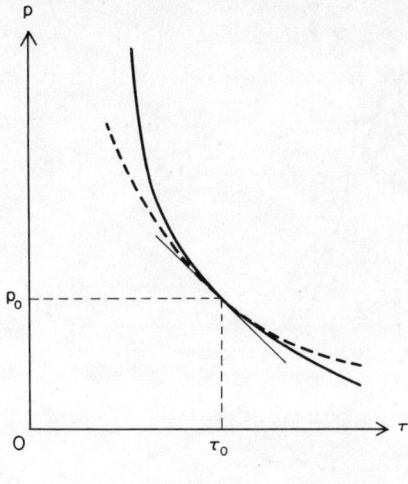

Fig. 2.6

increase of entropy $[S] \geq 0$ then follows. (For a proof in the general case, in which C_v is not constant, we refer, for example, to Courant and Friedrichs (1948) and to Jeffrey and Taniuti (1964)).

Let us now prove the important property that the flow is supersonic at the front of a shock and subsonic at the back. As shown in Eq. (2.4.6), the mass flow $\tilde{m} = \rho \tilde{u}$ remains constant at the front and the back of the shock, and by means of the conservation law for momentum, it is given by Eq. (2.4.8). On the other hand, $[p]/[\tau]$ is just the slope of the straight line connecting the two points (τ_0, p_0) and (τ_1, p_1), in Fig. 2.7. Also, using the definition of the sound speed, we find that the slope of the tangent at the point (τ_0, p_0) is $(dp/d\tau)_{(\tau_0,p_0)} = -a_0^2 \rho_0^2$. As may be seen from Fig. 2.7, the slope of the straight line connecting the two points (τ_0, p_0) and (τ_1, p_1) is larger than that of the tangent at (τ_0, p_1). Hence, we have $|[p]/[\tau]| > |dp/d\tau|_{(\tau_0,p_0)}$, provided $\tau_1 < \tau_0$. So, we arrive at the result,

$$|\tilde{u}_0| > a_0. \qquad (2.4.15a)$$

Furthermore, at the back of the shock, drawing the curve representing the Rankine–Hugoniot relation which passes through the point (τ_1, p_1), which is not taken to belong to the set of values on the front of the shock, and noticing that $|[p]/[\tau]| < |dp/d\tau|_{(\tau_1,p_1)}$ for $\tau_1 < \tau_0$, we have

$$|\tilde{u}_1| < a_1. \qquad (2.4.15b)$$

Conversely, the compressibility of a shock follows from the condition (2.4.15), and hence it turns out that this condition is equivalent to the law of increase of entropy. Thus, by means of condition (2.4.15), it is possible to select a unique physical solution from amongst the weak solutions

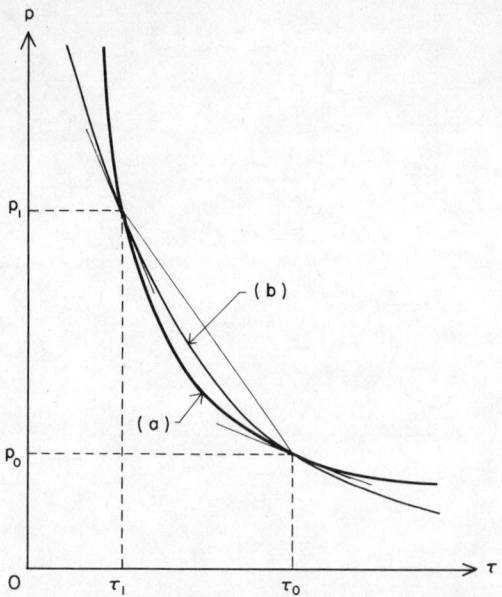

Fig. 2.7   The curve (a) is the R–H curve for τ_0, p_0 in front. The curve (b) is the R–H curve for τ_1, p_1 in front.

given by the shock condition (Rankine–Hugoniot relation). Also conditions (2.4.15) are to be regarded as an inequality condition to be satisfied by the shock speed $\tilde{\lambda}$, so that henceforth we shall call it an evolutionary condition, like condition (1) in Section 1.3.

Let us now explain more precisely what the evolutionary condition means, both in terms of its mathematical definition and its physical meaning. To do so, let us consider an equation with n components written in the conservation form

$$U_{,t} + F_{,x} = 0. \qquad (2.4.16)$$

In the case that the first term is the time derivative of $V(U)$, some function of U as in Eq. (2.4.2a), we consider F to be a function of V. Now, defining the components of the matrix A to be $A_{ji} \equiv \partial f_j / \partial x_i$ in terms of the components f_i of F, Eq. (2.4.16) reduces to

$$U_{,t} + A U_{,x} = 0. \qquad (2.4.17)$$

When Eq. (2.4.17) is hyperbolic, we call the original matrix equation (2.4.16) a hyperbolic system of conservation laws.

Suppose that two distinct constant states are separated by a certain discontinuity line, then the jump condition (shock condition) across it may be expressed as

$$-\tilde{\lambda}[U]+[F]=0. \tag{2.4.18}$$

When $\tilde{\lambda}$ does not correspond to a characteristic speed, we usually call Eq. (2.4.18) the generalized Rankine–Hugoniot relation. When $U = U_0$ is specified at $x \to -\infty$ and one component of U_1, say u_{11} (or $\tilde{\lambda}$), is given at $x \to \infty$, Eq. (2.4.18) yields n nonlinear algebraic equations for the n unknown quantities $\tilde{\lambda}, u_{21}, \ldots, u_{n1}$ (or U_1), as may be seen from the results for a gas shock. Since Eq. (2.4.18) is not linear, however, the solution is not determined uniquely.

For our selection principle we shall require the continuous dependence of the solution on the boundary conditions. This will enable us to select a physically meaningful solution from the class of solutions admitted by the conservation laws, that is, from the class of weak solutions. This requires that, with respect to a suitable norm, a small change of a boundary condition produces a correspondingly small change in the solution. This is a self-evident principle which is a natural requirement for physical solutions. Thus, for a state (solution) satisfying the generalized Rankine–Hugoniot relation to be physically significant, the time evolution of the system must be uniquely determined after a slight disturbance, at least for an infinitesimal time interval. That is, the initial value problem for that state must be well-posed. This obviously places a restriction on the weak solutions admitted by the generalized Rankine–Hugoniot relation, and thus may be expressed in the form of a set of inequalities for λ as we shall show below. We call it the evolutionary condition. Also, we shall hereafter call a discontinuous solution satisfying the evolutionary condition a generalized shock, or simply a shock. Since the evolutionary condition can also be interpreted as the condition for the stability of the system over an infinitesimally small time interval, it is possible that a shock which is unstable in the usual sense satisfies the evolutionary condition.

We can express the evolutionary condition in physical terms in the following way. When incoming waves of infinitesimal amplitude are incident upon a shock at infinity the shock surface is slightly disturbed, and at the same time disturbances propagate outwards in the form of outgoing waves at infinity. For any incoming waves the infinitesimal vibration of the shock surface and the out-going waves must be uniquely determined by the hyperbolic system of conservation laws given in Eq. (2.4.16). In order to represent such a physical condition in a concrete form, let $\bar{U}$ be the vector U satisfying the generalized Rankine–Hugoniot relation. Then transform the problem to a coordinate system moving with the shock by means of the transformation $x' = x - \tilde{\lambda}t$ and $t' = t$. Eq. (2.4.16) then becomes

$$U_{,t'} + (-\tilde{\lambda}U + F)_{,x'} = 0, \tag{2.4.19}$$

though the generalized Rankine–Hugoniot relation is unchanged. Suppose also that $\bar{U} = U_1$ for $x' > 0$ and $\bar{U} = U_0$ for $x' < 0$. Then, denoting the infinitesimal amplitude disturbance of $\bar{U}$ by δU, we have $U = \bar{U} + \delta U$. When we rewrite x' and t' in terms of x and t, the speed $\delta\lambda$ of the displacement δx of the shock surface ($x = 0$) due to the disturbance δU can be found by applying Gauss' theorem to the Eq. (2.4.19) at the oscillating shock surface. Namely, we find

$$\delta\lambda [U] = [-\tilde{\lambda} U + F]. \tag{2.4.20}$$

Linearizing this equation and using the generalized Rankine–Hugoniot relation for $\bar{U}$, we obtain at $x = 0$

$$\delta\lambda [\bar{U}] = [\tilde{A}\,\delta U], \qquad \tilde{A} \equiv A(\bar{U}) - \tilde{\lambda} I. \tag{2.4.21}$$

Since U satisfies Eq. (2.4.19) except on the shock surface, that is, it satisfies it for $x \neq 0$, the disturbance δU is governed for $x \neq 0$ by the equation

$$\delta U_{,t} + \tilde{A}\,\delta U_{,x} = 0. \tag{2.4.22}$$

When the disturbances incident at $x = \pm\infty$ (the incoming waves) oscillate in time with a certain fixed angular frequency ω, the change of speed $\delta\lambda$ of the vibration of the shock surface at $x = 0$ may be written

$$\delta\lambda = -\delta s \cdot e^{i\omega t}. \tag{2.4.23}$$

In addition, δU can be written in the form $\delta U \propto \exp(i\omega t - ik^{(\alpha)} x)$. We see that $k^{(\alpha)}$ is given by $k^{(\alpha)} = \omega/\tilde{\lambda}^{(\alpha)}$, where the $\tilde{\lambda}^{(\alpha)}$ are the eigenvalues of the matrix $\tilde{A}$. Therefore, a general disturbance δU may be expressed as the following linear combination:

$$\delta U_j = \sum_{\alpha=1}^{n} \delta a_j^{(\alpha)} \tilde{R}_j^{(\alpha)} \exp\left[i\omega\left(t - \frac{x}{\tilde{\lambda}_j^{(\alpha)}}\right)\right] \qquad (j = 0, 1), \tag{2.4.24}$$

where $\tilde{R}^{(\alpha)}$ is the right eigenvector of matrix $\tilde{A}$ corresponding to $\tilde{\lambda}^{(\alpha)}$, and we have written $j = 0$ and $j = 1$ to indicate the values in the regions $x < 0$ and $x > 0$, respectively. The reason the values in the two regions are different is because the constant states $\bar{U}$ for the two regions are different.

To find a solution which is valid for all space, we must connect δU_1 and δU_0 in such a way that the boundary conditions at $x = 0$, Eqns (2.4.21) and (2.4.23), are satisfied. Our problem is to find the condition under which outgoing waves are uniquely determined for given incoming waves. So, using Eq. (2.4.24), we express δU_j in terms of both the incoming and outgoing waves. For an incoming wave, we have $\tilde{\lambda}^{(\alpha)} < 0$ in the region ($j = 1$) corresponding to $x > 0$, and $\tilde{\lambda}_0^{(\alpha)} > 0$ in the one corresponding to

2.4 SHOCK WAVES IN GAS DYNAMICS

$x<0$. For an outgoing wave, we have $\tilde{\lambda}_1^{(\alpha)}>0$ for $x>0$, and $\tilde{\lambda}_1^{(\alpha)}<0$ for $x<0$. Thus, denoting the eigenvalues for the incoming and outgoing waves by $\tilde{\lambda}_{\text{in}}^{(\alpha)}$ and $\tilde{\lambda}_{\text{out}}^{(\alpha)}$, respectively, and also using the same notation for other quantities, δU_j can be written in the following way:

$$\delta U_j = \sum_{\alpha=1}^{f_j} \delta a_{\text{out}}^{(\alpha)} \tilde{R}_{\text{out}}^{(\alpha)} \exp\left[i\omega\left(t-\frac{x}{\tilde{\lambda}_{\text{out}}^{(\alpha)}}\right)\right]$$
$$+ \sum_{\alpha=1}^{n-f_j} \delta a_{\text{in}}^{(\alpha)} \tilde{R}_{\text{in}}^{(\alpha)} \exp\left[i\omega\left(t-\frac{x}{\tilde{\lambda}_{\text{in}}^{(\alpha)}}\right)\right], \qquad (2.4.25)$$

where f_j denotes the number of outgoing waves. Substituting this disturbance δU_j into the boundary condition (2.4.21) at $x=0$, and setting $f=f_0+f_1$, we obtain

$$\sum_{\alpha=1}^{f} |\tilde{\lambda}_{\text{out}}^{(\alpha)}|\, \tilde{R}_{\text{out}}^{(\alpha)}\, \delta a_{\text{out}}^{(\alpha)} + [\bar{U}]\delta s = \sum_{\alpha=1}^{2n-f} |\tilde{\lambda}_{\text{in}}^{(\alpha)}|\, \tilde{R}_{\text{in}}^{(\alpha)}\, \delta a_{\text{in}}^{(\alpha)}. \qquad (2.4.26)$$

When the incoming waves are given, Eqns (2.4.26) become algebraic equations for the $f+1$ unknown quantities $\delta U_{\text{out}}^{(\alpha)}$ ($\alpha=1, 2, \ldots, f$) and δs. If the solution of these equations is to be found, we must have $f=n-1$. That is, the number of outgoing waves must be $n-1$, and at the same time the vectors $R_{\text{out}}^{(\alpha)}$ and $[\bar{U}]$ must be linearly independent of each other. We call this the evolutionary condition (E). When we write Eq. (2.4.26) in matrix form, we have $T_{\text{out}}\psi_{\text{out}} = T_{\text{in}}\psi_{\text{in}}$, or $\psi_{\text{out}} = S\psi_{\text{in}}$. Hence, ψ_{out} and ψ_{in} are $(f+1)$-dimensional and $(2n-f)$-dimensional column vectors composed of $\delta a_{\text{out}}^{(\alpha)}$, δs, and $\delta a_{\text{in}}^{(\alpha)}$, respectively, while the matrices T_{out} and T_{in} are made up of the corresponding vectors $|\tilde{\lambda}_{\text{out}}^{(\alpha)}|\tilde{R}_{\text{out}}^{(\alpha)}$ and $[\bar{U}]$, and $|\tilde{\lambda}_{\text{in}}^{(\alpha)}|\tilde{R}_{\text{in}}^{(\alpha)}$, respectively, with $S = T_{\text{out}}^{-1}T_{\text{in}}$. Thus the evolutionary condition (E) may be regarded as the condition under which the matrix S is uniquely determined.

We shall consider Eq. (1.3.8) as the simplest example. Since $n=1$ and $\tilde{\lambda}^{(1)} = \bar{u}$ in this case, we must have $\bar{u}_1 < 0$ for $x>0$ and $\bar{u}_0 > 0$ for $x<0$ in order that $f=0$; that is, outgoing waves do not exist. We thus obtain the results of Section 1.3 in the limit as the dissipation coefficient tends to zero.

Next, we shall consider the equations of hydrodynamics. The eigenvalues of the matrix $\tilde{A}$ are

$$\lambda^{(1)} = \bar{u}, \qquad \lambda^{(2)} = \bar{u}+a, \quad \text{and} \quad \lambda^{(3)} = \bar{u}-a.$$

Since $\lambda^{(\alpha)} \lessgtr 0$, corresponding to $x \lessgtr 0$, for outgoing waves, if $\bar{u}_1 > a_1$ for $x>0$, the three waves corresponding to the eigenvalues $\lambda_1^{(1)} = \bar{u}_1 > 0$, $\lambda_1^{(2)} = \bar{u}_1 a_1 > 0$, and $\lambda_1^{(3)} = \bar{u}_1 - a_1 > 0$ are outgoing. Thus the number of outgoing waves is different according to the difference in the relative magnitudes of $\bar{u}$ and a, and the total number is shown in Fig. 2.8. In

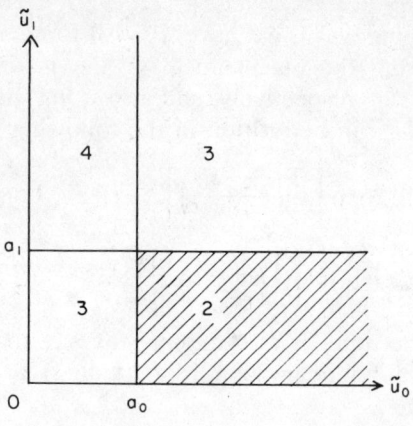

Fig. 2.8

order to satisfy the evolutionary condition (E), the number of outgoing waves is $f = n - 1 = 2$, for which we find $\tilde{u}_0 > a_0$ and $\tilde{u}_1 < a_1$. This is merely the result Eq. (2.4.15), as is clearly seen from Fig. 2.8. The evolutionary condition also requires the linear independence of $\tilde{R}_{\text{out}}^{(\alpha)}$ and $[\bar{U}]$. From the explicit form of these vectors it can easily be proved that this condition is also satisfied. (For the details we refer to Jeffrey and Taniuti (1964).)

In this case, we see that the evolutionary condition (E) amounts to the law of increase of entropy in hydrodynamics. Apparently, however, these two conditions are not equivalent. In fact, the evolutionary condition is not always satisfied for slow magnetoacoustic shocks in magnetohydrodynamics, even when the law of increase of entropy holds (Akhiezer et al. (1959); Polovin (1961); Jeffrey and Taniuti (1964)). In that case, it has been proved, however, that as long as the evolutionary condition is satisfied the law of increase of entropy is also true (Taniuti (1962)). In this sense, it is insufficient to use the law of increase of entropy as the condition for selecting a physical solution from amongst the possible weak ones, and it follows that we must adopt the evolutionary condition (E) instead.

The evolutionary condition given here, which enables us to select a physically relevant solution from amongst the others is not to be unique, and there will be alternative forms. Another condition, based on a generalization of the law of increase of entropy, has been proposed by Dafermos (1974) and Lax (1971). There are also methods for selecting physical solutions by virtue of finite difference schemes, which make it possible to obtain solutions, including shocks, by numerically integrating hyperbolic equations.

2.4 SHOCK WAVES IN GAS DYNAMICS

We will now examine whether the solution sought according to the evolutionary condition corresponds to the ones obtained in the limit by making the dissipation vanish. We first consider the case of gas dynamics, and use a coordinate system in which the shock wave is at rest, with the x-axis along the direction of flow. Then Eqns (2.2.1) become

$$(\rho u)_{,x} = 0, \tag{2.4.27a}$$

$$(\rho u^2 + p - \mu u_{,x})_{,x} = 0, \tag{2.4.27b}$$

$$\left\{\rho u\left(\frac{u^2}{2} + w\right) - \mu u u_{,x} - \chi T_{,x}\right\}_{,x} = 0, \tag{2.4.27c}$$

where w is the enthalpy per unit mass, so that $w = e + p\tau$, and also $\mu = (4/3)\zeta$. (For simplicity we set $\zeta' = 0$.) Integrating the above equations once, Eq. (2.4.27a) becomes Eq. (2.4.6), and by using Eq. (2.4.6), we obtain from Eqns (2.4.27b, c)

$$\tilde{m}u + p - \mu u_{,x} = P, \tag{2.4.28a}$$

$$\tilde{m}\left(\frac{u^2}{2} + w\right) - \mu u u_{,x} - \chi T_{,x} = E, \tag{2.4.28b}$$

where P and E are integration constants.

Suppose the state under consideration is connected to two uniform states as $x \to \pm\infty$, that is at the front and the rear of the shock wave, then the constants P and E are given by the values in these states. So we find the conservation laws

$$P = \tilde{m}u_0 + p_0 = \tilde{m}u_1 + p_1, \tag{2.4.29a}$$

$$E = \tilde{m}\left(\frac{u_0^2}{2} + w_0\right) = \tilde{m}\left(\frac{u_1^2}{2} + w_1\right). \tag{2.4.29b}$$

These are, of course, equivalent to Eqns (2.4.4). If we solve Eqns (2.4.28) for u and T (or p), the structure of the shock wave is obtained.

In general, though, the equations cannot be integrated analytically even when we assume that μ and χ are independent of T. However, in the special case that $\sigma \equiv \zeta C_p/\chi = \frac{3}{4}$, by means of the relations $\mu = (\frac{4}{3})\zeta$ and $w = C_p T$, Eq. (2.4.28b) becomes

$$\tilde{m}\left(\frac{u^2}{2} + w\right) - \mu\left(\frac{u^2}{2} + w\right)_{,x} = E. \tag{2.4.30}$$

(σ is the Prandtl number and nearly equal to $\frac{3}{4}$ for ages of diatomic molecules.) This can be integrated immediately to give $\tilde{m}(u^2/2 + w) = E$. Inserting this into Eq. (2.4.28a) to eliminate p, and also using the conservation laws Eq. (2.4.29), we get

$$\mu u u_{,x} = -\frac{\tilde{m}(\gamma+1)}{2\gamma}(u_0-u)(u-u_1)$$

so that, after an integration, we have

$$x - x_0 = \frac{D}{1-(u_1/u_0)}\left[\ln[1-(u/u_0)] - \frac{u_1}{u_0}\ln\{(u-u_1)/u_0\}\right.$$

$$\left. -\{1-(u_1/u_0)\}\ln\{(u_0-u_1)/(2u_0)\}\right]. \quad (2.4.31)$$

Here, x_0 is the integration constant and D is given by

$$D \equiv \frac{2}{\gamma+1}\frac{\mu}{\rho_0 u_0}, \qquad (2.4.32)$$

and it has the dimension of length. Denoting $D[1-(u_1/u_0)]^{-1}$ by d, Eq. (2.4.31) reduces to

$$(u_0-u)(u-u_1)^{-(u_1/u_0)}\{(u_0-u_1)/2\}^{(u_1/u_0)-1} = e^{(x-x_0)/d}. \quad (2.4.33)$$

Hence, as $d \to 0$, u approaches a step function in such a way that $u \to u_1$ for $x > x_0$ and $u \to u_0 > u_1$ for $x < x_0$. That is, u equals u_0 at $x \to -\infty$ (the front), rapidly decreases with increasing x in the neighbourhood of x_0, and passes into u_1 as $x \to +\infty$ (the rear).

Since, roughly speaking, we may regard d as the thickness of the shock, the thickness is inversely proportional to $u_0 - u_1$, that is, to the strength of the shock, and it is proportional to μ. Letting L be a typical length, it follows that D is inversely proportional to the Reynolds number $\mathcal{R}$ (see Eq. (2.4.1)). Accordingly, we may consider that the thickness of the shock depends upon $\mathcal{R}^{-1}$.

In particular, we can approximate the left-hand side of Eq. (2.4.33) by $(u_0-u)/(u-u_1)$ for a weak shock, in which $u_1/u_0 = 1-\varepsilon(\varepsilon \ll 1)$, and hence we obtain solutions of the same form (which are called the Taylor solutions) as the shock solutions Eq. (1.3.1) for the Burgers equation. In this case involving a weak shock the Taylor solutions are obtained for any Prandtl number. The structure of a shock may also be investigated in the general case when μ and χ depend on the temperature, though we may infer qualitative results from those mentioned above.

For a strong shock, however, the thickness, in fact, becomes small to the same order of magnitude as the mean free path of the gas molecules. Accordingly, the applicability of the equations (2.2.1) of hydrodynamics becomes doubtful, and we must then use the Boltzmann equation (for instance, cf. Grad (1949; 1952), Mott-Smith (1951)). On one hand, however, the approximation involving replacing a shock by a discon-

2.4 SHOCK WAVES IN GAS DYNAMICS

tinuity surface becomes better for stronger shocks. The variation of entropy follows from Eqns (2.4.27) and (2.4.3b), showing that

$$\tilde{m}TS_{,x} = (\chi T_{,x})_{,x} + \mu(u_{,x})^2 \tag{2.4.34}$$

holds. Supposing χ and μ to be constant, integration of this equation gives

$$\tilde{m}[S]_{-\infty}^{\infty} = \chi \int \frac{1}{T^2}\left(\frac{\partial T}{\partial x}\right)^2 dx + \mu \int \frac{1}{T}\left(\frac{\partial u}{\partial x}\right)^2 dx > 0, \tag{2.4.35}$$

and we obtain the law of increase of entropy. For a weak shock, we have $T_{,x} \sim u_{,x} \sim \varepsilon/d$, and, since $d \sim \varepsilon^{-1}$, $\Delta S \sim \varepsilon^2/d \sim \varepsilon^3$, so that the increment in the entropy is of the third order of smallness in the shock strength.

It can be inferred from the above results for the special case of gas dynamics that, in general, the solutions selected by the law of increase of entropy, and consequently by the evolutionary condition, correspond to the ones obtained in the limit by setting the dissipation coefficients equal to zero. Here, an important point is that since the x-axis is taken along the direction of the flow, u decreases downstream as can be seen from Eq. (2.4.27), and consequently the compressibility of a shock is derived automatically. This is the result of the (local) conservation laws Eq. (2.4.27), together with the law of increase of entropy Eq. (2.4.35). However, it should be noticed that these results do not follow from the (global) conservation laws Eqns (2.4.6) and (2.4.29a, b) derived by integration of the local ones.

As will be mentioned in the next chapter, a general system which becomes of hyperbolic type provided the dissipation is neglected, can be reduced to Burgers' equation in an asymptotic sense (cf. Section 3.2). Since the solution to Burgers' equation with small dissipation is approximated by the weak solution of the corresponding hyperbolic equation subject to the evolutionary condition, the solution of such a general system in the limit of small dissipation could be determined by the evolutionary condition.

The hyperbolic conservation laws Eq. (2.4.16), supplemented by the evolutionary condition, makes it possible to find the solutions of various kinds, subject to initial and boundary values comprising discontinuities. Though the original equation involving the effect of dissipation is, in general, a very complicated nonlinear evolution equation, we can still obtain its approximate solution. We mention here a simple case in gas dynamics. (For the details, we refer to Courant and Friedrichs (1948); see also Coulson and Jeffrey (1977).)

Suppose that a shock proceeds with speed $\tilde{\lambda}(>0)$ from the origin $x = 0$ at $t = 0$. The streamlines for $x > 0$ are shown schematically in Fig. 2.9(a), where the fluid elements are accelerated by the shock.

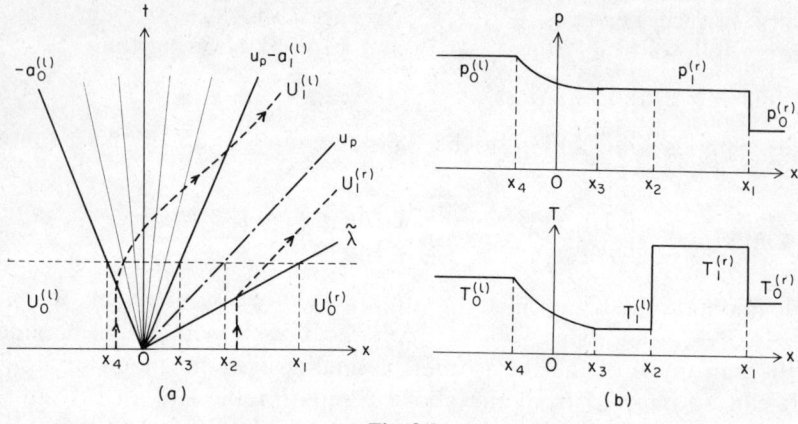

Fig. 2.9

Regarding the streamline issuing from the origin as a moving boundary line, we have the situation that the gas at rest for $x>0$ is suddenly pushed by a piston which is kept moving with a constant velocity. Since the velocity at the rear of the shock is given by that of the piston u_p, the shock is uniquely determined when the state $U_0^{(r)}$ at the front is specified. Next, suppose that at $t=0$ the gas in the state $U_0^{(l)}$ is at rest in $x<0$. We then have the problem which arises when we suddenly withdraw the piston. In this case a (centred) expansion wave $x-(u-a)t=0$ issues from the origin (Fig. 2.9(a)). The decrease in density across the wave is given by

$$[u]_0^{u_p} = -\int_{\rho_0^{(l)}}^{\rho_1^{(l)}} a/\rho \, d\rho.$$

The wave is adjacent to $U_0^{(l)}$ across the boundary (characteristic) line $x + a_0^{(l)} t = 0$, and to the constant state $U_1^{(l)}$ across the boundary (characteristic) line $x - (u_p - a_1^{(l)}) t = 0$. In this case, the conservation laws admit (an expansion) shock propagating to the left. However, this is forbidden by the evolutionary condition in the same way as in the case of Eq. (1.3.8).

Now, if at $t=0$ two constant states $U_0^{(r)}$ and $U_0^{(l)}$ are brought into contact at the origin, the solution can be obtained by joining the solutions to the above two problems. In this case, the piston is to be replaced by a contact surface. Its velocity u_p is determined by equating the pressure in the state $U_1^{(r)}(U_0^{(r)}, u_p)$ linked to $U_0^{(r)}$ across the shock, to that in the state $U_1^{(l)}(U_0^{(l)}, u_p)$ linked to $U_0^{(l)}$ across an adiabatic expansion wave; that is, by the equation $p_1^{(r)}(U_0^{(r)}, u_p) = p_1^{(l)}(U_0^{(l)}, u_p)$. Since we clearly have $p_1^{(r)} > p_0^{(r)}$ and $p_1^{(l)} < p_0^{(l)}$, we have $p_0^{(r)} < p_0^{(l)}$. The corresponding pressure p and temperature T variation is shown in Fig. 2.9(b). This provides the solution to the problem of a shock tube, in which two gases at different pressures are separated by a diaphagm, and a shock is formed by removing it instantaneously (Problem 2.4.2).

In general, the initial value problem for a discontinuous step function comprising such different constant states is called the Riemann problem. In particular, if the initial flow velocity is discontinuous, the resulting Riemann problem is related to the interaction of shocks. If, for instance, there are two shocks propagating in the same direction, the fast shock inevitably overtakes the slow one by virtue of the evolutionary condition. It is readily seen that after overtaking there must exist a contact surface between the two shocks. When, in general, two shocks collide, it follows that they split into two shocks and a contact surface (except in the case of a head-on collision betwen two shocks with the same speed). If the instant of collision is taken as the origin for the time, we have a Riemann problem involving a discontinuity in speed. The Riemann problem for a general hyperbolic conservation system of n components was solved by Lax (1957) in the genuinely nonlinear case, and it has been investigated in detail for magnetohydrodynamics by, for example, Jeffrey and Taniuti (1964).

Problem 2.4.1

Examine the propagation of waves when the velocity of a piston is given as a function of time. In particular, consider the process of concentrating several shocks.

Problem 2.4.2 Shock tube

There are two masses of gas at different pressures $p_0^{(r)}$ and $p_0^{(l)}$ which are separated by a diaphragm. Investigate the propagation of the resulting shock when the diaphragm is removed at $t=0$. In particular, find the jump in the pressure due to the shock, and the variation of the temperature at the contact surface (Zel'dovich and Raizer (1966)).
Hint cf. Fig. 2.9.

Problem 2.4.3

Show that the Rankine–Hugoniot relation for magnetohydrodynamic (magnetoacoustic) shocks is given by

$$[e+\langle p\rangle\tau] = -\left(\frac{\mu}{16\pi}\right)[\tau][\mathbf{H}_t]^2,$$

where $\tau = \rho^{-1}$, and $\mathbf{H}_t$ is the transverse magnetic field, so that $\mathbf{H}_t = \mathbf{H} - \mathbf{H}_x$ (Jeffrey and Taniuti (1964)).
Hint Use Eq. 2.3.17.

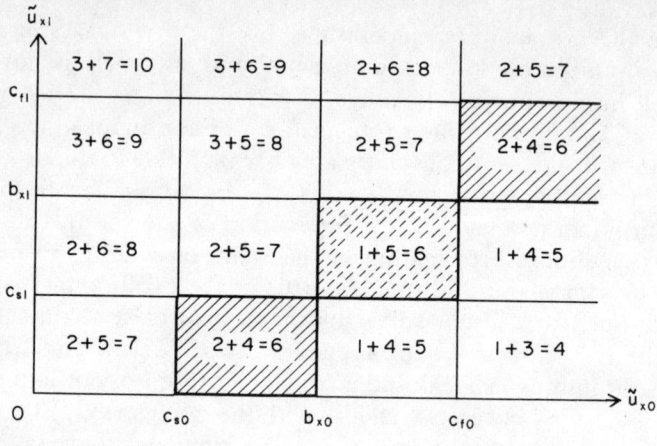

Fig. 2.10

Problem 2.4.4

Consider the slow magnetoacoustic shock wave, and show that the component of the magnetic field transverse to the direction of propagation of the shock decreases. Prove also that a shock for which the transverse component of the magnetic field is reversed is forbidden by the evolutionary condition (Jeffrey and Taniuti (1964)).
Hint Cf. Fig. 2.10.

2.5 Extension to the multi-dimensional case

The propagation of nonlinear waves in three-dimensional space has not been studied extensively. However, there is a special three-dimensional case to which we can easily extend the one-dimensional results. Let us first consider the following equation in the conservation form

$$\frac{\partial U}{\partial t} + \nabla \cdot \boldsymbol{F} + B = 0. \qquad (2.5.1)$$

Here, U is a column vector with n components $U = (u_1, u_2, \ldots, u_n)$, ∇ and $\boldsymbol{F}$ are the vector gradient operator and vector in three-dimensional space, $\nabla \equiv (\partial/\partial x_1, \partial/\partial x_2, \partial/\partial x_3)$ and $\boldsymbol{F} = (F_1, F_2, F_3)$, respectively, F_r ($r = 1, 2, 3$) and B are also column vectors with n components given by $(f_1^{(r)}, f_2^{(r)}, \ldots, f_n^{(r)})$ and $(b_1, b_2, \ldots, b_n)$, respectively, and these components are functions of x_r, t and u_i ($i = 1, 2, \ldots, n$) (we assume the necessary smoothness).

2.5 EXTENSION TO THE MULTI-DIMENSIONAL CASE

Now suppose that U changes discontinuously across the hypersurface $\varphi(x_1, x_2, x_3, t) = 0$ in four-dimensional space. By applying the four-dimensional Gauss theorem, we then obtain the generalized Rankine–Hugoniot relation

$$-\tilde{\lambda}[U] + [\boldsymbol{n} \cdot \boldsymbol{F}] = 0, \qquad (2.5.2)$$

where $\tilde{\lambda}$ and $\boldsymbol{n}$ are the speed of the surface and its vector normal in three-dimensional space, respectively, given by

$$\tilde{\lambda} = -\frac{\varphi_{,t}}{|\nabla \varphi|}, \qquad \boldsymbol{n} = \frac{\nabla \varphi}{|\nabla \varphi|}. \qquad (2.5.3)$$

For an infinitesimal jump δU about the stationary state $U_0(x)$ satisfying $\nabla \cdot \boldsymbol{F}(U_0) + B(U_0) = 0$, Eq. (2.5.2) becomes, as in the one-dimensional case,

$$(\boldsymbol{n} \cdot \mathbf{A}(U_0) - \tilde{\lambda}) \, \delta U = 0. \qquad (2.5.4)$$

Here $\mathbf{A}$ is the vector whose components are given by matrices $A_r\{(A_r)_{ij}\} \equiv \partial f_i^{(r)}/\partial u_j$ ($r = 1, 2, 3$). Thus, for an infinitesimal jump to exist (in which case λ may replace $\tilde{\lambda}$), the algebraic equation for λ

$$\det |-\lambda + \boldsymbol{n} \cdot \mathbf{A}(U_0)| = 0 \qquad (2.5.5)$$

must have real roots, at least for some range of $\boldsymbol{n}$. When the n-dimensional matrix $\boldsymbol{n} \cdot \mathbf{A}$ has n real eigenvalues for all $\boldsymbol{n}$ (in some range of U) and the coresponding eigenvectors are linearly independent, we say Eq. (2.5.1) is (totally) hyperbolic.

As the extension of Eq. (2.3.9) in Section 2.3, in general, let us consider a system of partial differential equations of the form

$$\sum_{\mu=0}^{3} A_\mu(U) U_{,\mu} + B(U) = 0, \qquad (2.5.6)$$

in which we take the four-dimensional coordinates x_μ ($\mu = 0, 1, 2, 3$) as the independent variables, and suppose that U is specified on the surface $\varphi(x_\mu) = 0$ ($U_{,\mu} \equiv \partial U/\partial x_\mu$). That is, in the neighbourhood of this surface, let us introduce the curvilinear coordinates $\{\varphi, \xi_s\}$ ($s = 1, 2, 3$), in which $\varphi = 0$ is embedded, and let $U(0, \xi_s)$ be given as sufficiently smooth function of ξ_s. Then Eq. (2.5.6) can be expressed in the form

$$(A_\mu \varphi_{,\mu}) U_{,\varphi} + \sum_{s=1}^{3} (A_\mu \xi_{s,\mu}) U_{,\xi_s} + B = 0. \qquad (2.5.7)$$

Here we use the convention that when a Greek index appears twice, it is to be understood that summation takes place with respect to this index from 0 to 3. Therefore, $U_{,\varphi}|_{\varphi=0}$ is not determined uniquely provided the condition

$$\det|A_\mu \varphi_{,\mu}| = 0 \qquad (2.5.8)$$

is satisfied. Accordingly, it follows that an infinitesimal jump δU of U across $\varphi = 0$ is permitted. Generally, we call equation (2.5.8) the characteristic equation of Eq. (2.5.6), and a curved surface $\varphi(x_\mu) = $ constant on which it is satisfied is called a characteristic surface. Condition (2.5.8) is a necessary one in order that the solution of Eq. (2.5.6) admits waves moving with finite speed.

If $\det|A_\mu \varphi_{,\mu}| \neq 0$, by successive differentiation of Eq. (2.5.7) we find that all of the higher order derivatives of U on $\varphi = 0$ are uniquely determined. Hence, a necessary condition for a solution satisfying a boundary condition on $\varphi = 0$ to exist uniquely in the neighbourhood of $\varphi = 0$ (the Cauchy problem) is that $\varphi = 0$ is nowhere tangential to a characteristic surface. In particular if $\det A_0 \neq 0$, without loss of generality, we may write Eq. (2.5.6) as

$$U_{,0} + \mathbf{A} \cdot \nabla U + B = 0. \qquad (2.5.9)$$

If Eq. (2.5.9) is (totally) hyperbolic, we may regard x_0 as the time and the characteristic equation (2.5.8) reduces to Eq. (2.5.5). If we consider the characteristic equation to be the condition for the indeterminacy of the higher order normal derivatives with respect to a characteristic surface (the derivatives in the direction normal to that surface), then the characteristic equation for the system of higher order derivatives can be derived in a similar way. For instance, let $A_{\mu\nu}$ and B be an $(n \times n)$ matrix and a column vector of n components, respectively, and let both of them be functions of U and $U_{,\mu}$. Now let us consider the equation

$$A_{\mu\nu} U_{,\mu\nu} + B = 0.$$

When U and $U_{,\varphi}$ are specified on $\varphi = 0$, we get

$$\det|A_{\mu\nu} \varphi_{,\mu} \varphi_{,\nu}| = 0 \qquad (2.5.8')$$

as the condition that $U_{,\varphi\varphi}$ is indeterminate. Also, let us consider a system of simultaneous equations of different orders

$$A_\mu^{(1)} U_{,\mu}^{(1)} + B^{(1)} = 0,$$
$$A_{\mu\nu}^{(2)} U_{,\mu\nu}^{(2)} + B^{(2)} = 0,$$

where $A_\mu^{(1)}$, $B^{(1)}$, $A_{\mu\nu}^{(2)}$, and $B^{(2)}$ are functions of $U^{(1)}$, $U^{(2)}$ and $U_{,\mu}^{(2)}$. When $U^{(1)}$, $U^{(2)}$, and $U_{,\varphi}^{(2)}$ are prescribed on $\varphi = 0$, we find without difficulty that

$$\det|A_\mu^{(1)} \varphi_{,\mu}| = 0, \qquad (2.5.8''a)$$

$$\det|A_{\mu\nu}^{(2)} \varphi_{,\mu} \varphi_{,\nu}| = 0, \qquad (2.5.8''b)$$

is the condition for the indeterminacy of $U_{,\varphi\varphi}^{(2)}$ and $U_{,\varphi}^{(1)}$.

2.5 EXTENSION TO THE MULTI-DIMENSIONAL CASE

Setting
$$p_\mu \equiv \varphi_{,\mu}, \tag{2.5.10}$$

Eq. (2.5.8) becomes
$$H(x_\mu, u_1, \ldots, u_n, p_\mu) = 0. \tag{2.5.11}$$

Since H is a homogeneous function of degree n in $p_0, p_1, p_2,$ and p_3, we have
$$p_\mu \frac{\partial H}{\partial p_\mu} = nH = 0, \tag{2.5.12}$$

so that, by using the four-dimensional vectors $\boldsymbol{p} = \{p_\mu\}$ and $\nabla_p H \equiv \{\partial H/\partial p_\mu\}$, Eq. (2.5.12) can be written in the form
$$\boldsymbol{p} \cdot \nabla_p H = 0. \tag{2.5.13}$$

By definition, the vector $\boldsymbol{p}$ is parallel to the four-dimensional normal vector $\boldsymbol{n}^{(4)}$ of the characteristic surface $\varphi = $ constant. Accordingly, it follows from the relation (2.5.13) that the vector $\nabla_p H$ is orthogonal to the four-dimensional normal vector $\boldsymbol{n}^{(4)}$, and consequently is tangent to the characteristic surface $\varphi = $ constant. Therefore, on the characteristic surface $\varphi = $ constant, we can choose the differential vector, which is defined by
$$d\boldsymbol{x}^{(4)} = \{dx_\mu\}, \tag{2.5.14}$$

so as to satisfy the conditions $d\varphi = (\partial\varphi/\partial x_\mu) \, dx_\mu = p_\mu \, dx_\mu = 0$, so that it is parallel to $\nabla_p H$. So, we have
$$\frac{dx_0}{(\partial H/\partial p_0)} = \frac{dx_1}{(\partial H/\partial p_1)} = \frac{dx_2}{(\partial H/\partial p_2)} = \frac{dx_3}{(\partial H/\partial p_3)} = ds, \tag{2.5.15}$$

where
$$ds = \frac{1}{|\nabla_p H|} \left(\sum_{\mu=0}^{3} (dx_\mu)^2 \right)^{1/2} \tag{2.5.16}$$

is a line element on the characteristic surface.

Equations (2.5.15) determine a family of curves lying on $\varphi = $ constant. We call these either characteristic curves or characteristic rays, or more simply still, rays. This shows that a characteristic surface may be considered as a surface generated by characteristic rays. Since Eq. (2.5.11) is satisfied on the characteristic surface, that variation of H on the surface gives
$$\delta H = \left[\frac{\partial H}{\partial p_\nu} \frac{\partial p_\nu}{\partial x_\mu} + \frac{\partial H}{\partial x_\mu} \right] \delta x_\mu = 0. \tag{2.5.17}$$

Furthermore, by virtue of the identity $\partial p_\nu/\partial x_\mu = \partial p_\mu/\partial x_\nu$, we find

$$\frac{\partial H}{\partial p_\nu}\frac{\partial p_\mu}{\partial x_\nu} = -\frac{\partial H}{\partial x_\mu}. \tag{2.5.18}$$

Now, from Eq. (2.5.15) we have

$$\frac{dx_\mu}{ds} = \frac{\partial H}{\partial p_\mu}. \tag{2.5.19}$$

Also, using this relation, Eq. (2.5.18) becomes

$$\frac{dp_\mu}{ds} = -\frac{\partial H}{\partial x_\mu}. \tag{2.5.20}$$

Solving Eqns (2.5.19) and (2.5.20) gives the characteristic curves $\{x_\mu(s)\}$.

Equations (2.5.19) and (2.5.10) have a form similar to that of the canonical equation of motion for a particle. Hence, to clarify the relationship, we solve Eq. (2.5.11) algebraically for p_0, when we obtain

$$H = \prod_{i=1}^{n} H^{(i)} = 0, \quad H^{(i)} \equiv p_0 + \mathcal{H}^{(i)}(p_r, x_r, t, u_1, \ldots, u_n) \quad (r = 1, 2, 3). \tag{2.5.21}$$

Since $p_0 = \varphi_{,t}$, this equation is exactly the Hamilton–Jacobi equation. The expressions p_0 so obtained as functions of p_r correspond to the n eigenvalues λ of Eq. (2.5.5). That is, there exists a p_0 corresponding, respectively, to each of the n possible modes. We also find that Eqns (2.5.19) and (2.5.20) can be written in the form of the canonical equations for a Hamiltonian $\mathcal{H}^{(i)}$:

$$\frac{dx_r}{dt} = \frac{\partial \mathcal{H}^{(i)}}{\partial p_r}, \quad \frac{dp_r}{dt} = -\frac{\partial \mathcal{H}^{(i)}}{\partial x_r} \quad (r = 1, 2, 3, \quad i = 1, 2, \ldots, n). \tag{2.5.22}$$

This shows that the characteristic curves for the waves of the n possible modes are equivalent to the motion of particles determined by the Hamiltonians $\mathcal{H}^{(i)}$ corresponding to their respective modes. We call $\boldsymbol{v}_r = d\boldsymbol{x}/dt$, which corresponds to the particle velocity, the ray velocity. The velocity (characteristic velocity) determined by writing Eq. (2.5.3) in terms of the momentum $\boldsymbol{p}(p_1, p_2, p_3)$ to give $\lambda = -p_0/|\boldsymbol{p}|$, $\boldsymbol{n} = \boldsymbol{p}/|\boldsymbol{p}|$ is called the normal velocity. As may be seen from Eq. (2.5.13), there is the relationship $\boldsymbol{n} \cdot \boldsymbol{v}_r = \lambda$ between the ray velocity and the normal velocity. This corresponds to the fact that, in a general dynamical system, the momentum and the velocity are not always in the same direction.

Since the Hamiltonian $\mathcal{H}^{(i)}$ is a function of $U(u_1, u_2, \ldots, u_n)$, we cannot determine the characteristic surface unless the solution U is

2.5 EXTENSION TO THE MULTI-DIMENSIONAL CASE

known. For a wavefront, however, we can solve Eq. (2.5.22) by using the value of U in the state ahead of the wavefront, say U_0.

As an example, let us consider the case that A_r does not depend on x_r and t explicitly. Suppose that the variation (disturbance) of U at $t = 0$ is confined to a closed surface $\psi(x_r) = 0$ enclosing the origin in three-dimensional space, and that U is constant, so that $U = U_0$ on and outside the surface. Then, on the surface, H becomes a function of $\boldsymbol{p}$ only, and the problem leads to an equation involving A_r with constant coefficients. Eq. (2.5.22) can immediately be integrated to give $p_r = $ constant, and the rays become straight lines. The momentum p_r^0 corresponding to a point x_r^0 on the initial wavefront is given by $\partial \psi / \partial x_r |_{x_r^0} = p_r^0$. Thus, for $t > 0$, we can determine the solution $\{x_r(t), p_r(t)\}$ of the canonical equations (2.5.22) satisfying the initial value $\{x_r^0, p_r^0\}$ at $t = 0$.

The curve in four-dimensional space described by the solution $x_r(t)$ is the (characteristic) ray, while the surface in four-dimensional space generated by the rays issuing from all the points on the initial surface of wavefront (wave surface) is the characteristic surface, $\varphi(x, t) = 0$. Also, the surface obtained by projecting the time section of this characteristic surface into three-dimensional space is the wavefront itself. Namely, the surface in three-dimensional space given by $\varphi(x_r, t_0) = 0$ gives the surface of the wavefront at $t = t_0$ (cf. Fig. 2.11).

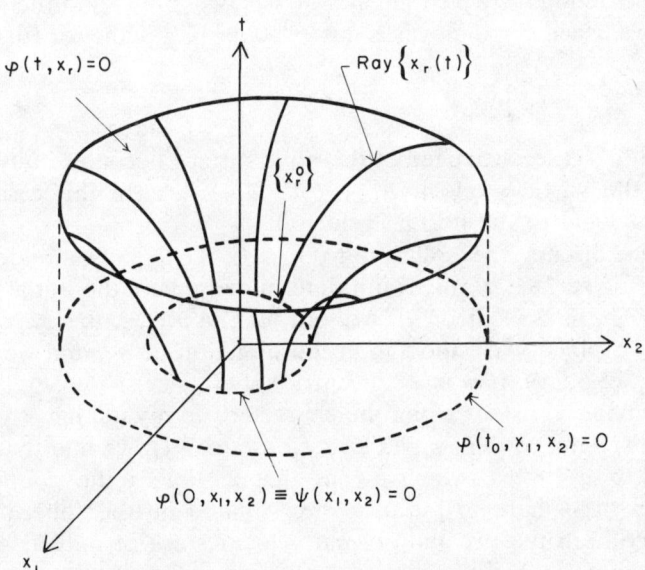

Fig. 2.11

Unlike the Hamilton–Jacobi equation in the classical mechanics of a point mass, however, H in Eq. (2.5.11) is homogeneous in p_μ ($\mu = 0, 1, 2, 3$). Consequently, we notice that the Hamiltonian $\mathcal{H}^{(i)}$ must be of degree one in p_r ($r = 1, 2, 3$). For example, when

$$H = p_0^2 - \boldsymbol{p}^2, \tag{2.5.23}$$

we have

$$\mathcal{H}^{(i)} = \pm \sqrt{\sum_{j=1}^{3} p_j^2}. \tag{2.5.24}$$

It should be noticed that this differs from classical mechanics, in which even the Hamiltonian for a free particle $\mathcal{H}_f \equiv \frac{1}{2}\boldsymbol{p}^2$ does not satisfy this condition. If the Hamiltonian $\mathcal{H}$ is not homogeneous, the characteristic curves determined by the canonical equations (2.5.22) do not lie on the characteristic surface. For, as is easily confirmed in the above Hamiltonian $\mathcal{H}_f$, we have

$$d\varphi = p_\mu \, dx_\mu = \left[p_0 + \sum_{j=1}^{3} (\partial \mathcal{H}^{(i)}/\partial p_j) p_j \right] dt \neq 0,$$

which does not lead to $\varphi = $ constant.

In fact, for $\mathcal{H}_f$, the particle velocity is not equal to the velocity with which the wave surface $\varphi = $ constant moves. To see this, let the particle velocity be denoted by $\boldsymbol{p}^0$, then $\varphi = -(\frac{1}{2})(\boldsymbol{p}^0)^2 t + \boldsymbol{p}^0 \cdot \boldsymbol{x}$, and the speed with which the surface $\varphi = 0$ moves is not $|\boldsymbol{p}^0|$ but $\frac{1}{2}|\boldsymbol{p}^0|$, while for Eq. (2.5.24)

$$\varphi = \mp \sqrt{(\boldsymbol{p}^0)^2} \, t + \boldsymbol{p}^0 \cdot \boldsymbol{x}. \tag{2.5.24'}$$

Hence, in this case, the speed of the wave surface becomes unity, which is equal to the particle velocity, $[\sum_k (\partial \mathcal{H}^{(i)}/\partial p_k)^2]^{1/2}$. (In this case, the ray velocity is equal to the normal velocity).

We now discuss the solution to Eq. (2.5.24) in some detail. The characteristic surface $\varphi(x, t)$ is not determined unless the initial condition $\varphi(x, 0) = \psi(x)$ is specified. If we put $\psi(x) = \boldsymbol{p}^0 \cdot \boldsymbol{x}$, the wave surface represents a plane wave moving in the direction of $\boldsymbol{p}^0$, and we naturally obtain Eq. (2.5.24'). If ψ is a concentric sphere $\psi(x) = r(\equiv \sqrt{\sum_{j=1}^{3} x_j^2})$, and $\boldsymbol{p} = \nabla \psi$ remains constant along the characteristic ray issuing from a point $\boldsymbol{x}^0$ on that sphere, so that, $\boldsymbol{p} = (\boldsymbol{x}^0/r^0)$. Consequently, we find that $x_j - x_j^0 = \pm(x_j^0/r^0)t$, or, $r = r^0 \pm t$, which means that at time t the points initially located on the sphere of radius r^0 are situated on the sphere of radius $r^0 \pm t$ (since both the ray and normal velocities are of unit magnitude in the radial direction). Therefore, $\varphi = r \pm t$ also becomes the characteristic surface.

2.5 EXTENSION TO THE MULTI-DIMENSIONAL CASE

It is easily seen that this result satisfies the Hamilton–Jacobi equation for the $\mathcal{H}^{(i)}$ in Eq. (2.5.24). Also, constructing a family of surfaces in terms of the parameter r^0, we find that $\varphi = r^0$ becomes a cone-like hypersurface enclosing the time axes, which reduces to the sphere of radius r^0 at $t = 0$, and it gives the wave surface comprising convergent and divergent waves if t is fixed. In particular, if we put $r^0 = 0$, we get hypercones enclosing the positive and negative parts of the time axis respectively. In what follows, using the terminology of the theory of relativity, we will call this the light cone. We may seek the light wave surface issuing from a sphere by using an envelope of plane waves solutions. For instance, denoting the normal velocity by $\lambda(\boldsymbol{n})$, the wave surface comprising waves originating from the tangent plane $\boldsymbol{n} \cdot (\boldsymbol{x} - \boldsymbol{x}^0) = 0$ at a point $\boldsymbol{x}^0 = (r^0 \boldsymbol{n})$ on the sphere of radius r^0 at $t = 0$ becomes $\boldsymbol{n} \cdot (\boldsymbol{x} - \boldsymbol{x}^0 - \lambda \boldsymbol{n} t) = 0$ at the time t. Generally, λ depends on $\boldsymbol{n}$. Let us suppose that λ is positive for all $\boldsymbol{n}$. Then, at each instant of time, the wave surface issuing from the sphere (r^0) is given by constructing the envelope of the planes formed as $\boldsymbol{x}^0$ covers the surface of the sphere (r^0). As $r^0 \to 0$, we of course obtain the wave surface generated by a point source. To obtain it geometrically, for a given $\boldsymbol{n}$, mark the point given by the vector $\lambda \boldsymbol{n}$ issuing from the origin. Then, as $\boldsymbol{n}$ ranges over all possible directions, construct the surface covered by the locus of these points. This is just the surface corresponding to the normal velocity.

The envelope of planes normal to the radial vectors drawn from the origin to arbitrary points on this surface gives the wave surface, at $t = 1$, of waves starting from the origin (point source). Since, in Eq. (2.5.23), we have $\lambda = 1$ independently of $\boldsymbol{n}$, the surface of the normal velocity becomes a sphere of unit radius, and so does the wave surface. In general, the wave surface of disturbances diverging from a point source (the time section of the light cone) may also be sought by constructing the envelope of the wave surfaces of plane waves radiating out from the origin in all possible directions, as above.

With this objective in mind, for a given $\boldsymbol{n}$, we first take a point whose distance from the origin is $\lambda(\boldsymbol{n})$, and then construct the surface of normal velocity obtained by allowing $\boldsymbol{n}$ to vary. Then, to complete the construction, we find the envelope of the planes normal to the radial vectors drawn from the origin to the points on that surface (the time section at $t = 1$).

The homogeneity of H in p_μ is due to the homogeneity in the time and the spatial derivatives. Though characteristic of a hyperbolic equation, in general, this is also related to the Lorentz invariance in a homogeneous isotropic field. In classical mechanics (the Hamilton–Jacobi equation), this homogeneity entirely breaks down, and this is also carried over to the Schrödinger equation. Explicitly, the Hamilton–Jacobi equation in classical mechanics is a geometrical-optics approximation to the (non-hyperbolic) Schrödinger equation. As will be mentioned below, the light cone becomes the characteristic surface and determines the domain of dependence and the range of influence in equations regarded as essentially describing the field of the d'Alembertian. For equations possessing the Lorentz invariance of the special theory of relativity, this is connected with the requirements of the invariance of the velocity of light and the finiteness of the velocity of propagation of action.

First of all, let us consider the following Maxwell equations

$$c^{-1}\boldsymbol{D}_{,t} - \nabla \times \boldsymbol{H} = 0, \tag{2.5.25a}$$

$$c^{-1}\boldsymbol{B}_{,t} + \nabla \times \boldsymbol{E} = 0, \tag{2.5.25b}$$

$$\nabla \cdot \boldsymbol{B} = 0, \tag{2.5.25c}$$

$$\nabla \cdot \boldsymbol{D} = 0, \tag{2.5.25d}$$

where $\boldsymbol{B}, \boldsymbol{D}, \boldsymbol{E}$, and $\boldsymbol{H}$ are the magnetic induction vector, the electric displacement, the electric field vector, and the magnetic field vector in Gaussian units, respectively, and c is the velocity of light in a vacuum.

It can easily be proved by means of Eqns (2.5.25a, b) that Eqns (2.5.25c, d) hold for all time if they are true at a certain initial time, and hence we may regard them as restrictions for the initial values. Consequently, the time evolution is determined according to Eqns (2.5.25a, b). However, these equations alone are, of course, not complete, and we need to specify the constitutive equations relating $\boldsymbol{D}$ and $\boldsymbol{H}$ to $\boldsymbol{B}$ and $\boldsymbol{E}$, respectively. For instance, in an isotropic homogeneous transparent medium, we can put

$$\boldsymbol{D} = \varepsilon \boldsymbol{E}, \qquad \boldsymbol{B} = \mu \boldsymbol{H}, \tag{2.5.26}$$

where ε and μ (both of which are constant) are the dielectric constant and magnetic permeability, respectively.

In general, when $\boldsymbol{D}$ and $\boldsymbol{H}$ are given as functions of $\boldsymbol{E}$ and $\boldsymbol{B}$ according to the constitutive equations, Maxwell's Eqns (2.5.25a, b) can be written in matrix form as $A_\alpha U_{,\alpha} = 0$, where U is a column vector of six components made up of $\boldsymbol{E}(E_i)$ and $\boldsymbol{B}(B_i)$. From this, we find the characteristic equation $\det |A_\alpha \, \partial \varphi / \partial x_\alpha| = 0$. Or, recalling that the equation for an infinitesimal jump δU of U may be derived immediately from the field equations by making the replacement $\partial/\partial t \to -\lambda \delta$, $\nabla \to \boldsymbol{n} \delta$, we obtain from Eqns (2.5.25)

$$\frac{\lambda}{c} \delta \boldsymbol{D} + \boldsymbol{n} \times \delta \boldsymbol{H} = 0, \tag{2.5.27a}$$

$$-\frac{\lambda}{c} \delta \boldsymbol{B} + \boldsymbol{n} \times \delta \boldsymbol{E} = 0, \tag{2.5.27b}$$

$$\boldsymbol{n} \cdot \delta \boldsymbol{D} = 0, \qquad \boldsymbol{n} \cdot \delta \boldsymbol{B} = 0. \tag{2.5.27c}$$

2.5 EXTENSION TO THE MULTI-DIMENSIONAL CASE

Using Eqns (2.5.26) as the constitutive equations we have the well known result

$$\lambda^2 = \frac{c^2}{\varepsilon\mu}. \tag{2.5.28}$$

Also, it easily follows that $\delta \boldsymbol{E}$, $\delta \boldsymbol{B}$, and $\boldsymbol{n}$ are mutually orthogonal, and represent a transverse wave. Applying the definition of λ to Eq. (2.5.28), and using the refractive index $n^2 = \varepsilon\mu$, H in Eq. (2.5.11) becomes

$$H = \frac{1}{2}\left(p_0^2 - \left(\frac{c}{n}\right)^2 \boldsymbol{p}^2\right), \tag{2.5.29}$$

so that the consequences of Eq. (2.5.23) can be applied straightforwardly.

In particular, the characteristic surface passing through any space-time point (x_p, t_p) becomes the light cone

$$(x - x_p)^2 - \left(\frac{c}{n}\right)^2 (t - t_p)^2 = 0. \tag{2.5.30}$$

As is well known, introducing the four-potential $\{A_\alpha\} = (\phi, \boldsymbol{A})$ through

$$\boldsymbol{E} = -c^{-1}\boldsymbol{A}_{,t} - \nabla\phi, \tag{2.5.31a}$$

$$\boldsymbol{B} = \nabla \times \boldsymbol{A}, \tag{2.5.31b}$$

Eqns (2.5.25b, c) of Maxwell's equations are always satisfied, and for the constitutive equations (2.5.26) together with the Lorentz condition $\nabla \cdot \boldsymbol{A} + (\varepsilon\mu/c)\phi_{,t} = 0$, Eqns (2.5.25a, d) reduce to

$$\nabla^2 A_\alpha - \left(\frac{\varepsilon\mu}{c^2}\right)\frac{\partial^2 A_\alpha}{\partial t^2} = 0 \quad (\alpha = 0, 1, 2, 3, 4).$$

The above equation then shows us immediately that

$$(\nabla\varphi)^2 - \frac{\varepsilon\mu}{c^2} \varphi_{,t}^2 = 0.$$

That is, the vanishing of Eq. (2.5.29) gives the characteristic equation (2.5.8′). Since the constitutive relation (2.5.26) is not relativistically invariant, it is not valid when the medium moves with a constant velocity v. However, Eqns (2.5.26) are still valid in a coordinate system moving together with the medium (the medium system). The components of $(\boldsymbol{D}, \boldsymbol{H})$ and $(\boldsymbol{E}, \boldsymbol{B})$ are transformed like the components of an antisymmetrical tensor of rank two. Hence, when the medium moves with velocity $\boldsymbol{v}$, the constitutive equations take the form

$$\boldsymbol{D} + \frac{1}{c}\boldsymbol{v} \times \boldsymbol{H} = \varepsilon\left(\boldsymbol{E} + \frac{1}{c}\boldsymbol{v} \times \boldsymbol{B}\right),$$

$$\boldsymbol{H} - \frac{1}{c}\boldsymbol{v} \times \boldsymbol{D} = \mu^{-1}\left(\boldsymbol{B} - \frac{1}{c}\boldsymbol{v} \times \boldsymbol{E}\right).$$

(Møller (1972)). Expressing $(\mathbf{D}, \mathbf{H})$ in terms of $(\mathbf{E}, \mathbf{B})$ by means of the above equations, and inserting the results into Eqns (2.5.27), we obtain the characteristic equation for a moving medium,

$$-c^2 + \lambda^2 + (\varepsilon\mu - 1)\gamma^2(-\lambda + v_n)^2 = 0 \quad \left(\gamma^2 \equiv \frac{1}{1 - v^2/c^2}\right), \quad (2.5.32)$$

where $v_n = \mathbf{v} \cdot \mathbf{n}$.

The above equation may be derived from physical considerations by transforming the equation $\nabla'^2 A_\alpha - (\varepsilon\mu/c^2)\partial^2 A'_\alpha/\partial t'^2 = 0$ in the medium system into the corresponding equation for the rest system. By virtue of the Lorentz transformation,

$$\mathbf{x} = \mathbf{x}' + \mathbf{v}[(\gamma - 1)\mathbf{x}' \cdot \mathbf{v}/v^2 + \gamma t'],$$
$$t = \gamma(t' + (\mathbf{v}/c^2) \cdot \mathbf{x}'),$$

the operator $\nabla'^2 - (\varepsilon\mu/c^2)\partial^2/\partial t'^2$ is transformed into

$$\nabla^2 - \frac{1}{c^2}\frac{\partial^2}{\partial t^2} + \frac{1 - \varepsilon\mu}{c^2}\gamma^2\left(\frac{\partial}{\partial t} + \mathbf{v}\cdot\nabla\right)^2. \quad (2.5.32')$$

It easily follows from this that the characteristic equation in the rest system reduces to Eq. (2.5.32). Accordingly, the light wave surface diverging from the origin according to the characteristic equation (2.5.32) may be derived from the light cone Eq. (2.5.30) $(x'_p = t'_p = 0)$ by means of the Lorentz transformation. From this, taking the direction of the velocity $\mathbf{v}$ as the x-axis, and putting $\bar{\mathbf{x}} \equiv \mathbf{x} - \mathbf{x}_p$ and $\bar{t} \equiv t - t_p$, the characteristic surface passing through the point $P(\mathbf{x}_p, t_p)$ becomes

$$\frac{\gamma^2\chi}{\varepsilon\mu}\left(\bar{x} - \frac{\varepsilon\mu - 1}{\chi}v\bar{t}\right)^2 + \bar{y}^2 + \bar{z}^2 - \frac{(c\bar{t})^2}{\gamma^2\chi} = 0 \quad \left(\chi \equiv \varepsilon\mu - \beta^2, \beta \equiv \frac{v}{c}\right). \quad (2.5.33)$$

The section of this light cone at any time $\bar{t}$ is an ellipsoid, and it turns out that its centre is at $\bar{x} = (\varepsilon\mu - 1)\chi^{-1}v\bar{t}$, $y = z = 0$, and that the semi-principal axis (in the x-direction) is $(1 - \beta^2)\sqrt{\varepsilon\mu}\, c\bar{t}\chi^{-1}$. Therefore, when the speed of light in the medium exceeds the speed of the medium, so $|v| < c/n$ $(n \equiv \sqrt{\varepsilon\mu})$, Eq. (2.5.33) represents a distorted light cone, enclosing the time axis, with its apex at the point P. Conversely, when $|v| > c/n$, so that the speed of the medium exceeds the speed of light in the medium, the light cone is shifted in the direction of the velocity in such a way that the time axis is always outside the light cone.

In this case, the projection of the light cone (a four-dimensional hypersurface) on to the plane $\bar{t} = 0$ is then a cone whose apex is at the point P in three-dimensional space. Consequently, a surface of discontinuity may also exist in three-dimensional space. This is merely the

2.5 EXTENSION TO THE MULTI-DIMENSIONAL CASE

envelope of the wave surface at each instant of time (the envelope is obtained in the usual manner by eliminating t from Eq. (2.5.33) and the equation obtained by differentiating Eq. (2.5.33) with respect to t), and hence the wave surface does not proceed in the direction of the normal to this surface. Namely we have $\lambda = -\varphi_{,t}/|\nabla\varphi| = 0$ and, from the mathematical point of view, the time-independent Maxwell equations become hyperbolic. In fact, putting $\partial/\partial t \equiv 0$ in Eq. (2.5.32′) brings it to the form

$$\frac{\partial^2}{\partial y^2} + \frac{\partial^2}{\partial z^2} + \gamma^2[1-(\beta n)^2]\frac{\partial^2}{\partial x^2}. \tag{2.5.34}$$

When $|v| > c/n$ we obtain a cone, encircling the x-axis, with its apex at the point P in three-dimensional space

$$\bar{y}^2 + \bar{z}^2 - \frac{1}{\gamma^2[(\beta n)^2 - 1]}\bar{x}^2 = 0 \tag{2.5.35}$$

as a solution of Eq. (2.5.32) for $\lambda = 0$. If we observe this in the medium reference system, then with the aid of the relation $\bar{x} = \gamma\bar{x}'$ (the Lorentz contraction), we find for an arbitrary constant time t' that

$$\bar{y}'^2 + \bar{z}'^2 - [(\beta n)^2 - 1]^{-1}\bar{x}'^2 = 0, \tag{2.5.36}$$

where $\bar{x}' = x' - x'_p$.

Since in the medium system a point source of light waves moves with the speed $|v|$ the above equation is just the well-known result obtained from the Huygens' principle. As the observed direction of propagation of the light waves is normal to the surface of the cone, if we denote by θ the angle between the normal and the x-axis, we find $\cos\theta = (\beta n)^{-1}$. When an electron passes through the medium with a velocity greater than the velocity of light in the medium, the light then observed is the well-known Čerenkov radiation (see, for example, Taniuti (1953)).

A characteristic surface does not always reduce to the Minkowski light cone even in a relativistically invariant theory (Boillat (1965); Taniuti (1959)). In general, a relativistically invariant equation can be derived from a Lagrangian $\mathscr{L}$ which is invariant under a Lorentz transformation. For an electromagnetic field there exist the two invariant quantities $Q \equiv (1/2)(\boldsymbol{E}^2 - \boldsymbol{B}^2)$ and $R \equiv (1/2)(\boldsymbol{E} \cdot \boldsymbol{B})^2$ which are formed from $\boldsymbol{E}$ and $\boldsymbol{B}$ satisfying Eqns (2.5.25b, c), and so the Maxwell equations are relativistically invariant provided $\mathscr{L}$ is a function of these quantities. Varying it with respect to $\boldsymbol{A}$ and ϕ, introduced through Eqns (2.5.31a, b), and deriving the Euler equation, we obtain Eqns (2.5.25a, d) by putting

$$\boldsymbol{D} = \mathscr{L}_{,Q}\boldsymbol{E} + \mathscr{L}_{,R}(\boldsymbol{E}\cdot\boldsymbol{B})\boldsymbol{B}, \tag{2.5.37a}$$

$$\boldsymbol{H} = \mathscr{L}_{,Q}\boldsymbol{B} - \mathscr{L}_{,R}(\boldsymbol{E}\cdot\boldsymbol{B})\boldsymbol{E}. \tag{2.5.37b}$$

Suppose, for simplicity, that $\mathscr{L}_{,R} = 0$. Then, by using Eqns (2.5.27b, c), Eqns (2.5.27a, c) become, respectively,

$$\delta\mathscr{L}_{,Q}E_n + \mathscr{L}_{,Q}\delta E_n = 0, \tag{2.5.38a}$$

$$\delta\mathscr{L}_{,Q}\left(\frac{\lambda}{c}\boldsymbol{E} + \boldsymbol{n}\times\boldsymbol{B}\right) + \frac{c}{\lambda}\mathscr{L}_{,Q}\left[\left(\frac{\lambda^2}{c^2}-1\right)\delta\boldsymbol{E} + \delta E_n\boldsymbol{n}\right] = 0. \tag{2.5.38b}$$

The following equation is also true:

$$\delta\mathscr{L}_{,Q} = \mathscr{L}_{,QQ}\left(\boldsymbol{E}\cdot\delta\boldsymbol{E} - \frac{c}{\lambda}\boldsymbol{B}\cdot\boldsymbol{n}\times\delta\boldsymbol{E}\right). \tag{2.5.38c}$$

Let us take the scalar product of Eq. (2.5.38b) with $\boldsymbol{E}$, and the vector product of Eq. (2.5.38b) with $\boldsymbol{n}$. Eliminating $\boldsymbol{E}\cdot\delta\boldsymbol{E}$, $\boldsymbol{n}\times\delta\boldsymbol{E}$, $\delta\mathscr{L}_{,Q}$ and δE_n from the resulting equations and Eq. (2.5.38a, c), we obtain the characteristic equation

$$(\mathscr{L}' + \mathscr{L}''E^2)\left(\frac{\lambda}{c}\right)^2 - 2(\boldsymbol{E}\times\boldsymbol{B})_n\mathscr{L}''\frac{\lambda}{c} + [B^2 - (E_n^2 + B_n^2)]\mathscr{L}'' - \mathscr{L}' = 0 \tag{2.5.39}$$

(where the prime denotes a derivative with respect to Q). Consequently, for waves propagating into a vacuum with $\boldsymbol{E} = \boldsymbol{B} = \boldsymbol{0}$, we have $(\lambda/c)^2 = 1$ provided $\partial\mathscr{L}/\partial Q|_{Q=0} \neq 0$, and the speed of the wavefront becomes the speed of light. However, as $\lambda/c = \pm 1$ are not normally the roots of Eq. (2.5.39), we conclude that the characteristic cone does not in general correspond to the Minkowski light cone.

For instance, the speed of the wavefront of a light wave moving into a region of a constant magnetic field $\boldsymbol{B}_0$ (and $\boldsymbol{E} = \boldsymbol{0}$) is determined by

$$\left(\frac{\lambda}{c}\right)^2 = 1 - [B_0^2 - (\boldsymbol{B}_0\cdot\boldsymbol{n})^2]\frac{\mathscr{L}_0''}{\mathscr{L}_0'}. \tag{2.5.40}$$

Choosing the z-axis along the direction of $\boldsymbol{B}_0$, and introducing $c' \equiv c(1 - B_0^2\mathscr{L}_0''/\mathscr{L}_0')^{1/2}$ and $\mathscr{L}_0' \equiv (\partial\mathscr{L}/\partial Q)_{Q=-B_0^2/2}$, the above equation reduces to $\varphi_{,t}^2 = c'^2(\varphi_{,x}^2 + \varphi_{,y}^2) + c^2\varphi_{,z}^2$. Thus, although the wave is propagated with the speed c along the magnetic field, it is propagated with the speed c' in a direction perpendicular to the magnetic field, and also with a speed lying between c and c' for propagation in an oblique direction.

In particular, if $\mathscr{L}_0''/\mathscr{L}_0' < 0$, the speed λ is superlight. If $\mathscr{L}_0''/\mathscr{L}_0' > B_0^{-2}$, λ becomes imaginary for some range of $\boldsymbol{n}$, and the original equation ceases to be (totally) hyperbolic. For instance, it becomes elliptic for a variation which is independent of z.

Setting $\boldsymbol{E}\cdot\boldsymbol{B} = 0$ in the constitutive equation (2.5.37), it is easily seen that there generally exists a solution representing a linearly polarized (nonlinear) plane wave. For instance, the plane wave with $\boldsymbol{E} = E\boldsymbol{e}_x$ and

2.5 EXTENSION TO THE MULTI-DIMENSIONAL CASE

$B = Be_z$, propagating in the x-direction, is described by

$$(\mathscr{L}'E)_{,t} + c(\mathscr{L}'B)_{,x} = 0,$$
$$B_{,t} - cE_{,x} = 0.$$

The remaining two Maxwell equations in (2.5.25) are satisfied automatically. The above equation is the same as Eq. (2.2.28) for a scalar field, so that the results obtained there can be applied straightforwardly.

As an example of a relativistically invariant electromagnetic field, we can consider the Heisenberg–Euler equation (Heisenberg and Euler (1936)). This equation provides an approximate description of the photon–photon scattering process due to the effect of virtual pair production-annihilation of electrons and positrons, and the Langrangian is

$$\mathscr{L} = Q + 4aQ^2 + bR,$$

where a and b are constants, $a = \hbar e^4/(360\pi^2 m^4)$, $b = 7\hbar e^4/(180\pi^2 m^4)$. This equation was derived on the basis of perturbation theory applied quantum field theory, and so it does not, of course, hold for very strong fields. The characteristic equation for a linearly polarized plane wave is obtained by setting $E_n = B_n = 0$ in Eq. (2.5.39).

From Problem 2.2.2 we have $\mathscr{L}_{,Q}|_{Q=0} = 1$, so that the wavefront of light propagating into a vacuum moves with the speed of light. In this case, a simple wave does not lead to steepening. When there is a constant magnetic field B_0, the wavefront does not travel with the speed of light, though its speed is then only slightly smaller than the speed of light because $aB_0^2 \ll 1$. However, after a finite time the solution fails to have a meaning because of the steepening which then occurs in the simple wave and, in addition, the classical approximation ceases to hold as the field quantity begins to vary rapidly. We must then take into account the quantum effect instead of the effect of dissipation (Lutzky and Toll (1959)).

As another example of a nonlinear electromagnetic field, we can take the Born–Infeld field (Born and Infeld (1934, 1935)). To remove the singularity of the Coulomb force, this was proposed under the assumption that Maxwell's equations cease to hold in close proximity to a point charge. By means of an appropriate normalization, the Lagrangian can be expressed as

$$\mathscr{L} = \{1 - 2(Q + R)\}^{1/2} - 1.$$

Setting $R = 0$, the present problem reduces to Problem 2.2.2, so that it corresponds to the completely exceptional case as long as the propagation of a polarized plane wave is concerned.

As another example of a relativistically invariant field equation we mention Dirac's equation,

$$\gamma_\mu \frac{\partial \psi}{\partial x_\mu} + \kappa \psi = 0. \qquad (2.5.41)$$

Here, $x_0 = ict$ and ψ is a complex quantity with four components, the γ_μ are 4×4 matrices with complex constant elements and are defined by the commutation relation $\gamma_\mu \gamma_\nu + \gamma_\nu \gamma_\mu = 2\delta_{\mu\nu}$, while κ is a constant. Let us here take κ to be a scalar quantity dependent on ψ as the general case. We remark that, to write Eq. (2.5.41) in terms of real quantities, it suffices to introduce a column vector U with eight components composed of Re ψ and Im ψ. Now, from Eq. (2.5.8), the characteristic equation for Eq. (2.5.41) is

$$\det |\gamma_\mu \varphi_{,\mu}| = 0. \qquad (2.5.41')$$

Squaring the above equation, and using the commutation relation for the γ_μ, we have immediately that $\varphi_{,\mu} \varphi_{,\mu} = 0$, which gives us the Minkowski light cone.

Next, let us consider the dynamical equations (2.2.3) of a perfect gas. By making the replacements $\partial/\partial t \to -\lambda \delta$, $\nabla \to \boldsymbol{n}\delta$, the characteristic equations become

$$-\lambda \, \delta\rho + \delta(\rho v_n) = 0, \qquad (2.5.42a)$$

$$-\lambda \, \delta\boldsymbol{v} + v_n \, \delta\boldsymbol{v} + \boldsymbol{n}\left[\left(\frac{a^2}{\rho}\right)\delta\rho + \frac{1}{\rho}\left(\frac{\partial p}{\partial S}\right)\delta S\right] = 0, \qquad (2.5.42b)$$

$$(-\lambda + v_n)\, \delta S = 0, \qquad (2.5.42c)$$

where $v_n = \boldsymbol{n} \cdot \boldsymbol{v}$. Suppose $\delta S \neq 0$. Then we have from Eq. (2.5.42c)

$$\lambda = v_n. \qquad (2.5.43)$$

We then find $\delta\rho = \delta v_n = 0$, so that it is clear that the wave corresponds to a contact surface (cf. Section 2.4). Denoting by θ the angle between $\boldsymbol{v}$ and $\boldsymbol{n}$, Eq. (2.5.43) may be written in the form $\lambda = v \cos \theta$, and hence, setting $\boldsymbol{v} = \boldsymbol{v}_0$ (constant), the surface of normal velocity represents a circle of diameter $|\boldsymbol{v}_0|$ whose centre is at $\frac{1}{2}\boldsymbol{v}_0$. Accordingly, the wave surface for entropy waves originating from a point source at the origin is located at the point $\boldsymbol{v}_0$ at $t = 1$. So, a point entropy disturbance moves along $\boldsymbol{v}_0$ without spreading out, which is the same as in the one-dimensional case.

Assuming $\lambda \neq v_n$, we have $\delta S = 0$, and from Eqns (2.5.42a, b) we find

$$\lambda = v_n \pm a, \qquad (2.5.44a)$$

$$\rho^{-1} \, \delta\rho = \pm a^{-1} \, \delta v_n. \qquad (2.5.44b)$$

Here, the $\pm$ signs in the above two equations correspond, respectively,

2.5 EXTENSION TO THE MULTI-DIMENSIONAL CASE

and also the transverse component of Eq. (2.5.42b) yields $\delta v_t = 0$. That is, this mode of propagation represents a longitudinal wave (sound wave). From Eq. (2.5.44) the Hamiltonian $\mathcal{H}^{(i)}$ becomes

$$\mathcal{H}^{(i)} = \boldsymbol{v} \cdot \boldsymbol{p} \pm a\sqrt{\boldsymbol{p}^2}. \tag{2.5.45}$$

For waves propagating into a constant state $(\rho_0, \boldsymbol{v}_0, S_0)$, $\boldsymbol{p}$ is constant and the ray velocity is given by

$$\frac{d\boldsymbol{x}}{dt} = \boldsymbol{v}_0 \pm a_0 \frac{\boldsymbol{p}}{\sqrt{\boldsymbol{p}^2}}. \tag{2.5.46}$$

Integrating this, we obtain

$$(\boldsymbol{x} - \boldsymbol{v}_0 t)^2 = (a_0 t)^2, \tag{2.5.47}$$

which is a sphere of radius $a_0 t$ whose centre is moving with the velocity $\boldsymbol{v}_0$. If v_0 exceeds a_0, the time axis is outside the charactersitic surface in the four-dimensional space. Therefore there exists a spatial discontinuity surface, as in the case of Čerenkov radiation. We call this discontinuity surface a Mach wave.

Like the case of Čerenkov radiation, we may find Eq. (2.5.47) by means of a transformation of the coordinate system. In the system moving with the fluid, an infinitesimal variation $\delta \rho$ of the density is governed by $\Box' \delta \rho = 0$ (where $\Box' \equiv \partial^2/\partial t'^2 - a_0^2 \nabla'^2$). Passing into the laboratory system through the Galilean transformation $\boldsymbol{x}' = \boldsymbol{x} - \boldsymbol{v}_0 t$, $t' = t$, the operator $\Box$ is not invariant, but changes into $(\partial/\partial t + \boldsymbol{v}_0 \cdot \nabla)^2 - a_0^2 \nabla^2$. From this, we obtain Eq. (2.5.47) immediately. The characteristic cone is, of course, derived from $\boldsymbol{x}'^2 - (a_0 t)^2 = 0$ according to a Galilean transformation.

When $v_0 > a_0$, the envelope of the wavefronts of waves which emerged from the origin at $t = 0$, is a sphere of radius $a_0 t$, and this corresponds to the wave surface of the Mach wave. Here t serves as a parameter, and the centre of the sphere is at $\boldsymbol{v}_0 t$. The wave surface does not spread itself in the direction of the normal to this wave surface, and hence $\varphi_{,t} = 0$. Hence a Mach wave is a characteristic surface in the steady state case. Putting $\varphi_{,t} = 0$ in Eq. (2.5.44a), the characteristic equation reduces to $(\boldsymbol{v}_0 \cdot \boldsymbol{n})^2 - a_0^2 = 0$ or, equivalently, to $\cos \theta_0 = \pm a_0/|\boldsymbol{v}_0|$. Consequently, there must be a real (vector) $\boldsymbol{n}$ satisfying the above equation in order that the characteristic surface (discontinuity surface) exists in the steady case as well. Thus we must have $|\boldsymbol{v}_0| > a_0$, which is the condition for Eq. (2.2.3) to be hyperbolic in the steady case.

From these considerations of the relationship between a steady spatial discontinuity surface and a wavefront in time-dependent propagation, we can derive the following important general rule. Let the system be invariant under a Galilean transformation, and also be homogeneous, so that the characteristic equation does not depend explicitly on x and t. In the laboratory system, construct a conoidal surface by drawing the tangents from the point $x = -v_0$ to the wavefront surface at $t = 1$. Then, if this is possible, it corresponds to a steady spatial discontinuity surface. In the hydrodynamic case considered here, the wavefront obtained for $t = 1$ and $v_0 = 0$ is a sphere of radius a_0. Hence it is obvious that such a tangential conoidal surface exists only for $|v_0| > a_0$; that is, for the supersonic case.

Problem 2.5.1

Find the wave surface of magnetohydrodynamic waves starting from a point source in a uniform state (Friedrichs and Kranzer (1958); Jeffrey and Taniuti (1964)).
Hint Assume $a \ll b_x$ (Problem 2.3.1).

3
Asymptotic methods

3.1 Far fields for hyperbolic equations

In this chapter we show that when the wave amplitude is small, the wave propagation in many physical systems can be described asymptotically by means of only a few simple equations, such as Burgers' equation, the Korteweg–de Vries equation, the nonlinear Schrödinger equation, and others. For that purpose, let us start by introducing the concept of a far field.

When we take the linear wave equation (2.1.1) as an example, the general solution can be expressed as the superposition of waves moving to the right and left. In general, these two waves are simultaneously excited by an arbitrary initial condition. However, if the initial disturbance is localized in a certain region ($|x| \leq x_0$) such as given by Eq. (2.1.27), that is for an initial disturbance with compact support, after the time $t = x_0$ the disturbance separates into two progressive waves propagating to the right and left, as may be seen from Fig. 2.2.

In general, for an initial disturbance with compact support, the field observed at a sufficiently great distance necessarily corresponds to a progressive wave moving to the right or left, and is not the linear superposition of two such waves (i.e. it corresponds to an eigenstate of the matrix A in Eq. (2.1.10)). Since these progressive waves are general solutions to the following first order equations of one fewer degrees of freedom

$$\phi_{,t} \pm \phi_{,x} = 0, \qquad (3.1.1)$$

we call the fields described by equations (3.1.1), the far fields of the original equation (2.1.1). It should be noticed that in the far field, half of the initial data is lost. Namely, the initial condition for ϕ of Eq. (3.1.1) is, in general, not equal to that for ϕ of Eq. (2.1.1). For instance, in Fig. 2.2, the initial condition for the far field in the domain 2 must be specified on the boundary line between 2 and 6. However, this boundary line is a characteristic of Eq. (2.1.1), and hence ϕ and $\phi_{,t}$ cannot be given arbitrarily on this line. This is the so-called characteristic initial value problem. For this reason, the number of degrees of freedom of the far field is less than that for the original equation.

We remark that, originally, the concept of a far field came from the idea of finding out properties inherent to a given evolution equation which do not depend in any sensitive manner on the details of the initial conditions. Thus such properties will be observed after a suitably long period of time for a wide class of initial conditions. In this sense, it is not really meaningful to investigate the initial value problem for the far field equation (3.1.1) in detail. Rather, it will be of greater significance for us to study the asymptotic behaviour which will be observed after a suitably long lapse of time. In fact, these far fields are merely the simple waves mentioned in Section 2.1.

With these ideas in mind, the concept of a far-field can easily be extended, not only to a general linear hyperbolic system of equations, but also to the nonlinear hyperbolic system of equations with n components, (2.2.21'), that is,

$$U_{,t} + A(U)U_{,x} = 0. \tag{3.1.2}$$

The far fields are described by Eqns (2.2.27c') which govern simple waves, that is, by the n equations

$$u_{1,t} + \lambda_j(u_1)u_{1,x} = 0 \qquad (j = 1, 2, \ldots, n). \tag{3.1.3}$$

For initial data with compact support, Friedrichs' theorem asserts that the domain adjacent to a constant state is necessarily a simple wave (until such time as a shock forms). It shows that, even in the nonlinear case, the signal arriving first at a sufficiently great distance is necessarily in the form of a simple wave, and it provides the justification for regarding a simple wave as a far field.

Now, when wave amplitudes are small, we may find the solution u_1 to Eq. (3.1.3) (which we shall hereafter denote by u) by means of a perturbation calculation. Let ε be a small parameter, and let us expand u about a constant value $u^{(0)}$ in powers of ε by writing

$$u = u^{(0)} + \varepsilon u^{(1)} + \varepsilon^2 u^{(2)} + \cdots. \tag{3.1.4}$$

Substituting this expression into Eq. (3.1.3), equating coefficients of corresponding powers of ε and using the fact that $u^{(0)} \equiv \text{const.}$, we arrive at the results

$$O(\varepsilon) \qquad u^{(1)}_{,t} + \lambda_0 u^{(1)}_{,x} = 0, \tag{3.1.5a}$$

$$O(\varepsilon^2) \qquad u^{(2)}_{,t} + \lambda_0 u^{(2)}_{,x} = -\lambda_{,u^{(0)}} u^{(1)} u^{(1)}_{,x}, \tag{3.1.5b}$$

where $\lambda_0 \equiv \lambda(u^{(0)})$, $\lambda_{,u^{(0)}} \equiv d\lambda/du|_{u=u^{(0)}}$, and the subscript j in λ_j has been omitted.

Transforming to a new moving system of coordinates x', t' by means of the equations

$$x' = x - \lambda_0 t, \qquad t' = t, \tag{3.1.6}$$

we find from Eq. (3.1.5a) that $u^{(1)} = f(x')$. Inserting this result into Eq. (3.1.5b), we get $u^{(2)} = -\lambda_{,u^{(0)}} f f_{,x'} \cdot t$. The solution will cease to have meaning for a time of the order of $t \sim 1/\varepsilon$, for then we have $|\varepsilon^2 u^{(2)}| \sim |\varepsilon u^{(1)}|$, so that terms with different orders (powers) of ε will be of the same magnitude.

A term proportional to t is called a secular term, and difficulty due to terms of this kind often appear in the perturbation treatment of nonlinear equations. As a result, the above simple perturbation method proves to be useless for the investigation of the long-time behaviour of a solution in a nonlinear far field.

An alternative method is to solve exactly the nonlinear equation which is obtained by substituting the expansion Eq. (3.1.4) for u into Eq. (3.1.3), and then to neglect $\varepsilon^2 u^{(2)}$ and the higher order terms. Making the transformation (3.1.6), we finally obtain $u^{(1)}_{,t} + \varepsilon \lambda_{,u^{(0)}} u^{(1)}_{,x'} = 0$, in which the orders of the first and second terms appear to be different. However, if we consider the behaviour of the system after a long time of the order ε^{-1}, we have $t' = \varepsilon t \sim 1$. Therefore, in place of Eq. (3.1.6), introducing a transformation to stretch the time scale,

$$x' = x - \lambda_0 t, \qquad t' = \varepsilon t, \tag{3.1.7}$$

we find that Eq. (3.1.3) may be approximated by

$$u^{(1)}_{,t'} + \lambda_{,u^{(0)}} u^{(1)} u^{(1)}_{,x'} = 0. \tag{3.1.8}$$

Since the equation is exactly solvable, its reduction to the nonlinear equation (3.1.8) is not so effective, though Eq. (3.1.8) is more easily solved than Eq. (3.1.3).

It should be remarked, however, that Eq. (3.1.8) gives rise to a class of approximate solutions to the complicated system (3.1.2). In what follows, we will derive Eq. (3.1.8) from Eq. (3.1.2) immediately, without proceeding via the above course of reduction; namely, Eq. (3.1.2) → Eq. (3.1.3) → Eq. (3.1.8). At the same time we will consider further the meaning of the transformation (3.1.7).

To begin with, let us expand the vector U with n components in equation (3.1.2) about a constant state $U^{(0)}$, using powers of a parameter ε which denotes the smallness of the amplitude

$$U = U^{(0)} + \varepsilon U^{(1)} + \varepsilon^2 U^2 + \cdots. \tag{3.1.9}$$

Since the amplitude of this wave is now small enough, this suggests that the solution of Eq. (3.1.2) approaches that of the linearized form of the equation. In the linear case, the families of characteristics are merely parallel straight lines. As time evolves, because of nonlinear effects, the true characteristics deviate from the linear ones and hence, after a sufficiently long time, the characteristics in any one family will cease to be parallel. Consequently, in a far field, nonlinear effects such as the crossing of characteristics cannot be neglected. To show this explicitly, let us expand the following characteristic equation in powers of ε

$$\frac{dx}{dt} = \lambda_j, \qquad (3.1.10)$$

and then retain only terms of the first order in ε, to get

$$\frac{dx'}{dt} = \varepsilon \sum_{k=1}^{n} u_k^{(1)} \left(\frac{\partial \lambda_j}{\partial u_k}\right)_{U=U^{(0)}} \equiv \varepsilon U^{(1)} \cdot \nabla_u \lambda_{j0}, \qquad (3.1.10')$$

where $x' = x - \lambda_{j0} t$ and $\lambda_{j0} \equiv \lambda_j(U^{(0)})$. Hereafter, the subscript 0 will be used to denote the value of any quantity when $U = U^{(0)}$.

Equation (3.1.10') means that the characteristics deviate from parallel straight lines by an amount of the order ε. If we measure the time using the scale εt, the phase velocity in the frame moving with the speed λ_{j0}, $(dx'/d(\varepsilon t))$, becomes of the order of unity. Thus, for small $t \ll \varepsilon^{-1}$, the phase velocity can be well approximated by the phase velocity in the linear case. However, for long times ($t \sim \varepsilon^{-1}$), the characteristics deviate appreciably from those in the linear case, even if the amplitude is very small ($\varepsilon \ll 1$). It may be inferred from this that a far field can be described by using the stretched time $t \sim \varepsilon^{-1}$. Consequently, from now on, we shall consider solutions at both a long-distance and a long-time which will evolve from a localized disturbance.

In Eq. (3.1.10'), $\nabla_u \lambda_{j0}$ is constant, and hence $U^{(1)}$ becomes a function of x' and εt. As a result the time derivative of $U^{(1)}$ is small and of the order ε. That is, along the linear parallel equi-phase lines $x' = \text{const.}$, $U^{(1)}$ varies slowly with time. As a result of these considerations we are led to introduce the transformation Eq. (3.1.7) quite naturally. Accordingly, using the relations $\partial/\partial x \equiv \partial/\partial x'$, $\partial/\partial t \equiv \varepsilon \partial/\partial t' - \lambda_{j0} \partial/\partial x'$, introducing the expansion Eq. (3.1.9) into Eq. (3.1.2) and expanding the matrix A as

$$A = A_0 + \varepsilon U^{(0)} \cdot \nabla_u A_0 + \cdots, \qquad (3.1.11)$$

we first find for the lowest order $O(\varepsilon)$ that

$$(\lambda_{j0} I - A_0) U^{(1)}_{,x'} = 0, \qquad (3.1.12)$$

and for the next order, $O(\varepsilon^2)$, that

$$(-\lambda_{j0} I + A_0) U^{(2)}_{,x'} + U^{(1)}_{,t'} + (U^{(1)} \cdot \nabla_u A_0) U^{(1)}_{,x'} = 0. \qquad (3.1.13)$$

3.1 FAR FIELDS FOR HYPERBOLIC EQUATIONS

Solving Eq. (3.1.12) algebraically for $U_{,x'}^{(1)}$ we have

$$U^{(1)} = \varphi_j^{(1)} R_{j0} + V_j^{(1)}, \tag{3.1.14}$$

where R_{j0} is the right eigenvector of the matrix A_0 corresponding to the eigenvalue λ_{j0}, and $\varphi_j^{(1)}$ is a function of x' and t'. Here $V_j^{(1)}$ is a vector arising as a result of the integration, and it is only a function of t'. It is to be determined by the boundary conditions. Moreover, $\varphi_j^{(1)}$ cannot be determined from this equation, and it must be determined later. Next, Eq. (3.1.13) may be regarded as an algebraic equation for $U_{,x'}^{(2)}$, but we have $\det|-\lambda_{j0}I + A_0| = 0$. Hence, if the left eigenvector of the matrix A_0 corresponding to the eigenvalue λ_{j0} is denoted by L_{j0}, the necessary and sufficient condition for the solution of this equation to exist becomes

$$L_{j0} \cdot [U_{,t'}^{(1)} + (U^{(1)} \cdot \nabla_u A_0) U_{,x'}^{(1)}] = 0. \tag{3.1.15}$$

Assuming, for simplicity, that $V_j^{(1)} = 0$ (hereafter, the subscript j will be omitted), and inserting Eq. (3.1.14) into the above condition (3.1.15), we obtain the nonlinear equation for $\varphi^{(1)}$:

$$\varphi_{,t'}^{(1)} + \alpha \varphi^{(1)} \varphi_{,x'}^{(1)} = 0. \tag{3.1.16}$$

Here, α is defined by the expression $\alpha = (L_0 \cdot (R_0 \cdot \nabla_u A_0) R_0)/(L_0 \cdot R_0)$, and after some manipulation it turns out that α may be expressed in the alternative form

$$\alpha = \nabla_u \lambda_0 \cdot R_0. \tag{3.1.17}$$

Since for a simple wave we have $\lambda_{,u_1} \equiv \sum_{k=1}^{n} (\partial \lambda / \partial u_k) du_k/du_1 = \nabla_u \lambda \cdot R$, we find that $\alpha = (\lambda_{,u_1})_0$. This expression for α can also be obtained directly from Eq. (3.1.10). The characteristic equation of (3.1.16) is $dx'/dt' = \alpha \varphi^{(1)}$, so that comparing this equation with Eq. (3.1.10') yields the result Eq. (3.1.17). So far we have assumed that $V^{(1)}$ in Eq. (3.1.15) is zero. However, we can generally derive Eq. (3.1.16) without this assumption (Taniuti (1974)).

Thus, in the asymptotic sense, the far field of the general hyperbolic system Eq. (3.1.2) can be described by means of the simple nonlinear equation (3.1.16) which is solvable exactly. We call the present asymptotic method the reductive perturbation method (RPM), in the sense that it enables us to reduce the complicated nonlinear system of equations (3.1.2) to the tractable nonlinear equation (3.1.6) of reduced rank.

There is an exceptional case in which α in Eq. (3.1.17) becomes zero. This usually arises in the following two cases. As already stated in Section 2.3, one is the case that Eq. (3.1.2) is exceptional (i.e. not genuinely nonlinear), so that $\nabla_u \lambda \cdot R = 0$. Examples of this are the entropy wave in hydrodynamics and the Alfvén wave in magnetohydrodynamics (Problem 2.4.4), etc. In these cases, the characteristics do not cross, and the steepening of a wave leading to the shock formation process does not occur. The other case arises when $\nabla_u \lambda_0 = 0$. In this case, a different stretching of the time scale must be introduced to account for the nonlinear effects. For example, putting $\lambda = u_1^2$ in Eq. (3.1.3), and considering the expansion about $u_1 = 0$, we arrive at the result $\nabla_u \lambda_0 = 0$. Setting $u_1 = \varepsilon u$, Eq. (3.1.3) then reduces to $\partial u/\partial t + \varepsilon^2 u^2 \, \partial u/\partial x = 0$, and we obtain the transformation $t' = \varepsilon^2 t$.

The transformation Eq. (3.1.7) is closely related to the fact that Eq. (3.1.16) is invariant under the following scale changes $\varphi^{(1)} \to \varepsilon \varphi^{(1)}$, $x' \to x'$, $t' \to \varepsilon^{-1} t'$. However, a transformation of this kind (a similarity law) is not unique. That is, Eq. (3.1.16) is also invariant under the following scaling:

$$\varphi^{(1)} \to \varepsilon \varphi^{(1)}, \qquad x' \to \varepsilon^{-\gamma} x', \qquad t' \to \varepsilon^{-(\gamma+1)} t', \tag{3.1.18}$$

where γ is an arbitrary constant. The above scaling may be found from Eq. (3.1.10'), and also the corresponding transformation

$$\xi = \varepsilon^\gamma (x - \lambda_0 t), \qquad \tau = \varepsilon^{\gamma+1} t, \tag{3.1.19}$$

similarily leads to Eq. (3.1.16). A transformation of this kind, representing a similarity law for a nonlinear wave equation, was introduced by Gardner and Morikawa (1960) in the case of a solitary wave in a plasma propagated normally across a magnetic field. For this reason, hereafter we shall call Eq. (3.1.19) the Gardner–Morikawa transformation, or more simply, the G–M transformation.

Since γ is arbitrary, for large values of γ, solutions vary more slowly over space and time than they do for smaller values. This implies that, for hyperbolic equations, the far field is not determined uniquely. The breaking time t_B after which smooth solutions cease to exist is determined by the spatial behaviour of the initial conditions, so that, in the case of hyperbolic equations, the time scale for the far-field must be less than the breaking time t_B. Accordingly, the concept of a far field in the hyperbolic case is not meaningful unless t_B is large enough. For instance, it is possible that a shock is formed before an initial disturbance breaks up into simple waves (Jeffrey (1976)).

Thus, for nonlinear hyperbolic equations, the γ which determines the space and time scales depends on the smoothness of the spatial change of the initial data, and cannot be determined uniquely from the properties inherent in the equations, such as the similarity law. In any case, for hyperbolic equations, within the domain of smooth solutions it is not meaningful to consider any really long time evolution of a solution. In order to introduce the concept of a far field, under these circumstances we must extend solutions to the class of weak solutions, since these admit discontinuities such as shocks.

In the next section, instead of considering weak solutions, we apply the above asymptotic method (reductive perturbation method) to physical systems with dissipative terms. In this case, generally speaking, smooth solutions exist for all time, so that $t_B = \infty$, unlike the case of hyperbolic equations. Thus we can quite sensibly consider an arbitrarily long time evolution of a solution. It can also be proved that the γ in Eq. (3.1.18) is then uniquely determined by the similarity law which is involved.

3.2 Dissipative systems

As already mentioned, generally, in hyperbolic equations, a smooth wave breaks at a finite time, so that hyperbolic equations are not adequate to describe the long time behaviour of physical systems. In any actual physical system, when a wave begins to break, some physical effect acts so as to prevent the breaking process. For example, in the case of a perfect gas, the viscosity begins to work effectively as the sound wave steepens and these two effects, the nonlinearity leading to breaking and the dissipation which causes smoothing, then balance one another. As a result, a shock wave is formed and propagated as already described in Section 2.4.

Such a physical phenomenon may be described by the following model equation:

$$u_{,t} + a(u)u_{,x} - \mu u_{,xx} = 0, \qquad (3.2.1)$$

where the third term corresponds to the dissipative effect provided μ is a positive constant. In particular, when $a(u)$ is proportional to u, Eq. (3.2.1) reduces to the Burgers equation (1.2.17) which was mentioned previously:

$$u_{,t} + \alpha u u_{,x} - \mu u_{,xx} = 0.$$

Unlike the case of nonlinear hyperbolic equations, this equation admits a solution in the form of a wave moving with a constant velocity without change of shape. The form of this progressive wave solution, or stationary solution, was given in Eq. (1.3.1).

It was shown in Section 2.4 that this solution represents a weak shock wave (shock) in gas dynamics, so let us now investigate the similarity law for this shock. The shock width is inversely proportional to the strength $|u'_\infty|$. In other words, the stationary solution Eq. (1.3.1) remains invariant when $u'(u'_\infty)$ is multiplied by a factor ε and at the same time x' is multiplied by ε^{-1}. Expressing this transformation in terms of the variables in the original coordinate system (laboratory system), we have

$$u \to \varepsilon u, \qquad \lambda \to \varepsilon \lambda, \qquad x \to \varepsilon^{-1} x, \qquad t \to \varepsilon^{-2} t. \qquad (3.2.2)$$

Under these transformations, the time derivatives u_t the nonlinear term uu_x, and the dissipative term u_{xx} are each multiplied by ε^3, so that the original equation (1.2.17) is invariant. In this sense, we may call the shock Eq. (1.3.1) the invariant far field.

Thus the similarity law for the Burgers equation uniquely determines the shock structure. This similarity law for the Burgers equation is different from the similarity law Eq. (3.1.18) for the nonlinear hyperbolic equation obtained by putting $\mu = 0$. We see that the uncertainty in the similarity law Eq. (3.1.18) for an hyperbolic equation does not appear in the similarity law Eq. (3.2.2) for the Burgers equation. The similarity law Eq. (3.1.18) becomes the similarity law Eq. (3.2.2) of the shock only when the arbitrary constant γ in Eq. (3.1.18) equals unity. Actually, when the transformation Eq. (3.1.18) is assumed, we have $\partial/\partial t \sim \varepsilon^{\gamma+1}$, $u\,\partial/\partial x \sim \varepsilon^{\gamma+1}$ and $\partial^2/\partial x^2 \sim \varepsilon^{2\gamma}$, and we find $\gamma = 1$ from the condition $2\gamma = \gamma + 1$.

We thus see that the relationship between the shock width and the strength in the shock solution of the Burgers equation is due to the similarity law Eq. (3.2.2) of the Burgers equation. The crucial point for this similarity law is that ε is connected with the amplitude as well as with the space-time scaling. This may be exhibited by rewriting the shock solution (for $\gamma = 1$) Eq. (1.3.1) in the following form:

$$u = u_\infty + \varepsilon\left(1 - \tanh\left[\frac{1}{2\mu}\{\varepsilon(x - u_\infty t) - \varepsilon^2 t\}\right]\right), \tag{3.2.3}$$

in which u tends asymptotically to u_∞ as $x \to \infty$ and the linear phase velocity λ_0 is given by u_∞, while the shock velocity λ becomes $\lambda = u_\infty + \varepsilon$. This form shows explicitly the G–M transformation for $\gamma = 1$,

$$\xi = \varepsilon(x - \lambda_0 t), \qquad \tau = \varepsilon^2 t. \tag{3.2.4}$$

The above ordering can also be derived without explicitly solving the Burgers equation; that is, without using Eq. (3.2.3). To see this, let us consider a wave proceeding to the right with constant velocity. Suppose that the solution decays exponentially to a constant value u_∞ as $x \to \infty$. For simplicity, let u_∞ be zero, then for $x \gg 1$ let us assume u to be of the form

$$u \propto e^{-\kappa(x - \lambda t)} \qquad (\kappa > 0, \lambda > 0),$$

which becomes sufficiently small as $x \to +\infty$. Hence, in such a region, the nonlinear term in the Burgers equation can be neglected, so that the linear approximation may be used to give

$$\lambda\kappa = \mu\kappa^2, \quad \text{or} \quad \lambda = \mu\kappa. \tag{3.2.4'}$$

On the other hand, when u approaches a constant $u_{-\infty} \sim \varepsilon$ as $x \to -\infty$, integrating Burgers' equation once with respect to x from $-\infty$ to $+\infty$ gives

3.2 DISSIPATIVE SYSTEMS

the shock condition Eq. (1.3.7). That is, $\lambda = \frac{1}{2}u_{-\infty}$, which leads to the ordering $\lambda \sim \varepsilon$ and consequently, from Eq. (3.2.4′), to the ordering $\kappa \sim \varepsilon$. If κ is replaced by ik, the asymptotic form of u becomes a plane wave, when Eq. (3.2.4′) is then the linear dispersion relation. Hence the orderings can be derived from the linear dispersion relation. Although Eq. (1.3.7) is an essentially nonlinear relationship by means of which we are led to the ordering $\kappa \sim u$, this can also be deduced from the linear dispersion relation, as will be shown below, Eq. (3.2.9).

We now extend Eq. (3.2.1) to the case of a vector U with n components, and consider a general dissipative system which becomes hyperbolic in the non-dissipative limit as, for example, occurs in the equation:

$$U_{,t} + AU_{,x} + K_1(K_2 U_{,x})_{,x} = 0, \tag{3.2.5}$$

where K_1 and K_2 are $n \times n$ matrices which are functions of the vector U. For this system we may combine the results of the reductive perturbation method applied to the hyperbolic equation (3.1.2) and the similarity law of the Burgers equation.

Let λ_0 a specific non-degenerate eigenvalue of the matrix A_0 appropriate to the constant state $U^{(0)}$, and assume that the dispersion relation for small wave number k of the linearized system about $U^{(0)}$ is given by

$$\omega = \lambda_0 k + i\mu k^2 + \cdots \qquad (\mu < 0).$$

Then it is easily anticipated that Eq. (3.2.5) may be reduced to Burgers' equation by means of the G–M transformation Eq. (3.2.4) with $\gamma = 1$. The G–M transformation suggests that, in the frame moving with the phase velocity λ_0, the nonlinear equation (3.2.5) admits approximate solutions which vary slowly over space and more slowly with time (i.e. with $\tau = \varepsilon^2 t$) because of the weakly nonlinear and weakly dissipative effects. In fact, introducing the above transformation into Eq. (3.2.5), and using $\partial/\partial x \equiv \varepsilon\, \partial/\partial \xi$, $\partial/\partial t \equiv -\varepsilon\lambda_0\, \partial/\partial \xi + \varepsilon^2\, \partial/\partial \tau$, equating the lowest order terms $O(\varepsilon^2)$ yields

$$(\lambda_0 I - A_0) U^{(1)}_{,\xi} = 0,$$

so that we have

$$U^{(1)} = \varphi^{(1)}(\xi, \tau) R_0, \tag{3.2.6}$$

where we have assumed the boundary condition $U^{(1)} \to 0$ as $\xi \to \infty$. The dissipative term does not appear to this order, and we get the same result as in the case of a hyperbolic equation.

To the order $O(\varepsilon^3)$, we have

$$(-\lambda_0 I + A_0) U^{(2)}_{,\xi} + U^{(1)}_{,\tau} + U^{(1)} \nabla_u A_0 U^{(1)}_{,\xi} + K_0 U^{(1)}_{,\xi\xi} = 0, \tag{3.2.7}$$

where $K_0 = (K_1 K_2)_{U=U^{(0)}}$. As the necessary and sufficient condition for the existence of solutions of the above algebraic equation for $U^{(2)}_{,\xi}$, the same procedure as the one we used for the hyperbolic system gives

$$\varphi^{(1)}_{,\tau} + \alpha \varphi^{(1)} \varphi^{(1)}_{,\xi} + \mu \varphi^{(1)}_{,\xi\xi} = 0, \tag{3.2.8}$$

which is the equation determining $\varphi^{(1)}$, and it becomes Burgers' equation if $\mu < 0$. Here, α and μ are, respectively,

$$\alpha = \nabla_u \lambda_0 \cdot R_0, \qquad \mu = \frac{L_0 K_0 R_0}{L_0 \cdot R_0}. \tag{3.2.9}$$

Of course, the α is the same as that in the case of the hyperbolic system. Also, μ is found from the linear dispersion relation. This can be proved as follows. We expand the linear dispersion relation of system Eq. (3.2.5), in terms of k as follows:

$$(-i\omega I + ikA_0 - k^2 K_0) \delta U = 0.$$

Because of the fact that $\omega = \varepsilon \omega_0 + \varepsilon^2 \omega_1 + \cdots$, $k = \varepsilon k_0$, and $\delta U = \varepsilon \delta U^{(1)} + \varepsilon^2 \delta U^{(2)} + \cdots$, to the order $O(\varepsilon^2)$ we find

$$(-i\omega_0 I + ik_0 A_0) \delta U^{(1)} = 0,$$

and to the order $O(\varepsilon^3)$, we have

$$(-i\omega_0 I + ik_0 A_0) \delta U^{(2)} + (-i\omega_1) \delta U^{(1)} - k_0^2 K_0 \delta U^{(1)} = 0,$$

and hence we arrive at the results

$$\omega_0 = \lambda_0 k_0, \qquad \omega_1 = ik_0^2 \frac{L_0 K_0 R_0}{L_0 \cdot R_0}.$$

Thus, without using Eq. (3.2.9), μ can be obtained as the coefficient of the second order term in the expansion of the dispersion relation with respect to k.

However, there are two exceptional cases given by $\alpha = 0$ and $\mu = 0$. The former case occurs under the same conditions we mentioned in connection with the nonlinear hyperbolic equation (3.1.2), so that we will call this case the nonlinearly exceptional case. The latter case, which occurs when $L_0 K_0 R_0 = 0$, is due to a property of the system of the linearized equations. Therefore, we will call this case the linearly exceptional case. In this case, a system of equations including the second derivatives of U, such as given by Eq. (3.2.5), does not necessarily exhibit a dissipative effect. This occurs, for instance, when the two-fluid equations for a collisionless plasma are approximated in the case of long wavelengths (i.e. the magnetohydrodynamic approximation is used). In this case, the system of equations takes a form similar to that of Eq. (3.2.5). However, $K_0 R_0$ is then orthogonal to L_0, and the term containing the second derivative of U leads to a dispersive effect instead of to dissipation. The system of equations cannot then be reduced to Burgers' equation, but to the Korteweg–de Vries equation which will be discussed in the next section (Taniuti and Wei (1968); Kakutani *et al.* (1968); Karpman (1975)).

Finally we notice that in the higher order approximations to give $U^{(n)}$ ($n \geq 2$) secular terms appear. (This is readily checked by means of the one-shock solution in gas dynamics.) In order to eliminate secular terms in higher order approximations Kodama and Taniuti (1979a), using a renormalization technique, showed that when the dispersion relation is expanded in the form

$$\omega = \lambda_0 k + \sum_{n=1}^{\infty} i^n \mu^{(n)} k^{n+1} \qquad (\mu^{(1)} \equiv \mu < 0)$$

Burgers' equation (3.2.8) should be modified to

$$\tilde{\varphi}_{,\tau}^{(1)} + \alpha \tilde{\varphi}^{(1)} \tilde{\varphi}_{,\xi}^{(1)} + \mu \tilde{\varphi}_{,\xi\xi}^{(1)} + \frac{2\mu}{\alpha} \sum_{m \geq 3}^{\infty} a_{m-1} \frac{\partial}{\partial \xi} C_m = 0,$$

where

$$a_m = \varepsilon^{m-1} \mu^{(m)}$$

and

$$C_m = -\left(\frac{\partial^m \psi}{\partial \xi^m}\right) \Big/ \psi,$$

in which ψ is given by the equation

$$\psi_{,\tau} = -\sum_{m \geq 2}^{\infty} a_{m-1} \frac{\partial^m \psi}{\partial \xi^m}$$

and is linked with $\tilde{\varphi}^{(1)}$ through the Cole–Hopf transformation (1.2.18) which takes the form $\tilde{\varphi}^{(1)} = (2/\alpha) \mu \psi_{,x}/\psi$.

Consequently, the above nonlinear equation is equivalent to a linear equation for ψ and so can be solved exactly. C_m may be obtained by means of the formula

$$C_{n+1} = C_{n,\xi} - \tilde{\varphi}^{(1)} C_n$$

where $C_0 = -1$, $C_1 = \tilde{\varphi}^{(1)}$, $C_2 = -(\tilde{\varphi}^{(1)})^2 + \tilde{\varphi}_{,\xi}^{(1)}$ and $C_3 = \tilde{\varphi}_{,\xi\xi}^{(1)} - 3\tilde{\varphi}^{(1)} \tilde{\varphi}_{,\xi}^{(1)} + \tilde{\varphi}^{(1)})^3$, etc.

3.3 Weakly dispersive systems

So far we have mentioned that the steepening due to the nonlinearity is smeared by the dissipation of a medium and, consequently, a smooth wave is propagated. However, a nonlinear wave can also propagate itself without breaking by virtue of the dispersion that may be present in a medium.

Generally speaking, dispersion means that the (phase) velocities of waves vary with the wavelength. If the wave amplitude is small, so that the variation of the phase velocity due to the nonlinearity can be neglected, the dispersion is determined by the dispersion relation of the linearized equation. In particular, when the group velocity is almost constant in some range of wavelengths, so that $|d\omega/dk| \gg |d^2\omega/dk^2|$, the dispersion is said to be weak; on the other hand, when the above inequality does not hold, the dispersion is said to be strong.

For example, in the case of the transmission of light through a transparent medium, the refractive index is almost constant for long wavelengths, and hence the dispersion is weak, though it becomes strong as the wavelength approaches that of the resonance absorption. We shall deal here with situations in which the dispersion is weak for long wavelengths and becomes strong for short wavelengths (but far from the resonance wavelengths).

Let a wave be smooth enough at the outset, so that the dispersion is weak. Then such a wave can become steep on account of a nonlinear effect. As a direct consequence, the short wavelength components are then excited, and their phase velocities vary appreciably with wavelength. Hence the various components of the wave giving rise to the steepening will be dispersed with different velocities, and a smooth wave will propagate. For instance, let us consider the case that the phase velocities become smaller for shorter wavelengths. Then it is easily predicted that, as the wave steepens because of the increasing of the local phase velocity due to the nonlinearity, this increase is balanced by a decrease owing to the dispersion, so that the wave is propagated without any overtaking.

As was mentioned in Section 1.2, we can cite the well-known Korteweg–de Vries (KdV) equation (1.2.19),

$$u_{,t} + uu_{,x} + \mu u_{,xxx} = 0,$$

as a model equation describing nonlinear waves in which the nonlinearity and the dispersion are balanced one against the other. Here μ is a positive or negative constant characterizing the dispersive effect. We take note of the fact that, unlike the Burgers equation, the last term (dispersive term) in the KdV equation involves a third-order spatial derivative. This difference is extremely important since the properties of the corresponding solutions are entirely different.

First of all, linearizing the solution about a constant value u_0, we obtain the dispersion relation

$$\omega = u_0 k - \mu k^3, \tag{3.3.1}$$

from which it follows that, unlike the Burgers equation, the frequency ω is always real and the wave never decays exponentially even for sufficiently short wavelengths. Since the phase velocity $\lambda \ (= \omega/k)$ is

$$\lambda = u_0 - \mu k^2, \tag{3.3.2}$$

the dispersion is weak in the range of long wavelengths. Now, as in the Burgers equation, the KdV equation admits a solution in the form of a persistent progressive wave, that is, a steady profile wave moving with a constant speed. The derivation of the solution is similar to that of the shock solution Eq. (1.3.1) for Burgers' equation. In particular, when seeking a solution which approaches a constant state at infinity for the case when μ is positive, we obtain

$$u = u_\infty + \varepsilon \operatorname{sech}^2 \left[\frac{1}{(12\mu)^{1/2}} \{ \varepsilon^{1/2}(x - u_\infty t) - \tfrac{1}{3}\varepsilon^{3/2} t \} \right]. \tag{3.3.3}$$

A remarkable feature of this solution is that u approaches the same value u_∞ as $x \to \pm\infty$. Namely, unlike the shock wave Eq. (3.2.3), the wave profile is symmetric with respect to its centre and a wave packet moves with the constant speed $\lambda = (u_\infty + \varepsilon/3)$. This wave is called a solitary wave. However, it has come to be called a soliton since Zabusky and Kruskal (1965) discovered, by means of numerical calculations, that these waves behave like particles involved in mutual collisions.

Zabusky and Kruskal investigated the time evolution of the solutions to the KdV equation (1.2.19) for the initial condition $u(x, 0) = \cos \pi x$, under the assumption of periodic boundary conditions (period $L = 2$), where the coefficient of the dispersive term was taken to be $\sqrt{\mu} = 0.022$. Since μ was small, the initial contribution from the dispersive term was negligibly small compared to the nonlinear term ($\{\max |\mu \, \partial u^3/\partial x^3|/\max |u \, \partial u/\partial x|\} \approx 0.005$). When the dispersive term is neglected, the formal solution of the KdV equation reduces to the implicit solution $u = \cos \pi(x - ut)$. That is, steepening then occurs on account of the nonlinear effect, and for $t = t_B = 1/\pi$, the solution becomes discontinuous at $x = \tfrac{1}{2}$. However, as the solution steepens, there comes a time beyond which the dispersive term can no longer be neglected. In fact, when considering the full KdV equation, at the time $t = t_B$, there appears in the vicinity of $x = \tfrac{1}{2}$, an oscillation due to the dispersive effect which is interacting with the nonlinearity. This may be seen quite clearly in the curve B of Fig. 3.1. For $t = 3.6 t_B$, a row of eight solitons can be seen.

It is easily confirmed that the amplitudes, widths, and propagation speeds of these solitons found by means of numerical calculation satisfy the similarity law Eq. (3.3.4) for the stationary solution Eq. (3.3.3). The paths followed by these solitons are shown in Fig. 3.2. The solitons alter their speeds only when they collide with each other (in this case they are accelerated). However, after the collision, they again propagate with the

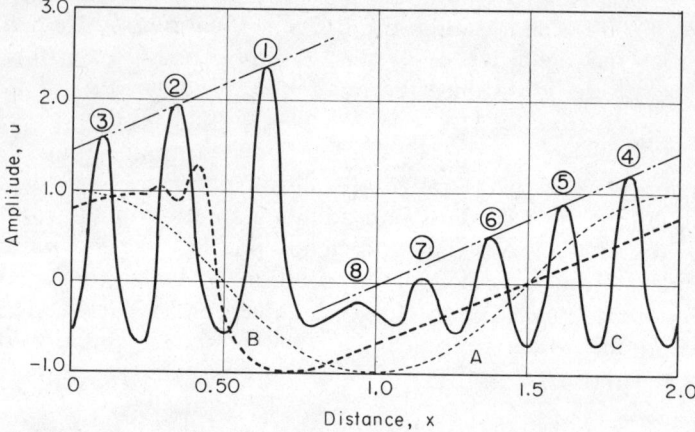

Fig. 3.1 Numerical solution to the KdV equation (1.2.19) ($\sqrt{\mu}=0.022$) for the initial condition $u(x, 0) = \cos \pi x$. A: $t = 0$, B: $t = 1/\pi$, C: $t = 3.6/\pi$ (Zabusky and Kruskal (1965)).

same speeds as before. Thus a soliton preserves its identity throughout any number of collisions. Also, although its amplitude decreases during a collision (the extreme right-hand trace in Fig. 3.2), this recovers after the collision. This property is one which differs crucially from the results which apply to the overtaking of linear pulses.

A further surprising observation is that at time $t = 0.5T_R$ ($T_R \approx 30.4 t_B$), odd numbered solitons and even numbered solitons gather together at different places, at which time the solution u reduces to the second harmonic of the initial condition ($\cos 2\pi x$). It was also observed that at $t = T_R$, all the solitons gather together at one point and the solution u returns to the initial condition ($\cos \pi x$). On account of this behaviour the time T_R is called the recurrence time. As will be shown in the next chapter, the persistence of solitons in collisions found by numerical calculation can be proved in an analytically rigorous way. This was first accomplished by Gardner et al. (1967).

It can be seen from Eq. (3.3.3) that the width of a soliton is proportional to $\varepsilon^{-1/2}$, and also that its propagation speed is larger than the characteristic speed u_∞ by an amount of order ε. Here, ε is a positive parameter denoting the magnitude of the amplitude. Therefore, we may say that Eq. (3.3.3) represents a supersonic, compressible soliton if we regard the characteristic speed as the sound speed. Unlike a shock wave, however, a soliton is not always compressible. If μ is negative, we have $\varepsilon < 0$ and then we obtain an expansive soliton. In this case the propagation speed is smaller than u_∞. This is connected with the reversibility of the system. Hereafter we will consider only the case $\mu > 0$ unless otherwise stated. In any case, in the limit $\varepsilon \to 0$, the width increases like $\varepsilon^{-1/2}$,

3.3 WEAKLY DISPERSIVE SYSTEMS

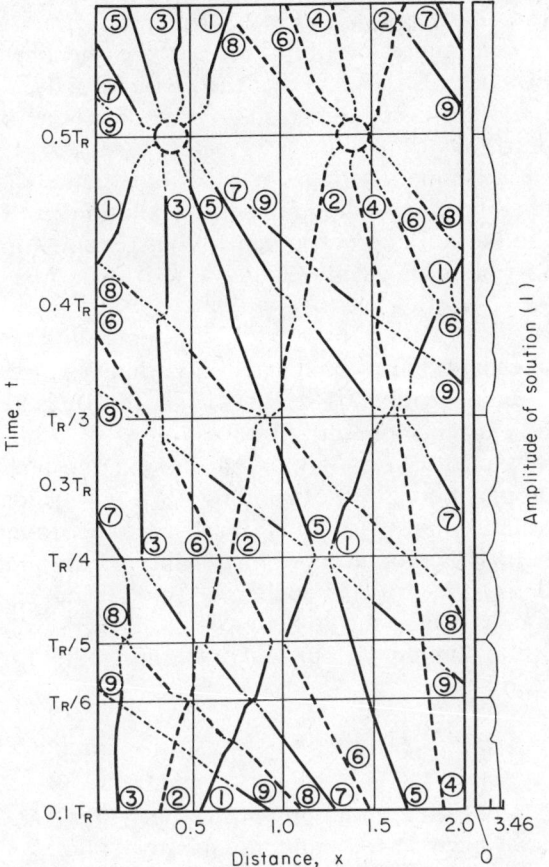

Fig. 3.2 Trajectories of eight solitons. The extreme right figure shows the time variation of the amplitude of soliton (1) (Zabusky and Kruskal (1965)).

and the soliton flattens, approaching the constant state u_∞, while its propagation speed decreases (or increases ($\mu < 0$)) proportionally with ε and approaches u_∞.

Setting $x - u_\infty t = x'$, it is readily seen that the soliton satisfies the following similarity law:

$$|u - u_\infty| \to \varepsilon |u - u_\infty|, \quad x' \to \varepsilon^{-1/2} x', \quad t \to \varepsilon^{-3/2} t. \quad (3.3.4)$$

That is, the widths and the propagation speeds of all the KdV solitons change so as to satisfy the above relations, regardless of their amplitudes. Since the KdV equation is invariant under the transformation Eq. (3.3.4), we may call the soliton the invariant far field.

The transformation Eq. (3.3.4) is equivalent to the similarity law $\omega \sim \varepsilon^{3/2}$ for $k \sim \varepsilon^{1/2}$, and it can be derived from the linear dispersion relation. It follows from Eq. (3.3.2) that the phase velocity is smaller than u_∞ (when $\mu > 0$). However, this result contradicts the fact that the propagation speed of the soliton is greater than u_∞. This apparent contradiction can be immediately understood if we notice that k must be purely imaginary, since, expanding the soliton solution Eq. (3.3.3) about u_∞ at infinity, it becomes an exponentially increasing (or decreasing) function. Thus, for a given phase velocity λ, k is determined by

$$k^2 = \mu^{-1}(u_\infty - \lambda) \tag{3.3.2'}$$

and, if $\mu > 0$, k becomes purely imaginary only when λ is greater than u_∞. Consequently, the inequality $\lambda > u_\infty$ ($\lambda < u_\infty$ if $\mu < 0$) is the necessary condition for the existence of solitary waves.

It may also be seen from Eq. (3.3.2') that the relationship between the width and the propagation speed of solitons is $\lambda - u_\infty \sim \varepsilon$, for $k \sim \varepsilon^{1/2}$. In this way, the relationships Eq. (3.3.4) can also be found from the dispersion relation. However, the proportionality of the amplitude to ε originates in the nonlinearity (though this is deduced also from the dispersion relation, as will be shown below, Eq. (3.4.10)). Namely without loss of generality, assuming $u_\infty = 0$, and integrating the KdV equation (1.2.19) twice yields

$$\frac{\mu}{2} \dot{u}^2 = \frac{\lambda}{2} u^2 - \tfrac{1}{6} u^3,$$

where the dot denotes differentiation with respect to $x - \lambda t$. Now at the top of the pulse we have $\dot{u} = 0$, and consequently $\lambda = u_{\max}/3 = \varepsilon/3$.

The propagation speed λ of a soliton is greater than u_∞, while the phase velocity ω/k of a plane wave is smaller than u_∞. Hence, in this respect, solitons and plane waves are essentially different from each other. For the case of a solution involving solitons and plane waves, we can refer to the wave train solution. The general progressive wave solution of the KdV equation is expressible by means of three parameters, u_0, ε, and Δ, in the form of the following oscillatory wavetrain solution:

$$u = u_0 + \varepsilon \, \text{cn}^2 \left[\frac{(\Delta + \varepsilon)^{1/2}}{(12\mu)^{1/2}} \left\{ x - \left(u_0 - \frac{\Delta}{3} + \frac{\varepsilon}{3}\right)t \right\}, m^2 \right],$$

$$m^2 = \frac{\varepsilon}{\Delta + \varepsilon} \quad (\Delta > 0). \tag{3.3.5}$$

In the special case in which $\Delta \to 0$, we have $m^2 = 1$, and Jacobi's elliptic function cn reduces to the hyperbolic function sech, so that the above oscillatory solution becomes the soliton solution Eq. (3.3.3) mentioned previously. Now, if Δ is sufficiently large, and when $\varepsilon \ll 1$, it follows that $m^2 \ll 1$, and the above wavetrain solution becomes a plane wave. In addition, in the limit $\Delta \to 0$, we have $\omega/k = u_0 + \varepsilon/3 > u_0$, while for sufficiently large Δ and when $\varepsilon \ll 1$, we have $\omega/k \simeq u_0 - \Delta/3 \ll u_0$. Since the wave number k is proportional to $(\Delta + \varepsilon)^{1/2}$, for a plane wave in which Δ is large enough, as $\varepsilon \to 0$ the wave number k becomes approximately proportional to $\Delta^{1/2}$ and does not change appreciably as ε changes. This situation is different from the case of a soliton whose width is inversely proportional to $\varepsilon^{1/2}$. Namely, as $\varepsilon \to 0$, the soliton becomes smoother and smoother and its amplitude approaches zero, while in the case of an approximate plane wave solution, the amplitude approaches zero without a change of the wave number.

The similarity law Eq. (3.3.4) of the KdV equation corresponds to $\gamma = 1/2$ in the general G–M transformation Eq. (3.1.18). Accordingly, by the use of

$$U = U^{(0)} + \varepsilon U^{(1)} + \varepsilon^2 U^{(2)} + \cdots,$$
$$\xi = \varepsilon^{1/2}(x - \lambda_0 t), \qquad \tau = \varepsilon^{3/2} t, \tag{3.3.6}$$

the reductive perturbation method will enable us to reduce a general dispersive system to the KdV equation. For example, let us consider the equation

$$U_{,t} + A U_{,x} + K_1 [K_2 (K_3 U_{,x})_{,x}]_{,x} = 0. \tag{3.3.7}$$

Then, following a procedure analogous to that of the previous section, it then follows that Eq. (3.3.7) may be reduced to

$$\varphi^{(1)}_{,\tau} + \alpha \varphi^{(1)} \varphi^{(1)}_{,\xi} + \mu \varphi^{(1)}_{,\xi\xi\xi} = 0. \tag{3.3.8}$$

Here, α is identical to the result given in Eq. (3.2.9), while μ is given by Eq. (3.2.9) with $K_0 = K_{10} K_{20} K_{30}$, and can be obtained from the linear dispersion relation when expanded in terms of $k \sim \varepsilon^{1/2}$ as

$$\omega = \lambda_0 k - \mu k^3 + \cdots. \tag{3.3.9}$$

As was shown in the previous section, the exceptional cases are given by $\alpha = 0$, corresponding to $\nabla_u \lambda_0 \cdot R_0 = 0$, or by $\mu = 0$, corresponding to $L_0 \cdot K_0 \cdot R_0 = 0$. As an example of the former case (the nonlinearly exceptional case), we can refer to solutions about $u = 0$ of the modified KdV equation:

$$u_{,t} + u^r u_{,x} + \mu u_{,xxx} = 0 \qquad (r = 2, 3, \ldots). \tag{3.3.10}$$

In this case, putting $u_\infty = 0$, the solitary wave solutions are given by

$$u = \varepsilon \left[\mathrm{sech}\left(\frac{x - ct}{\Gamma}\right) \right]^{2/r}, \tag{3.3.11}$$

where $c = 2\varepsilon^r/(r+1)(r+2)$ and $\Gamma = [2\mu(r+1)(r+2)/\varepsilon^r r^2]^{1/2}$. Therefore these solutions satisfy the similarity law $u \to \varepsilon u$, $x \to \varepsilon^{-r/2} x$, $t \to \varepsilon^{-3r/2} t$, and it is also easily seen that Eq. (3.3.10) is invariant under this transformation. In addition, let the general nonlinear dispersive equation for the scalar u be given by

$$u_{,t} + l(u) u_{,x} + \mu u_{,xxx} = 0,$$

where at $u = u_0$, $l(u)$ satisfies the conditions $(l_{,u})_0 = (l_{,uu})_0 = \cdots = (d^{r-1} l/du^{r-1})_0 = 0$, and $(d^r l/du^r)_0 \neq 0$. Then, by means of the G-M transformation

$$u = u_0 + \varepsilon u_1 + \cdots, \qquad \xi = \varepsilon^{r/2}(x - l_0 t), \qquad \tau = \varepsilon^{3r/2} t, \qquad (3.3.12)$$

this equation can be reduced to the following modified KdV equation:

$$u_{,t} + \alpha u^r u_{,x} + \mu u_{,xxx} = 0, \qquad \alpha = \frac{1}{r!} \left(\frac{d^r l}{du^r} \right)_0. \qquad (3.3.13)$$

In general, solutions of higher order equations to give $U^{(n)}$ ($n \geq 2$) involve secular terms. Namely, $U^{(2)}$ may be expressed as $\varphi^{(2)} R_0 + \frac{1}{2} R_0 \cdot \nabla_u R_0 (\varphi^{(1)})^2 + F \varphi^{(1)}_{,\xi\xi}$ where F is a solution of $(-\lambda_0 I + A_0) F = (\mu I - K_0) R_0$; $\varphi^{(2)}$ is determined in the next order by the linearized KdV equation with a source term given by $\varphi^{(1)}$, which admits secular solutions. When the dispersion relation is expanded for small k as

$$\omega = \sum_{n=0}^{\infty} \mu^{(n)} k^{2n+1}, \qquad \mu \equiv \lambda_0, \qquad \mu^{(1)} \equiv -\mu,$$

it was shown by Kodama and Taniuti (1978a, b) (1979b) by means of a renormalization technique that the secular terms in N-soliton solutions can be removed by modifying the KdV equation (3.3.8) to

$$\tilde{\varphi}^{(1)}_{,\tau} + \alpha \tilde{\varphi}^{(1)} \tilde{\varphi}^{(1)}_{,\xi} + \mu \tilde{\varphi}^{(1)}_{,\xi\xi\xi} + \sum_{n=2}^{\infty} a^{(n)} \sigma_{n+1,\xi} = 0.$$

Here σ_n ($n = 3, 4, \ldots$) are the higher order conserved densities which will be given by Eq. (4.2.13) in Section 4.2, and $a^{(n)}$ ($n \geq 2$) are constants determined by the dispersion relation so that $a^{(n)} \propto \mu^{(n)}$. The equation is called the generalized KdV equation and it is completely integrable (Lax (1971)). (See also Section 4.7.) ($U^{(n)}$ ($n \geq 2$) of Kodama and Taniuti (1978a) should be corrected as above.)

For weak nonlinear waves of the order ε considered here, we may assume that all the eigenvalues of A differ from each other by magnitudes of the order of unity. The speeds of the solitons associated with each eigenvalue of A then differ from the corresponding eigenvalue by an amount of the order of ε. Consequently, when an initial disturbance which has compact support at $t = 0$ splits into several solitons as $t \to \infty$, solitons with different modes (eigenvalues λ_{j0}) will not overlap. That is, we will then be able to express a solution in terms of a linear combination of various solitons in the form

$$U \approx U^{(0)} + \varepsilon \sum_{j=1}^{n} \varphi_j^{(1)} R_{j0}, \tag{3.3.14}$$

where $\varphi_j^{(1)}$ corresponds to the soliton solution of the jth mode.

Because of the time reversal phenomenon in dispersive systems, this implies that a converging series of solitons at $t = -\infty$ tends to a diverging series of solitons at $t = +\infty$. This was clarified by the Oikawa–Yajima theory (1973), which is mentioned in the next section by way of an extension of the reductive perturbation method.

3.4 The reductive perturbation method for long waves and its extension

The above-mentioned reductive perturbation method was first established by Taniuti and Wei (1968) in a more general form which is applicable to both weakly dissipative and weakly dispersive systems. The basic system of equations considered by them is of an extremely general type of the form

$$\frac{\partial U}{\partial t} + A \frac{\partial U}{\partial x} + \left\{ \sum_{\beta=1}^{s} \prod_{\alpha=1}^{p} \left(H_\alpha^\beta \frac{\partial}{\partial t} + K_\alpha^\beta \frac{\partial}{\partial x} \right) \right\} U = 0 \qquad (p \geq 2), \tag{3.4.1}$$

where H_α^β and K_α^β are $n \times n$ matrices, all of which are functions of U. An expansion about a constant solution $U^{(0)}$ in terms of a small parameter ε is assumed.

By means of the G–M transformation

$$U = U^{(0)} + \varepsilon U^{(1)} + \cdots, \tag{3.4.2}$$

$$\xi = \varepsilon^a (x - \lambda_0 t), \qquad \tau = \varepsilon^{a+1} t \left(a = \frac{1}{p-1} \right), \tag{3.4.3}$$

Eq. (3.4.1) can be reduced, in an asymptotic sense, to

$$\frac{\partial \varphi^{(1)}}{\partial \tau} + \alpha \varphi^{(1)} \frac{\partial \varphi^{(1)}}{\partial \xi} + \mu \frac{\partial^p \varphi^{(1)}}{\partial \xi^p} = 0. \tag{3.4.4}$$

Here we have used the boundary condition as $x \to \infty$ that $U^{(1)} \to 0$ and, consequently, that $\varphi^{(1)} \to 0$, while $U^{(1)}$ is given by

$$U^{(1)} = \varphi^{(1)} R_0, \tag{3.4.5}$$

with R_0 the right eigenvector of A_0 corresponding to λ_0, so that

$$(A_0 - \lambda_0 I) R_0 = 0. \qquad (3.4.6)$$

The coefficients α and μ are given by

$$\alpha = \frac{L_0 (R_0 \cdot \nabla_u A_0) R_0}{L_0 \cdot R_0}, \qquad (3.4.7)$$

$$\mu = \frac{L_0 K_0 R_0}{L_0 \cdot R_0}, \qquad (3.4.8)$$

where L_0 denotes the left eigenvector of A_0 corresponding to λ_0, and

$$K_0 = \sum_{\beta=1}^{s} \prod_{\alpha=1}^{p} (-\lambda_0 (H_\alpha^\beta)_0 + (K_\alpha^\beta)_0). \qquad (3.4.9)$$

The proof is omitted, but the results may readily be deduced from the results of the previous section. As before, α is rewritten in the form

$$\alpha = \nabla_u \lambda_0 \cdot R_0. \qquad (3.4.7')$$

The quantity μ can be obtained from the linear dispersion relation

$$\omega = \lambda_0 k + i^{p-1} \mu k^p + O(k^{2p-1}), \qquad (3.4.10)$$

where we have used the fact that $k \sim \varepsilon^a$ in order to find the dispersion relation. That is, in the linearized equation

$$\left\{ -i\omega I + ik A_0 + \sum_{\beta=1}^{s} \prod_{\alpha=1}^{p} (-i\omega (H_\alpha^\beta)_0 + ik (K_\alpha^\beta)_0) \right\} \delta U = 0,$$

let us expand ω and U as

$$\omega = \varepsilon^a (\omega_0 + \varepsilon \omega_1 + \cdots), \qquad \delta U = \delta U_0 + \varepsilon \delta U_1 + \cdots.$$

We then get from the terms to $O(\varepsilon^a)$

$$\omega_0 = \lambda_0 k_0 \ (k = \varepsilon^a k_0), \qquad \delta U_0 = R_0,$$

and from the terms to $O(\varepsilon^{a+1})$

$$(-i\omega_0 I + ik_0 A_0) \delta U_1 - i\omega_1 \delta U_0 + (ik_0)^p K_0 \delta U_0 = 0,$$

which yields

$$-i\omega_1 + (ik_0)^p (L_0 K_0 R_0)/(L_0 \cdot R_0) = 0.$$

Under a general boundary condition at $x = \infty$ we may have $U^{(1)} = R^{(1)} + V$, instead of Eq. (3.4.5), where the vector V is a function of τ. However, we can eliminate V by making an appropriate transformation, so that Eq. (3.4.1) can be reduced to an equation of the form of Eq. (3.4.4) (Taniuti and Wei (1968)).

3.4 REDUCTIVE PERTURBATION METHOD

Example 3.4.1 The hydrodynamic equation

For a perfect gas, the conservation laws Eq. (2.2.1) of mass, momentum, and energy can be expressed in the following form

$$\rho_{,t} + (\rho u)_{,x} = 0,$$

$$u_{,t} + uu_{,x} + \frac{p_{,x}}{\rho} - \nu \frac{u_{,xx}}{\rho} = 0,$$

$$p_{,t} + \gamma p u_{,x} + u p_{,x} - \frac{\gamma-1}{R} \chi \left(\frac{p_{,x}}{\rho} - \frac{p}{\rho^2} \rho_{,x} \right)_{,x} + \nu(\gamma-1)(uu_{,xx} - (uu_{,x})_{,x}) = 0,$$

where $\nu = (\frac{4}{3})\zeta$ and we have also assumed $\zeta' = 0$.

We now consider orders of magnitude and normalization. The viscosity ν and the thermal conductivity χ can be made unity by normalizing the velocity relative to the sound speed a_0 (or the thermal velocity), the density relative to the background density, and the length relative to the molecular mean free path l_{mfp} (and consequently the time relative to l_{mfp}/a_0). In this normalization, the scales of the experimental length and time become sufficiently large that stretching by the Gardner–Morikawa transformation is required. It should be understood that in what follows all the quantities are normalized in this way.

The system of equations can be written in the matrix form Eq. (3.4.1) with $p = s = 2$, if U, A, and K are given as follows:

$$U = \begin{pmatrix} \rho \\ u \\ p \end{pmatrix}, \qquad A = \begin{pmatrix} u & \rho & 0 \\ 0 & u & \rho^{-1} \\ 0 & \gamma p & u \end{pmatrix},$$

$$K_1^1 = \begin{pmatrix} 0 & 0 & 0 \\ 0 & 0 & 0 \\ \chi(\gamma-1)R^{-1} & -\nu(\gamma-1) & -\chi(\gamma-1)R^{-1} \end{pmatrix},$$

$$K_2^1 = \begin{pmatrix} p/\rho^2 & 0 & 0 \\ 0 & u & 0 \\ 0 & 0 & \rho^{-1} \end{pmatrix}, \qquad K_1^2 = \begin{pmatrix} 0 & 0 & 0 \\ 0 & -\nu/\rho & 0 \\ 0 & \nu(\gamma-1)u & 0 \end{pmatrix}$$

$$K_2^2 = \begin{pmatrix} 1 & 0 & 0 \\ 0 & 1 & 0 \\ 0 & 0 & 1 \end{pmatrix}.$$

The eigenvalues of A are $u \pm (\gamma p/\rho)^{1/2}$, u, so that for the constant state $U^{(0)} = (1, 0, \gamma^{-1})$ they reduce to ± 1 and 0. For a wave with speed $\lambda_0 = 1$, the eigenvectors R_0 and L_0 are given by $R_0 = (1, 1, 1)$ and $L_0 = (0, 1, 1)$, and we have the Burgers equation

$$\rho_{,\tau}^{(1)} + \alpha \rho^{(1)} \rho_{,\xi}^{(1)} + \mu \rho_{,\xi\xi}^{(1)} = 0,$$

with

$$\alpha = \left(\frac{\gamma+1}{2}\right), \quad \mu = -\tfrac{1}{2}(R^{-1}\chi(\gamma-1)^2\gamma^{-1}+\nu).$$

Here $\rho^{(1)}$ is the first-order term of the expansion of ρ in terms of ε, so that $\rho = 1 + \varepsilon\rho^{(1)} + \cdots$, and the boundary condition $U \to U^{(0)}$ as $x \to \infty$ has been assumed. (The way to approximate the hydrodynamic equations by means of Burgers' equation was found by Lighthill (1956) and Khokhlov (cf. Rudenko and Soluyan).

Example 3.4.2 Vibration of dispersive string

We mentioned a nonlinear string in Problem 2.2.1. Let us now consider the vibration of such a string with dispersive effects. The equation of motion for the displacement y is given by

$$y_{,tt} - c^2 y_{,xx} - \sigma^2 y_{,xxxx} = 0, \qquad (3.4.11)$$

where $c^2 = (1+y_{,x}^2)^{-3/2}$ (see Problem 2.2.1) and σ is a real constant. Letting $y_{,t} = u$, $y_{,x} = v$, as before, the equation of motion can be reduced to Eq. (3.3.7) with U, A, and K given by

$$U = \begin{pmatrix} u \\ v \end{pmatrix}, \quad A = \begin{pmatrix} 0 & -c^2 \\ -1 & 0 \end{pmatrix}, \quad K = \begin{pmatrix} 0 & -\sigma^2 \\ 0 & 0 \end{pmatrix}.$$

For the eigenvalues $\lambda_{0\pm} = \pm c_0 = \pm 1$, we have $L_{0\pm} = (\pm 1, 1)$, $R_{0\pm} = \begin{pmatrix} 1 \\ \mp 1 \end{pmatrix}$, and $(\nabla_u \lambda_\pm)_0 = (0, 0)$, which gives the nonlinearly exceptional case. Hence, Eq. (3.4.11) reduces to the modified KdV equation (3.3.13), with $r = 2$ and $\alpha = -\tfrac{3}{4}$, $\mu = \sigma^2/c_0 = \sigma^2/2$. μ can also be derived from the linear dispersion relation of Eq. (3.4.11), namely $\omega^2 = c_0^2 k^2 - \sigma^2 k^4$, so that $\omega \simeq c_0 k - (\sigma^2/2c_0)k^3$.

Example 3.4.3 Ion-sound wave

The motion of the cold ions in a hot electron gas with the Boltzmann distribution is governed by Eqns (A.4), (A.5) in the Appendix. We now normalize the relevant quantities as follows: $n_i \to n_i/n_0$, $u_i \to u_i/c_s$, $E \to eE/(m_e T_e)^{1/2}\omega_{pe}$ (or $\phi \to e\phi/T_e$), $x \to x/\lambda_{De}$, $t \to \omega_{pi} t$, where λ_{De} and ω_{pi} are the electron Debye length $(T_e/m_e)^{1/2}/\omega_{pe}$ and ion plasma frequency $(m_e/m_i)^{1/2}\omega_{pe}$ ($\omega_{pe}^2 = 4\pi e^2 n_0/m_e$), respectively, and c_s is the product of them, $(T_e/m_i)^{1/2}$. Then Eqns (A.5) reduce to

$$\frac{\partial n_i}{\partial t} + \frac{\partial (n_i u_i)}{\partial x} = 0, \qquad (3.4.12a)$$

3.4 REDUCTIVE PERTURBATION METHOD

$$\frac{\partial u_i}{\partial t} + u_i \frac{\partial u_i}{\partial x} = -\frac{\partial \phi}{\partial x}, \tag{3.4.12b}$$

$$\frac{\partial^2 \phi}{\partial x^2} = \exp(\phi) - n_i. \tag{3.4.12c}$$

Putting $n_i = 1 + n_i' e^{i(kx-\omega t)}$, $u_i = u_e' e^{i(kx-\omega t)}$, and $\phi = \phi' e^{i(kx-\omega t)}$, the linear dispersion relation for the normalized k and ω is obtained as

$$\omega^2 = \frac{k^2}{1+k^2}.$$

That is, for long wavelength ranges ($k^2 \ll 1$), we have

$$\omega \simeq k - \tfrac{1}{2}k^3.$$

Employing the assumption of quasi-neutrality, $n_i = n_e \equiv \exp(\phi)$, the equation of motion becomes

$$\frac{\partial u_i}{\partial t} + u_i \frac{\partial u_i}{\partial x} = -\frac{1}{n_i} \frac{\partial n_i}{\partial x}.$$

This equation is identical to the equation of motion for a gas with unit sound speed c_s, and it implies that the electrons exert a pressure on the ions through the electric field $E = -\partial \phi/\partial x$. In this sense, we call the wave described by the above dispersion relation the ion-sound wave. As the wave steepens by the inertia term, $\partial^2 \phi/\partial x^2$ ($= n_i - n_e$) can no longer be neglected. Consequently a soliton is formed by the dispersion associated with the deviation from the quasi-neutrality (the second term in the dispersion relation).

To begin with, let us seek a stationary solution of Eqns (3.4.12) (which is a function of $\xi' = x - \lambda' t$). Assuming the boundary conditions $n_i \to 1$, $u_i \to 0$, $\phi \to 0$, $\phi_{,\xi'}' \to 0$ as $|\xi'| \to \infty$, we get

$$\tfrac{1}{2}(\phi_{,\xi'})^2 = \exp(\phi) + \lambda'(\lambda'^2 - 2\phi)^{1/2} - (\lambda'^2 + 1). \tag{3.4.13}$$

Denoting the value of ϕ at $\phi_{,\xi'} = 0$ by ϕ_M, we have

$$\lambda'^2 = \frac{[\exp(\phi_M) - 1]^2}{2[\exp(\phi_M) - 1 - \phi_M]}. \tag{3.4.14}$$

Assuming that $\phi_M \ll 1$, $\lambda' - 1 = \delta\lambda \ll 1$, Eq. (3.4.13) reduces to

$$(\phi_{,\xi'})^2 = \tfrac{2}{3}\phi^2(3\delta\lambda - \phi),$$

and after an integration we get a soliton solution of the form

$$\phi = 3\delta\lambda \, \text{sech}^2 \left[\left(\frac{\delta\lambda}{2}\right)^{1/2} (x - \lambda' t) \right]. \tag{3.4.15}$$

Thus we have a KdV soliton when the amplitude is finite but small. As the amplitude of the potential approaches $\phi_M = \lambda'^2/2$ (i.e., $e\phi_M = m_i \lambda'^2/2$), the ions cannot cross the potential and so will be reflected. Then the intersection of ion orbits occurs, and as a consequence the cold ion approximation ceases to hold (cf. Section 1.2).

Accordingly, the maximum amplitude of the soliton is given by $\phi_M = \lambda'^2/2$, and from Eq. (3.4.14) we have $\phi_M \simeq 1.3$, and hence $\lambda' \simeq 1.6$. Namely, the maximum Mach number of the soliton is 1.6 (Sagdeev (1966)). In order to apply the reductive perturbation method to the system of ion waves we eliminate n_i and E from Eqns (A.4) and (A.5) to give

$$\frac{\partial n_e}{\partial t} + \frac{\partial (n_e u_i)}{\partial x} - \frac{\partial}{\partial x}\left(\frac{\partial}{\partial t} + u_i \frac{\partial}{\partial x}\right)\left[\left(\frac{1}{n_e}\right)\frac{\partial n_e}{\partial x}\right] = 0,$$

$$\frac{\partial u_i}{\partial t} + u_i \frac{\partial u_i}{\partial x} + \frac{1}{n_e}\frac{\partial n_e}{\partial x} = 0.$$

This system of equations corresponds to system Eq. (3.4.1) with $p = 3$, $s = 1$ and U, A, H, and K given by

$$U = \begin{pmatrix} n_e \\ u_i \end{pmatrix}, \quad A = \begin{pmatrix} u_i & n_e \\ 1/n_e & u_i \end{pmatrix},$$

$$H_1 = 0, \quad K_1 = -\begin{pmatrix} 1 & 0 \\ 0 & 0 \end{pmatrix}, \quad H_2 = \begin{pmatrix} 1 & 0 \\ 0 & 0 \end{pmatrix},$$

$$K_2 = \begin{pmatrix} u_i & 0 \\ 0 & 0 \end{pmatrix}, \quad H_3 = 0, \quad K_3 = \begin{pmatrix} 1/n_e & 0 \\ 0 & 0 \end{pmatrix}.$$

For $U^{(0)} = \begin{pmatrix} 1 \\ 0 \end{pmatrix}$, the eigenvalues of A_0 become $\lambda_0 = \pm 1$. For example, considering the case of $\lambda_0 = 1$, we have

$$R_0 = \begin{pmatrix} 1 \\ 1 \end{pmatrix}, \quad L_0 = (1, 1), \quad \nabla_u \lambda_0 = (0, 1),$$

so that we find $\alpha = 1$, $\mu = \frac{1}{2}$,

$$\frac{\partial n_e^{(1)}}{\partial \tau} + n_e^{(1)} \frac{\partial n_e^{(1)}}{\partial \xi} + \frac{1}{2}\frac{\partial^3 n_e^{(1)}}{\partial \xi^3} = 0. \tag{3.4.16}$$

Since electrons obey the Boltzmann distribution, we obtain $n_e^{(1)} = \phi^{(1)}$ by expanding e^ϕ, and so the soliton solution of the KdV equation coincides with Eq. (3.4.15).

Problem 3.4.1 Dispersive shallow water waves

The equations of shallow water waves obtained by adding the dispersive effect to the Eqns (2.2.35) in the Example 2.2.2 can be written, according to Boussinesq (1872), as follows (Jeffrey and Kakutani (1972)):

$$\frac{\partial \bar{y}}{\partial t} + \frac{\partial (\bar{y}u)}{\partial x} = 0,$$

$$\frac{\partial u}{\partial t} + u \frac{\partial u}{\partial x} = -\frac{\partial \bar{y}}{\partial x} - \frac{1}{3}\frac{\partial^3 \bar{y}}{\partial t^2 \partial x},$$

where we have normalized the relevant quantities as follows: $\bar{y} \to \bar{y}/y_0$, $u \to u/(gy_0)^{1/2}$, $x \to x/y_0$, $t \to t(g/y_0)^{1/2}$. Show that the propagation of this shallow water wave can be described by the KdV equation (Oikawa and Yajima (1973)).

So far we have supposed that only a wave involving one mode, corresponding to one of the eigenvalues of A, is excited in the lowest order. However, we can extend the reductive perturbation method to the case that waves with different modes coexist and interact with each other, and it can be proved that Eq. (3.4.4) holds for each mode (Oikawa and Yajima (1973, 1974)). As an extension of Eq. (3.4.5), let

$$U^{(1)} = \sum_j \varphi_j^{(1)} R_{j0}, \tag{3.4.17}$$

then the equation for the jth characteristics becomes $dx/dt \simeq \lambda_{j0} + \varepsilon U^{(1)} \nabla_u \lambda_{j0}$, and the deviation from a straight line is caused by the contribution resulting from self-interaction and that due to interaction with other modes.

Since the former is of the same order as that in the case of a single mode, that is, of the order $\varepsilon^{-a}\xi_j$, we have

$$x - \lambda_{j0} t = \varepsilon^{-a}\xi_j + \varepsilon \int \sum_{k \neq j} \varphi_k^{(1)} R_{k0} \cdot \nabla_u \lambda_{j0} \, dt$$

where, for the jth characteristic, the last term represents the deviation from a straight line due to interaction with other modes. This term is the product of the amplitude and the interaction time $dt \sim \lambda_{j0}^{-1} dx \sim \varepsilon^{-a}$ (dx is the width of the wave and $\sim \varepsilon^{-a}$). That is, it is of the order ε^{-a+1}. Consequently, the corresponding transformation reduces to the generalized G–M transformation

$$\xi_j = \varepsilon^a(x - \lambda_{j0}t - \varepsilon^{1-a}\zeta_j), \qquad \tau = \varepsilon^{a+1}t, \tag{3.4.18}$$

where ζ_j is a function of x and t which is to be determined later on so as to include the distortion of the characteristic caused by the interaction.

Inserting these into Eq. (3.4.1), it follows from the terms in that equation involving the lowest order of ε that we may assume $\varphi_j^{(1)}$ to be a function of ξ_j and τ. From terms involving the next order of ε, assuming that $U^{(2)} = \sum_{j=1}^{n} \varphi_j^{(2)} R_j$, we find the equation for $\varphi_j^{(2)}$:

$$-\sum_{l=1} \Lambda_{jl} \frac{\partial \varphi_j^{(2)}}{\partial \xi_l} + \sum_{l \neq j} \left[\sum_m \alpha_{jml} \varphi_m^{(1)} \frac{\partial \varphi_l^{(1)}}{\partial \xi_l} + \mu_{jl} \frac{\partial^p \varphi_l^{(1)}}{\partial \xi_l^p} \right] + \frac{\partial \varphi_j^{(1)}}{\partial \tau}$$

$$+ \alpha_{jjj} \varphi_j^{(1)} \frac{\partial \varphi_j^{(1)}}{\partial \xi_j} + \mu_{jj} \frac{\partial^p \varphi_j^{(1)}}{\partial \xi_j^p} + \sum_{l \neq j} \left\{ \Lambda_{jl} \frac{\partial \zeta_j}{\partial \xi_l} + \alpha_{jlj} \varphi_l^{(1)} \right\} \frac{\partial \varphi_j^{(1)}}{\partial \xi_j} = 0,$$

where $\Lambda_{jl} \equiv \lambda_{l0} - \lambda_{j0}$, the eigenvectors are normalized so that $(L_{j0}, R_{k0}) = \delta_{jk}$, and when $l = m = j$ the coefficients α_{jml} and μ_{jl} reduce to α and μ in the case of a single mode, respectively, so that,

$$\alpha_{jml} \equiv L_{j0} \cdot (R_{m0} \nabla_u A_0) R_{l0}, \qquad \mu_{jl} \equiv L_{j0} \sum_{\beta=1}^{s} \prod_{\alpha=1}^{p} ((K_\alpha^\beta)_0 - \lambda_l (H_\alpha^\beta)_0) R_{l0}.$$

The last term on the left-hand side gives the distortion of the characteristics of the lth mode caused by interactions with other modes, and we determine ζ_l so as to make this term zero (thus ξ_l is chosen to absorb this distortion). Integrating with respect to s_j by means of

$$\sum_l \Lambda_{jl} \frac{\partial}{\partial \xi_l} = \frac{d}{ds_j},$$

$\varphi_j^{(2)}$ may be found. Since $d\xi_j/ds_j = 0$ in this integration, a function of ξ_j alone is regarded as a constant, and for each mode we arrive at an equation of the form (3.4.4) from the condition that the terms proportional to s_j (the secular terms) do not appear.

As the simplest examples, we cite the interaction of two solitons propagated in opposite directions in ion-sound waves and in shallow water waves (Oikawa and Yajima (1973)), and also the interaction of two shocks of different modes in sound waves (Tatsumi and Tokunaga (1974)). These involve systems comprising two components which admit two wave modes, one propagated to the right and the other to the left.

3.5 A far field for linear wave modulation—the Schrödinger field

As seen from an argument used for the KdV equation, solutions for large wave numbers differ remarkably from those for long waves. For example, the wave train solution given by Eq. (3.3.5) can reduce, for large Δ and when $\varepsilon \ll 1$, to a plane wave, whose amplitude and phase are modulated

3.5 A FAR FIELD FOR LINEAR WAVE MODULATION

slowly, so that

$$u \approx u_0 + \frac{\varepsilon}{2} \cos\left[(kx - \omega t) + \frac{\varepsilon}{24\mu k}\left(x - \frac{d\omega}{dk}t\right)\right]. \tag{3.5.1}$$

We now examine the following general problem. When does there exist a plane wave of short wavelength whose amplitude and phase are modulated slowly over a certain localized region in space. How will such modulations propagate themselves, and what sort of far field will be observed for such waves. Towards this end, let us first consider the propagation of a quasi-monochromatic plane wave of the form

$$u = \int u(k) e^{i(kx - \omega t)} \, dk. \tag{3.5.2}$$

Here, we assume that the spectrum $u(k)$ has a peak at k_0 with a narrow wave number width $\delta k \sim \varepsilon$, and also that the angular frequency is a function of the wave number k. Then $\omega(k)$ can be expanded about k_0 as follows:

$$\omega(k) = \omega_0 + \omega_0' \, \delta k + \tfrac{1}{2}\omega_0'' \, \delta k^2 + \cdots, \tag{3.5.3}$$

where the subscript 0 denotes the value for k_0, and the prime denotes differentiation with respect to k. In what follows, let ω_0' and ω_0'' be the order of unity, that is, we assume strong dispersion. The substitution of Eq. (3.5.3) into Eq. (3.5.2) then yields

$$u \approx e^{i(k_0 x - \omega_0 t)} \int u(k_0 + \delta k) \exp\left[i\,\delta k\left(x' - \frac{\omega_0'' \, \delta k t}{2}\right)\right] d(\delta k), \tag{3.5.4}$$

where $x' = x - \omega_0' t$.

The integral with respect to δk is a slowly varying function of x' and t, and hence Eq. (3.5.4) expresses the slow modulation of the plane carrier wave. Denoting this modulation by φ, so that we may rewrite u as

$$u \approx \varphi e^{i(k_0 x - \omega_0 t)} \tag{3.5.5}$$

and, moreover, introducing ξ and τ by

$$\xi = \varepsilon x', \qquad \tau = \varepsilon^2 t, \tag{3.5.6}$$

it can easily be proved that the modulating function φ is a solution of the following Schrödinger equation,

$$i\varphi_{,\tau} + \frac{\omega_0''}{2} \varphi_{,\xi\xi} = 0. \tag{3.5.7}$$

In addition, letting $u(k_0 + \delta k) = u_0 e^{-\delta k^2/\varepsilon^2}$ in Eq. (3.5.4), φ takes the form

$$\varphi = \frac{\varepsilon u_0 \sqrt{\pi}}{(1+i\omega_0'' \varepsilon^2 t/2)^{1/2}} \exp\left[\frac{-(\varepsilon x')^2}{4(1+i\omega_0'' \varepsilon^2 t/2)}\right].$$

So, in the limit $t \to 0$, we have $\varphi \to \varepsilon u_0 \sqrt{\pi} \exp[-(\varepsilon x')^2/4]$, from which it follows that the spatial width of the packet is of the order $(1/\varepsilon)$, while in the limit $t \to \infty$, we find

$$\varphi \to [\varepsilon u_0 \sqrt{2\pi}/(i\omega_0'' \varepsilon^2 t)^{1/2}] \exp[-\{\varepsilon x'/(\omega_0'' \varepsilon^2 t)\}^2 + i(\varepsilon x')^2/(2\omega_0'' \varepsilon^2 t)],$$

which shows that in a frame moving with the group velocity, the spatial width of the packet is given by $|x'| \sim \varepsilon \omega_0'' t$ and the spatial variation of the phase may be estimated as $\mathrm{d}\{ix'^2/(2\omega_0'' t)\}/\mathrm{d}x' \sim \varepsilon$. From these results it follows that, after a sufficiently long time, modulations of n plane waves of independent modes propagate as n packets.

To see this, suppose that there exist n linearly independent modes, ω_j, k_j ($j = 1, 2, \ldots, n$), and let u be given by

$$u = \sum_{j=1}^{n} \varphi^{(j)} e^{i(k_j x - \omega_j t)},$$

where

$$\varphi^{(j)} = \int u^{(j)}(k_j + \delta k) \exp\left[i\left(\delta k x_j' - \frac{\omega_j'' \delta k^2 t}{2}\right)\right] \mathrm{d}(\delta k).$$

Then for $t = 0$, these modulations spatially overlap each other, because each of them has a spatial width of the order $1/\varepsilon$. In the limit $t \to \infty$, the widths become of the order $\varepsilon \omega_j'' t$, while the centres of the packets moving with the group velocities are spaced apart by the distances $(\omega_j' - \omega_{j-1}')t$, which are much greater than the widths, provided $(\omega_j' - \omega_{j-1}') \sim 1$. Therefore, after a sufficiently long time, these modulations propagate without overlapping one another. It follows from the above that in a wave field with strong dispersion, a far field for the modulation of plane waves can be described by the Schrödinger equation.

3.6 Strongly dispersive systems

In the previous section it was found that, for a linear field, the modulation of a plane wave is described by the Schrödinger equation. Let us now show that in a similar sense a strongly dispersive nonlinear field can also be described by the Schrödinger equation with a nonlinear term (the nonlinear Schrödinger equation):

$$i\varphi_{,\tau} + \frac{\omega_0''}{2} \varphi_{,\xi\xi} + Q |\varphi|^2 \varphi = 0. \tag{3.6.1}$$

3.6 STRONGLY DISPERSIVE SYSTEMS

As will be remarked in the next chapter, it was shown by Zakharov and Shabat (1972) that equation (3.6.1) may be solved analytically (the details will be mentioned in Section 4.5), and hence we may say that in an asymptotic sense we are able to find the solutions to a wide class of nonlinear wave systems with strong dispersion.

As a general nonlinear system, let us consider the following system which may be regarded as a generalization of equation (3.4.1):

$$\frac{\partial U}{\partial t} + A\frac{\partial U}{\partial x} + B + \sum_{\beta=1}^{s} \prod_{\alpha=1}^{p} \left(H_\alpha^\beta \frac{\partial}{\partial t} + K_\alpha^\beta \frac{\partial}{\partial x} \right) U = 0, \qquad (3.6.2)$$

where B is a vector with n components and also a function of U. A crucial assumption is that the linear dispersion relation of this system admits strong dispersion. Any systems may be considered as long as this assumption is satisfied, though of course the necessary analytic properties in the U-space, such as we assumed for weakly dispersive systems, will be required. In fact the system may be an integro-partial differential equation in nonlinear optics (Taniuti and Yajima (1973)), or the system may be multi-dimensional so that solutions have a dependence on a direction transverse to that of propagation, such as appears in the problem of a wave guide (Example 3.6.2). Moreover, the system may be weakly unstable and/or dissipative (Asano (1974a)). In all of these cases, the essential procedure of the method mentioned below is the same, and this has been confirmed by applications involving many specific examples.

In order to avoid unnecessary complications, we consider the following equation in place of Eq. (3.6.2):

$$U_{,t} + AU_{,x} + B = 0. \qquad (3.6.3)$$

Many physically interesting problems involving strong dispersion are described by this equation (cf. Example 3.6.1). Now the uniform state is given by a constant solution $U^{(0)}$ satisfying the algebraic equation

$$B(U^{(0)}) = 0. \qquad (3.6.4)$$

We assume here that there exists a physically meaningful constant solution, and consider a plane wave of infinitesimal amplitude imposed on the constant state, so that

$$U = U^{(0)} + \delta U_k \exp[i(kx - \omega t)] + \text{c.c.},$$

where c.c. denotes the complex conjugate of the preceding expression. Then the linear dispersion relation for this plane wave is given by

$$|\mp i\omega I \pm ikA_0 + \nabla_u B_0| = 0, \qquad (3.6.5)$$

which determines $\omega(k)$ as a function of k. Now we assume that there exists at least one real ω for a given k, such that $\omega(k)$ is a smooth function of k in some wave number domain containing the given k.

Moreover, we assume that for any integer l which is not equal to ± 1,

$$|\det W_l| \gg O(\varepsilon), \tag{3.6.6}$$

where W_l is the matrix defined by

$$W_l \equiv -il\omega I + ilkA_0 + \nabla_u B_0 \tag{3.6.7}$$

and ε is a small parameter.

The condition Eq. (3.6.6) is obviously invalid if ω is a linear function of k. That is, when ε is taken as the spectral width of a quasi-plane wave, the condition Eq. (3.6.6) corresponds to the condition for strong dispersion.

As shown in the previous section, the modulation of a linear plane wave is governed by the Schrödinger equation (3.5.7), provided the independent variables ξ, τ are introduced as follows:

$$\xi = \varepsilon(x - \lambda t), \qquad \tau = \varepsilon^2 t, \tag{3.6.8}$$

where λ is the group velocity $\partial \omega / \partial k$. We now study how such a quasi-plane wave is modulated further by nonlinear effects.

As is well known, nonlinear self-interaction of a plane wave gives rise to higher and lower harmonics, which appear first in the second order of the perturbation as the second harmonic mode and the (slowly varying) zeroth harmonic mode. They then couple with the fundamental mode to give nonlinear modulation in the third order. For instance, this modulation produces a nonlinear frequency shift proportional to the square of the amplitude, $\Delta\omega \sim |\delta U|^2$, which is often renormalized to the frequency in the linear dispersion relation. Hence, in order that this nonlinear self-modulation should be in balance with the other terms in the Schrödinger equation (3.6.1) already mentioned, the magnitude of the amplitude must be of the order ε. This is because the first two terms in Eq. (3.6.1) are of the order ε^2 and represent the slow variation, as may be seen clearly from Eq. (2.6.8), and hence this nonlinear term is balanced against the linear terms provided $|\delta U|^2$ is of the order ε^2.

On account of these considerations, let us assume the solution of Eq. (3.6.3) to be in the following form:

$$U = U^{(0)} + \sum_{\alpha=1}^{\infty} \varepsilon^{\alpha} U^{(\alpha)}, \tag{3.6.9a}$$

$$U^{(\alpha)} = \sum_{l=-\infty}^{\infty} U_l^{(\alpha)}(\xi, \tau) e^{il(kx-\omega t)}. \tag{3.6.9b}$$

Here, the $U^{(\alpha)}$ are real, so that the condition $U_l^{(\alpha)} = U_{-l}^{(\alpha)*}$ must be required, with $*$ denoting the complex conjugate operation. Since we consider modulation of the plane carrier wave, with the frequency ω and the wave number k, we may put $U_l^{(1)} = 0$ for $l \neq \pm 1$.

This can also be derived easily from condition (3.6.6) as follows. Substituting Eq. (3.6.9) into Eq. (3.6.3) and using Eq. (3.6.8), to the first order in ε we have

$$W_l U_l^{(1)} = 0. \tag{3.6.10}$$

So that, from the condition (3.6.6), we find for $|l| \neq 1$,

$$U_l^{(1)} = 0, \tag{3.6.11a}$$

while for $|l| = 1$,

$$U_1^{(1)} = \varphi^{(1)} R, \qquad U_{-1}^{(1)} = \varphi^{(1)*} R^*, \tag{3.6.11b}$$

in which R is a right eigenvector of W_1. Hence

$$W_1 R = 0, \tag{3.6.11c}$$

and $\varphi^{(1)}$ is a function of the variables ξ, τ which varies slowly with time and over space because of the modulation. We will now determine the equation governing $\varphi^{(1)}$. (The normalization of R is arbitrary, and this ambiguity is transferred to $\varphi^{(1)}$.)

Now the terms of order ε^2 in Eq. (3.6.3) can be written out in the following way:

$$W_l U_l^{(2)} + (-\lambda I + A_0) \frac{\partial U_l^{(1)}}{\partial \xi}$$

$$+ \left\langle U^{(1)} \cdot \nabla_u A_0 \sum_{l'=-\infty}^{\infty} il'k U_{l'}^{(1)} \exp[il'(kx - \omega t)] \right\rangle_l$$

$$+ \tfrac{1}{2} \nabla_u \nabla_u B_0 : \langle U^{(1)} U^{(1)} \rangle_l = 0, \tag{3.6.12}$$

where $\langle \ \rangle_l$ denotes the coefficient of the lth harmonic. The third and fourth terms in the above equation result from the self-interaction of the fundamental mode, and so those terms are non-vanishing only for $|l| = 2$ and $l = 0$. That is, this shows that the second harmonic and the slow mode appear to the second order of ε. Hence, for $l = 1$, Eq. (3.6.12) becomes

$$W_1 U_1^{(2)} + (-\lambda I + A_0) R \frac{\partial \varphi^{(1)}}{\partial \xi} = 0. \tag{3.6.13}$$

Since $\det W_1 = 0$, in order that this algebraic equation for $U_1^{(2)}$ is solvable, the following compatibility condition must be required:

$$L(-\lambda I + A_0) R = 0, \tag{3.6.14}$$

where L is the left eigenvector corresponding to R, so that

$$L W_1 = 0. \tag{3.6.15}$$

It is easily seen that this compatibility condition is satisfied automatically, for differentiating Eq. (3.6.11c) with respect to k gives

$$i(-\lambda I + A_0)R + W_1 \frac{\partial R}{\partial k} = 0, \qquad (3.6.16)$$

which multiplied by L from the left reduces to Eq. (3.6.14). Also, from Eq. (3.6.16), it follows that

$$U_1^{(2)} = \varphi^{(2)} R - i \left(\frac{\partial \varphi^{(1)}}{\partial \xi}\right)\left(\frac{\partial R}{\partial k}\right). \qquad (3.6.17)$$

Solving Eq. (3.6.12) for $|l| = 2$ and $l = 0$, $U_0^{(2)}$ and $U_{\pm 2}^{(2)}$ can be obtained, respectively, as

$$U_0^{(2)} = -W_0^{-1}[ik\{(\nabla_u A_0 \cdot R^*)R - \text{c.c.}\} + \nabla_u \nabla_u B_0 : R^* R]|\varphi^{(1)}|^2, \qquad (3.6.18a)$$

$$U_2^{(2)} = -W_2^{-1}\{ik(\nabla_u A_0 \cdot R)R + \tfrac{1}{2}\nabla_u \nabla_u B_0 : RR\}\{\varphi^{(1)}\}^2. \qquad (3.6.18b)$$

To the third order in ε, Eq. (3.6.3) gives

$$W_l U_l^{(3)} + (-\lambda I + A_0)\frac{\partial U_l^{(2)}}{\partial \xi} + \frac{\partial U_l^{(1)}}{\partial \tau} + \left\langle U^{(1)} \cdot \nabla_u A_0 \left(\frac{\partial U^{(1)}}{\partial \xi}\right)\right\rangle_l$$

$$+ \left\langle U^{(1)} \cdot \nabla_u A_0 \sum_{l'} il'k U_{l'}^{(2)} Z_{l'}\right\rangle_l + \tfrac{1}{2}\langle U^{(1)} U^{(1)} : \nabla_u \nabla_u A_0 \sum_{l'} il'k U_{l'}^{(1)} Z_{l'}\rangle_l$$

$$+ \langle U^{(2)} \cdot \nabla_u A_0 \sum_{l'} il'k U_{l'}^{(1)} Z_{l'}\rangle_l + \nabla_u \nabla_u B_0 : \langle U^{(1)} U^{(2)}\rangle_l$$

$$+ \tfrac{1}{6} \nabla_u \nabla_u \nabla_u B_0 : \langle U^{(1)} U^{(1)} U^{(1)}\rangle_l = 0, \qquad (3.6.19)$$

where $Z_l \equiv e^{il(kx-\omega t)}$. The non-vanishing terms in the summations over l' are only $U_{\pm 1}^{(1)}$, $U_{\pm 1}^{(2)}$, $U_0^{(2)}$, and $U_{\pm 2}^{(2)}$. Moreover, for $l = 1$, the contributions from $U_{\pm 1}^{(2)}$ vanish because

$$\langle Z_1 Z_{l'}\rangle_1 = \delta(l'), \qquad \langle Z_{-1} Z_{l'}\rangle_1 = \delta(l'-2),$$

and hence we notice that for $|l| = 1$ the nonlinear terms in Eq. (3.6.19) do not contain $\varphi^{(2)}$.

From Eq. (3.6.19), we can determine the equation governing $\varphi^{(1)}$ as follows. Let us multiply Eq. (3.6.19) for $l = 1$ by L from the left, and insert Eqns (3.6.11), (3.6.17), and (3.6.18) into the equation so obtained. Then, by means of Eq. (3.6.15), the first terms disappears, and thereby the compatibility condition Eq. (3.6.14) eliminates the part of the second term which is proportional to $\partial\varphi^{(2)}/\partial\xi$. Consequently all the terms containing $\varphi^{(2)}$ vanish, and hence we arrive at the equation for $\varphi^{(1)}$:

$$i\varphi_{,\tau}^{(1)} + P\varphi_{,\xi\xi}^{(1)} + Q|\varphi^{(1)}|^2 \varphi^{(1)} = 0. \qquad (3.6.20)$$

3.6 STRONGLY DISPERSIVE SYSTEMS

Here the constants P and Q are given by the following equations:

$$P = \frac{L(-\lambda I + A_0)}{L \cdot R} \frac{\partial R}{\partial k}, \tag{3.6.21}$$

$$Q = \frac{iL}{L \cdot R} [ik\{2(\nabla_u A_0 \cdot R^*)R_2^{(2)} - (\nabla_u A_0 \cdot R_2^{(2)})R^* + (\nabla_u A_0 \cdot R_0^{(2)})R$$
$$+ (\nabla_u \nabla_u A_0 : RR^*)R - \tfrac{1}{2}(\nabla_u \nabla_u A_0 : RR)R^*\}$$
$$+ \nabla_u \nabla_u B_0 \vdots (RR_0^{(2)} + R^* R_2^{(2)}) + \tfrac{1}{2}\nabla_u \nabla_u \nabla_u B_0 \vdots RR^*R], \tag{3.6.22}$$

where $R_2^{(2)}$ and $R_0^{(1)}$ are vectors introduced through the equations

$$U_2^{(2)} = (\varphi^{(1)})^2 R_2^{(2)}, \quad U_0^{(2)} = |\varphi^{(1)}|^2 R_0^{(2)}, \tag{3.6.23}$$

and they are given, respectively, by

$$R_2^{(2)} = -W_2^{-1}[ik(\nabla_u A_0 \cdot R)R + \tfrac{1}{2}\nabla_u \nabla_u B_0 : RR], \tag{3.6.24a}$$
$$R_0^{(2)} = -W_0^{-1}[ik\{(\nabla_u A_0 \cdot R^*)R - \text{c.c.}\} + \tfrac{1}{2}(\nabla_u \nabla_u B_0 : R^*R + \text{c.c.})]. \tag{3.6.24b}$$

The coefficient P is equal to $\omega''/2$, which can be proved as follows. Differentiating Eq. (3.6.16) with respect to k, we get

$$-i\omega'' R + 2i(-\lambda I + A_0) \frac{\partial R}{\partial k} + W_1 \frac{\partial^2 R}{\partial k^2} = 0,$$

and then multiplying this equation by L from the left, we have

$$-i\omega''(L \cdot R) + 2iL(-\lambda I + A_0) \frac{\partial R}{\partial k} = 0.$$

We thus obtain the generalized nonlinear Schrödinger equation

$$i\varphi_{,\tau} + \frac{\omega''}{2} \varphi_{,\xi\xi} + Q|\varphi|^2 \varphi = 0, \tag{3.6.25}$$

where Q is not necessarily real.

In particular, when Q is real, we call equation (3.6.25) the nonlinear Schrödinger equation. The extension of this equation to the case of three-dimensional modulation can easily be accomplished. Here, we give the most general form obtained by Asano (1974a),

$$i\frac{\partial \varphi}{\partial \tau} + \frac{1}{2}\sum_{i,j} \left(\frac{\partial^2 \omega}{\partial k_i \partial k_j}\right)\left(\frac{\partial^2 \varphi}{\partial \xi_i \partial \xi_j}\right) + Q|\varphi|^2 \varphi + i\delta\varphi = 0, \tag{3.6.26}$$

where k_i ($i = 1, 2, 3$) and ω denote the components of the wave vector and the frequency of the plane carrier wave, respectively. The coordinates ξ_i are given by $\varepsilon\{x_i - (\partial\omega/\partial k_i)t\}$, and the last linear term is added so the result will be applicable to weakly dissipative and/or unstable cases.

Simple solutions of the nonlinear Schrödinger equation can be found easily. For instance, when ω'' and Q take the same sign, the solution which tends to zero as $|\xi| \to \infty$ is a solitary wave,

$$\varphi = \left(-\frac{2\nu}{Q}\right)^{1/2} \operatorname{sech}\left\{\left(-\frac{2\nu}{\omega''}\right)^{1/2}\xi\right\} e^{-i\nu\tau}. \tag{3.6.27}$$

In addition, it follows from the Galilean invariance of the Schrödinger equation that Eq. (3.6.25) is invariant under the following transformation:

$$\xi' = \xi - V\tau, \qquad \tau' = \tau,$$
$$\varphi' = \varphi \exp\left[-i\frac{V}{\omega''}(\xi - \tfrac{1}{2}V\tau)\right]. \tag{3.6.28}$$

Hence, Eq. (3.6.25) also admits a solution of the form

$$\varphi' = \left(-\frac{2\nu}{Q}\right)^{1/2} \operatorname{sech}\left\{\left(-\frac{2\nu}{\omega''}\right)^{1/2}(\xi - V\tau)\right\} \exp\left[i\frac{V}{\omega''}\xi - \frac{i}{2}\left\{\left(\frac{V^2}{\omega''}\right) + 2\nu\right\}\tau\right], \tag{3.6.29}$$

where the parameters ν and V are independent of each other. These solutions decrease rapidly as $|x| \to \infty$, so that by analogy with the argument for the KdV equation, they may be considered to represent solitons. Unlike the KdV equation, however, the nonlinear Schrödinger equation does not admit a soliton solution corresponding to a steady wave moving with a constant velocity (a permanent progressive wave). In fact the plane-wave part represented by the exponential function, and the amplitude of the hyperbolic secant distribution, move with different velocities. In this sense, the solution Eq. (3.6.29) is called an envelope soliton. It was first proved numerically by Yajima and Outi (1971) that an envelope soliton behaves like a particle, as does the KdV soliton. This was shown analytically by Zakharov and Shabat (1972) by means of the inverse scattering method. The details will be given in the next chapter.

If φ approaches a constant value φ_0 at infinity, we have the plane wave solution

$$\varphi = \varphi_0 e^{i(\mu\xi - E\tau)},$$
$$E = \frac{\omega''}{2}\mu^2 - Q\varphi_0^2. \tag{3.6.30}$$

Let us notice, however, that as shown below, this solution is not stable, but is subject to a modulational instability. As in the case of Eq. (1.2.12), introduce the real functions ρ and u through the expressions

$$\varphi = \rho^{1/2} \exp\left(\frac{i}{\omega''}S\right), \qquad u = \frac{\partial S}{\partial \xi}. \tag{3.6.31}$$

3.6 STRONGLY DISPERSIVE SYSTEMS

Then, substituting this expression into Eq. (3.6.25), we have

$$\rho_{,\tau} + (\rho u)_{,\xi} = 0 \tag{3.6.32a}$$

$$u_{,\tau} + uu_{,\xi} = \omega'' Q \rho_{,\xi} + \left(\frac{\omega''}{2}\right)^2 [\rho^{-1/2}(\rho^{-1/2}\rho_{,\xi})_{,\xi}]_{,\xi}. \tag{3.6.32b}$$

If $\omega'' Q > 0$, in the long wavelength limit Eqns (3.6.32a, b) are equivalent to a hydrodynamic system with negative pressure. (The system becomes elliptic.) Considering the perturbation, $\rho_0 + \delta\rho \exp[i(K\xi - \Omega\tau)]$ and $u_0 + \delta u \exp[i(K\xi - \Omega\tau)]$, about a constant amplitude ρ_0 and phase u_0, we have $\Omega = u_0 K \pm (-\omega'' Q \rho_0)^{1/2} K + O(K^3)$, from which it follows that the perturbation grows for small K.

On the other hand, if ω'' and Q take opposite signs, the plane wave is stable. In this case, when Eq. (3.6.32) is written in the matrix form Eq. (3.3.7), the eigenvalues of A are real and non-degenerate, and hence the system can be reduced to the KdV equation (Asano, Taniuti, and Yajima (1973)). Namely, neglecting the dispersion, Eqns (3.6.32) correspond to the equations for a fluid whose sound speed a is given by $(|\omega'' Q| \rho)^{1/2}$, and also, in long wavelength ranges, the dispersion is such that the dispersive term of the order K^3 becomes $(\omega'')^2/(8a_0)$ $(a_0 \equiv a(\rho_0))$, so that Eqns (3.6.32) reduce to the KdV equation with $\mu < 0$. Consequently a soliton exists when $u_0 < a_0$ at infinity and it is expansive ($\rho \le \rho_0$).

It is also possible to obtain the exact solitary wave solution of Eqns (3.6.32) which has similar properties. That is, seeking solutions of Eqns (3.6.32) which are dependent only on ξ and which approach $\rho = \rho_0$, $u = u_0 < 0$ as $|\xi| \to \infty$, we get

$$\rho = \rho_0(1 - b^2 \operatorname{sech}^2 \theta), \qquad u = u_0(1 - b^2 \operatorname{sech}^2 \theta)^{-1}$$

$$b = (1 - M^2)^{1/2}, \qquad \theta = \frac{a_0 b}{\omega''} \xi,$$

where $M = |u_0|/a_0$. From these expressions, S is given by

$$S = u_0 \xi - \omega'' \left[\tan^{-1} \frac{b}{\sqrt{1-b^2}} \tanh \theta \right] - \nu\tau.$$

In addition the constant ν is assumed to be $-(1 - M^2/2)a_0^2$ since, letting $\xi \to \infty$ in Eq. (3.6.25), we have $\varphi''/\varphi \to -(u_0/\omega'')^2$. From the Galilean invariance, solutions obtained through the transformation $\xi \to \xi - V\tau$ $u \to u + V$ are also solutions of Eq. (3.6.32). Accordingly, the envelope soliton solutions of Eq. (3.6.25) for any V may be found by means of Eq. (3.6.28).

In contrast with the case that ω'' and Q take the same sign, such solutions are of an expansive type in which $|\varphi|^2$ decreases from a finite value at infinity and subsequently returns to the same value. This is often called the envelope cavity or dark pulse. The bright pulse arises when $|\varphi|^2$ increases from a finite value at infinity and subsequently returns to the same value. The degree of darkness is determined by the parameter b. In particular, if $b \to 1$ ($u_0 \to 0$), we have $\rho_{\min} \to 0$. Letting f be a real function, and substituting $\varphi = f \exp(-i\nu\tau)$ directly into Eq. (3.6.25), for $u_0 = 0$ we find a solution of the shock type instead of a soliton, namely

$$f = \left(-\frac{\nu}{Q}\right)^{1/2} \tanh\left[\left(\frac{\nu}{\omega''}\right)^{1/2} \xi\right].$$

This is called a kink solution and we have found this solution by an intuitive approach. For a systematic derivation of the envelope soliton solutions, the reader is referred, for example, to Scott *et al.* (1973) and Hasegawa (1975).

So far we have assumed the validity of the condition Eq. (3.6.6). In physical systems, this condition is valid for $|l| \geqslant 2$. However, for $l = 0$ it is not satisfied in most cases and we then encounter the difficulty that det $W_0 = \det |\nabla B_0| = 0$. As a result, $U_0^{(2)}$ cannot be determined uniquely, and a straightforward application of the above-mentioned method is no longer possible. Even in this case, however, we can still determine $U_0^{(2)}$ under a fairly general condition by using an expansion of higher order in ε (Inoue *et al.* (1974)).

Moreover, as shown in the following examples, in many cases $U_0^{(2)}$ can be determined by means of subsidiary conditions arising from the initial and boundary conditions. However, the coupling between the slow mode ($l = 0$) and the fast mode ($|l| \geqslant 2$) is a physically interesting problem in itself. For instance, as was pointed out by Hasegawa (1970), a resonant singularity appears in components such as $U_0^{(2)}$ for the mode in which $l = 0$. Thus, when a dispersive system takes the form Eq. (3.4.1), so that for long waves ω reduces to $\lambda_p k$ with λ_p equal to one of the eigenvalues of A_0, $U_0^{(2)}$ often involves the term $(\lambda - \lambda_p)^{-1}$. Consequently, when the group velocity of the carrier (fast) wave becomes close to the phase velocity of a slow mode, the two modes couple strongly, and the ε-expansion technique used here is no longer valid. The same phenomenon also occurs in the vicinity of $k \approx 0$ for a carrier wave with large ω. (Nishikawa *et al.* (1974); Ikezi *et al.* (1974).) In this case, we need another perturbation expansion, and we then obtain simultaneous equations involving the nonlinear Schrödinger equation and the KdV equation. (See also Zakharov (1972).)

The nonlinear Schrödinger equation (3.6.25) arising from a dispersive system is derived by other authors by means of different methods. One formal method is to make use of the nonlinear dispersion relation. In Eq. (3.6.25), the solution independent of ξ yields the frequency shift

3.6 STRONGLY DISPERSIVE SYSTEMS

$\Omega = Q |\varphi_0|^2$ and hence, renormalizing the frequency of the carrier wave by this amount, the nonlinear dispersion relation is obtained. Therefore, if this relation is known in some way in advance, this can be used to find the coefficient Q of the nonlinear term.

In general, let the nonlinear dispersion relation be $D_{NL}(\omega, k, |\varphi_k|^2) = 0$; that is, let $\omega = \omega(k, |\varphi|^2)$ be known. Then let us expand ω about $\omega = \omega_0(k_0)$, $k = k_0$ and $|\varphi|^2 = 0$ to obtain

$$\omega - \omega_0 = \frac{\partial \omega}{\partial k_0}(k - k_0) + \frac{1}{2}\frac{\partial^2 \omega}{\partial k_0^2}(k - k_0)^2 + \left(\frac{\partial \omega}{\partial |\varphi|^2}\right)_{|\varphi|^2=0} |\varphi|^2, \quad (3.6.33)$$

and make the replacements, $\omega - \omega_0 \to i\, \partial/\partial t$, $k - k_0 \to -i\, \partial/\partial x$. When the resulting operator operates on φ, thereby transforming the situation to a frame moving with the group velocity, we immediately obtain the nonlinear Schrödinger equation (3.6.25). Theoretical grounds for this operation were given by Karpman and Krushkal (1969) and their idea is based on Whitham's theory (Whitham (1965a, b), (1974)).

They consider a slowly modulated plane wave given by $\varphi \exp(i\theta)$, in which the phase θ is assumed to be of the form

$$\theta = k_0 x - \omega_0 t + \sigma,$$

where σ is the slowly varying part of the phase. In Whitham's theory, ω and k are introduced through the equations $\omega = -\theta_{,t} = \omega_0 - \sigma_{,t}$ and $k = \theta_{,x} = k_0 + \sigma_{,x}$, while the nonlinear dispersion relation is expanded as

$$\omega \simeq \omega^0(k^2) + (\partial\omega/\partial|\varphi|^2)_0 |\varphi|^2,$$

where $\omega^0(k^2)$ determines the linear dispersion relation. That is, $\omega^0(k^2) \simeq \omega_0 + \omega_0'(k - k_0) + \frac{1}{2}\omega_0''(k - k_0)^2$, where

$$\omega_0' \equiv \left(\frac{\partial \omega}{\partial k}\right)_{\substack{k=k_0 \\ \varphi=0}}, \quad \omega_0'' \equiv \left(\frac{\partial^2 \omega}{\partial k^2}\right)_{\substack{k=k_0 \\ \varphi=0}}$$

and

$$\left(\frac{\partial \omega}{\partial |\varphi|^2}\right)_0 \equiv \left(\frac{\partial \omega}{\partial |\varphi|^2}\right)_{\substack{k=k_0 \\ \varphi=0}}.$$

Hence we arrive at the result

$$\sigma_{,t} + \omega_0' \sigma_{,x} + \tfrac{1}{2}\omega_0''(\sigma_{,x})^2 + (\partial\omega/\partial|\varphi|^2)_0 |\varphi|^2 = 0.$$

Since $u \sim \sigma_{,x}$ and $\rho \sim |\varphi|^2$, it is easily seen that this equation is equivalent to Eq. (3.6.32b), provided the last term on the right-hand side is neglected. On the other hand in Whitham's theory (Whitham 1974, Section 14.2), Eq. (3.6.32a) is the equation of energy transport, which can be expressed in the conservation form $|\varphi|^2_{,t} + (v_E |\varphi|^2)_{,x} = 0$, where $v_E(k, |\varphi|^2)$ is the rate of energy transport. In the linear approximation it reduces to the group velocity, that is, to

$$v_E(k, |\varphi|^2) \approx v_E(k, 0) = \frac{\partial \omega(k, 0)}{\partial k} \approx \omega'_0 + \omega''_0(k - k_0),$$

and consequently, $(v_E |\varphi|^2)_{,x} \approx \omega'_0 |\varphi|^2_{,x} + \omega''_0 (\sigma_{,x} |\varphi|^2)_{,x}$.

Since the last term in Eq. (3.6.32b) comes from the (linear) Schrödinger equation, that is, from the linear dispersion (cf. Eqns (1.2.12), (1.2.14)), by means of an heuristic argument incorporating the linear result and Whitham's approximation, Karpman and Krushkal (1969) were the first to derive the nonlinear Schrödinger equation. Here it may be worth noticing that Eq. (3.6.32b), without dispersion, is the eikonal equation in nonlinear geometrical optics and, with Eq. (3.6.32a), constitutes the fundamental set of equations in nonlinear geometrical optics. Since the set of equations is a coupled system of first order partial differential equations of normal form, it can be linearized by means of a Legendre transformation and it permits various analogies with hydrodynamics. In this way a number of interesting solutions have been obtained by Shvartsburg (1980). As was noted already, the exact system Eq. (3.6.32), that is the nonlinear Schrödinger equation, can be solved by means of the inverse-scattering method, and the comparison of the solutions of the two systems has been carried out by Shvartsburg (1980). Here we note that nonlinear geometrical optics was first considered by Ostrovskii (1964). He derived a system of equations with dispersion, which is equivalent to Eqns (3.6.32) (Ostrovskii (1967)), and he then obtained the solitary wave as the stationary solution.

As examples of other methods for the derivation of the nonlinear Schrödinger equation, we can cite the derivative expansion method in which the orderings of the amplitude and the space-time scales are determined successively, in such a way that the secular terms do not appear in any order (Kawahara (1973), Jeffrey and Kawahara (1982)), and Kakutani's method which may be regarded as an extension of Bogoliubov's method for nonlinear oscillations to nonlinear waves (Kakutani and Sugimoto (1974), Jeffrey and Kawahara (1982)).

So far we have considered the self-modulation of a single mode belonging to the linear dispersion relation Eq. (3.6.5). However, when different modes with amplitudes of the same order (ε) coexist and undergo mutual interactions, as well as self-modulation, we can split the system of equations into the independent nonlinear Schrödinger equations just as in the case of long waves (Oikawa and Yajima (1974)). Consequently envelope solitons associated with different modes, propagating separately at the beginning, conserve their identities and still propagate as envelope solitons, even after the mutual interactions.

Nevertheless, we should mention the following point. Usually there are mutual interactions of the order of ε^2 taking place amongst waves of different modes of order ε, and when the resonance condition for the wave numbers and the frequencies is satisfied, strong resonant wave–wave interactions can occur. On the other hand, the self-modulation accompanying the envelope solitons is of the order of ε^3, and so it appears as a higher order effect. We discuss this matter in detail in Section 3.8. (A higher order approximation to the nonlinear Schrödinger equation in the reductive perturbation method was given by Kodama (1978).)

Example 3.6.1 The electron plasma wave

Eqns (A.3a, b, d) can be reduced to Eq. (3.6.3) (Poisson's equation (A.3c) may be regarded as a condition for the initial values) when we introduce

$$U = \begin{pmatrix} n \\ u \\ E \end{pmatrix}, \quad A = \begin{pmatrix} u & n & 0 \\ T/mn & u & 0 \\ 0 & 0 & u \end{pmatrix}, \quad B = \begin{pmatrix} 0 \\ (e/m)E \\ -4\pi e n_0 u \end{pmatrix}.$$

For the constant state $U^{(0)}(n_0, 0, 0)$, we have

$$W_1 \equiv \begin{pmatrix} -il\omega & ilkn_0 & 0 \\ ilkv_T^2/n_0 & -il\omega & e/m \\ 0 & -4\pi e n_0 & -il\omega \end{pmatrix},$$

where $v_T = (T/m)^{1/2}$. From $\det W_1 = 0$, we find the linear dispersion relation (strong dispersion),

$$\omega^2 = \omega_p^2 + k^2 v_T^2.$$

The right and left eigenvectors of the matrix W_1 are

$$R = \begin{pmatrix} 1 \\ \omega/kn_0 \\ i4\pi e/k \end{pmatrix}, \quad L = (v_T^2/n_0, \omega/k, -ie/mk),$$

so that from Eq. (3.6.21), P in Eq. (3.6.20) becomes

$$P = \frac{v_T^2 \omega_p^2}{2\omega^3},$$

which can also be found easily from $P = (\tfrac{1}{2}\partial^2 \omega/\partial k^2)$ through the linear dispersion relation.

Since $\det W_2 \neq 0$, Eq. (3.6.24a) gives

$$R_2^{(2)} = \frac{1}{3n_0 \omega_p^2} \begin{pmatrix} 4\omega^2 + 2\omega_p^2 \\ \omega(4\omega^2 - \omega_p^2)/kn_0 \\ i4\pi e(2\omega^2 + \omega_p^2)/k \end{pmatrix}.$$

However, since det $W_0 = 0$, we cannot obtain $R_0^{(2)}$ from Eq. (3.6.24b), though it can be found by means of Eq. (A.3c). Since $n_i = n_0$ (constant), the expansion of Eq. (A.3c) in powers of ε yields

$$\varepsilon^2 \frac{\partial E_0^{(1)}}{\partial \xi} + \varepsilon^3 \frac{\partial E_0^{(2)}}{\partial \xi} + \cdots = -4\pi e(\varepsilon n_0^{(1)} + \varepsilon^2 n_0^{(2)} + \cdots),$$

so that $n_0^{(1)} = 0$, and also from $W_0 U_0^{(1)} = 0$ it follows that $u_0^{(1)} = E_0^{(1)} = 0$, and that hence $U_0^{(1)} = 0$. Since $E_0^{(1)} = 0$, the above expansion of Eq. (A.3c) yields $n_0^{(2)}$. Consequently, Eq. (3.6.12) for $l = 0$ becomes

$$W_0 U_0^{(2)} + ik[(\nabla A_0 \cdot R^*)R - (\nabla A_0 \cdot R)R^*]|\varphi^{(1)}|^2 = 0,$$

which can be solved for the $U_0^{(2)}$ and $E_0^{(2)}$ components of $U_0^{(2)}$. As a result, we get

$$R_0^{(2)} = -\frac{2\omega}{kn_0^2}\begin{pmatrix} 0 \\ 1 \\ 0 \end{pmatrix}.$$

Hence, from Eq. (3.6.22), we have

$$Q = -\frac{k^2 v_T^2(8\omega^2 + \omega_p^2)}{6n_0^2 \omega \omega_p^2}.$$

Example 3.6.2 Nonlinear Klein–Gordon equation

We may consider a travelling wave propagated in a wave guide as an extension of the multi-dimensional case. (For a plasma wave, this problem was solved by VanDam (1973).) We here adopt the nonlinear Klein–Gordon equation as a model equation:

$$\frac{\partial^2 \psi}{\partial t^2} - \frac{\partial^2 \psi}{\partial x^2} - \frac{\partial^2 \psi}{\partial y^2} + m^2 \psi + \kappa \psi^3 = 0,$$

and the boundary condition we shall use is $\psi = 0$ for $y = 0$ and $y = a$. In the linear case, we may assume $\psi(x, y, t) = \tilde{\psi}(y)e^{i(kx-\omega t)}$ to find

$$\tilde{\psi}(y) = A \sin(\eta y) + B \cos(\eta y),$$

$$\eta^2 \equiv \omega^2 - k^2 - m^2.$$

Since $B = 0$, from the boundary condition we obtain

$$\eta^2 \equiv \omega^2 - k^2 - m^2 = \left(\frac{n\pi}{a}\right)^2 \qquad (n = 1, 2, \ldots),$$

so that in the dispersion relation $\omega(k, n)$, there exist various modes dependent on n.

3.6 STRONGLY DISPERSIVE SYSTEMS

When n is given, ω (or k) is determined through the dispersion relation for a fixed k (or ω). Now for the wave with $n=1$, let us consider the slow variation of amplitude caused by the nonlinear effect. Namely, using the ξ and τ defined by Eq. (3.6.8), we shall assume the following form of solution:

$$\psi = \sum_{\alpha=1}^{\infty} \varepsilon^{\alpha} \psi^{(\alpha)},$$

$$\psi^{(\alpha)} = \sum_{l=-\infty}^{\infty} \psi_l^{(\alpha)}(\xi, \tau, y) e^{il(kx-\omega t)}.$$

Now, from the dispersion relation, we have

$$\lambda \equiv \frac{\partial \omega}{\partial k} = \frac{k}{\omega}.$$

Retaining only terms of the first order in ε, we find

$$\psi_1^{(1)} = \varphi(\xi, \tau) \sin(\eta y), \qquad \psi_l^{(1)} = 0 \quad (|l| \neq 1),$$

and the terms of second order in ε yield

$$2il(k-\lambda\omega)\frac{\partial \psi_l^{(1)}}{\partial \xi} + \frac{\partial^2 \psi_l^{(2)}}{\partial^2 y} + (l^2\omega^2 - l^2 k^2 - m^2)\psi_l^{(2)} = 0,$$

where the first term vanishes because $\lambda = k/\omega$, so that $\psi_l^{(2)} = 0$ for $|l| \neq 1$. For $l = 1$, the terms of third order in ε give

$$\frac{\partial^2 \psi_1^{(3)}}{\partial y^2} + \eta^2 \psi_1^{(3)} + \left[2i\omega \frac{\partial \varphi}{\partial \tau} + (1-\lambda^2)\frac{\partial^2 \varphi}{\partial \xi^2}\right] \sin \eta y - 3\kappa |\varphi|^2 \varphi \sin^3 \eta y = 0.$$

Because of the compatibility condition, multiplying the above equation by $\sin \eta y$, followed by integration with respect to y from 0 to a, causes the first and second terms to vanish, and we obtain

$$i\varphi_{,\tau} + \frac{\omega^2 - k^2}{2\omega^3} \varphi_{,\xi\xi} - \frac{9\kappa}{8\omega} |\varphi|^2 \varphi = 0.$$

Since $\omega^2 - k^2 > 0$, a modulational instability results if $\kappa < 0$. Also, letting the amplitude be A, the envelope soliton solution for $\kappa < 0$ becomes

$$\psi_1^{(1)} = A \operatorname{sech}\left\{\left(-\frac{9\kappa\omega^2 A^2}{8(\omega^2 - k^2)}\right)^{1/2} \xi\right\} \sin \eta y \exp\left(i\frac{9\kappa A^2}{16\omega}\tau\right).$$

It is, in general, possible to apply the reductive perturbation method to a weakly nonlinear wave propagating in a channel, so that the system can be reduced to the nonlinear Schrödinger equation (in the strongly dispersive case) and to the KdV equation or the modified KdV equation (in the weakly dispersive case). It may also be worth mentioning that, in a similar way, a weakly unstable fluid flow through a pipe can be reduced to the generalized nonlinear Schrödinger equation (Stewartson and Stuart (1971)).

In a general multi-dimensional problem, however, the initial value problem for the nonlinear Schrödinger equation is not always valid for all time, as will be mentioned in the next section (Zakharov (1972)). Moreover, the envelope soliton is unstable against transverse perturbations (Yajima (1974)).

3.7 Self-focusing of waves

In the previous section we have seen that wave modulation occurring in the direction of wave propagation is described by the nonlinear Schrödinger equation. Let us illustrate here that wave modulation in a plane perpendicular to the direction of propagation is also described by the nonlinear Schrödinger equation.

As is well known in nonlinear optics, there are some phenomena in which the refractive index changes in proportion to the intensity of the light. For instance, we refer to the optical Kerr effect, in which anisotropy of the refractive index results from the total rotation of each molecule with anisotropy under the action of the light, and the electrostriction effect, in which the pressure varies due to the electric field, and the associated density variation then causes a change of the refractive index. The variations of the refractive index caused by these two effects are proportional to the square of the modulus of the electric field. That is, the refractive index is given by $n = n_0 + \Delta n = n_0 + (n_2^2/2n_0)|\boldsymbol{E}|^2$ (we assume $\Delta n > 0$), where n_0 is the linear refractive index and n_2 is the coefficient of the nonlinear refractive index. Moreover, we assume, for simplicity, that n_0 and n_2 are real constants independent of the frequency, etc.

Then Maxwell's equations reduce to

$$\nabla^2 \boldsymbol{E} - \left(\frac{n_0}{c}\right)^2 \boldsymbol{E}_{,tt} = \left(\frac{n_2}{c}\right)^2 (|\boldsymbol{E}|^2 \boldsymbol{E})_{,tt}. \qquad (3.7.1)$$

Let us consider a linearly polarized, quasi-plane, electromagnetic wave, with frequency ω and wave number k, propagating in the x-direction. The wave intensity will be assumed to vary over the plane ((y, z)-plane) normal to the direction of propagation. In addition, the spatial variation of the intensity will be assumed to be suitably slow compared with the wavelength. Furthermore, we shall consider steady propagation in which the spatial variation of the wave amplitude is slower in the x-direction of propagation than normal to it.

3.7 SELF-FOCUSING OF WAVES

We therefore assume a solution $E(r, t)$ of Eq. (3.7.1) in the form

$$E(r, t) = \varepsilon \varphi(\xi, \eta, \zeta) i e^{i(kx-\omega t)}, \qquad (3.7.2)$$
$$\xi = \varepsilon^2 x, \qquad \eta = \varepsilon y, \qquad \zeta = \varepsilon z,$$

where i is the unit vector along the direction of polarization of the electric field. Inserting Eq. (3.7.2) into Eq. (3.7.1), we obtain the linear dispersion relation $k^2 = (n_0 \omega/c)^2$ from the first order terms in ε. The second order terms do not exist, and from the third order terms in ε, we get the nonlinear Schrödinger equation

$$i\varphi_{,\xi} + \frac{1}{2k} \nabla^2 \varphi + \frac{k}{2}\left(\frac{n_2}{n_0}\right)^2 |\varphi|^2 \varphi = 0, \qquad (3.7.3)$$

where $\nabla^2 \equiv \partial^2/\partial \eta^2 + \partial^2/\partial \zeta^2$.

In the event that the amplitude does not change in the y and z-directions, the solution of Eq. (3.7.3) becomes

$$\varphi = \varphi_0 \exp\left[i\left(\frac{k}{2}\right)\left(\frac{n_2}{n_0}\right)^2 |\varphi_0|^2 \xi\right]. \qquad (3.7.4)$$

Thus, the change of phase, which is proportional to the intensity, results from the nonlinear effect. Since the effective wave number reduces to $k' = k[1 + (\frac{1}{2})(\varepsilon n_2/n_0)^2 |\varphi_0|^2]$, the phase velocity ω/k' becomes smaller in the regions of larger amplitude, and consequently the wave converges to the place of largest amplitude.

Let us suppose that the above relationship is also valid for the case in which the amplitude varies in the y and z-directions. Then, when the amplitude attains a maximum at the centre of the light beam, the wave focuses on its centre. This is called the self-focusing effect. However, diffusion also occurs in waves due to diffraction (which is represented by the second term in Eq. (3.7.3)), so that there exists a threshold value of the amplitude for self-focusing to occur. For example, let us consider the self-focusing of a cylindrical light beam whose radius is a. Since the refractive index inside the beam is larger than that outside (given by n_0), the rays issuing from points on the central axis are totally reflected, provided the angles between the rays and the axis are smaller than a critical angle $\theta_c \propto \cos^{-1}(k/k')$, and form a trapped beam. (The light propagates in a waveguide made by itself.)

On the other hand, a beam of radius a tends to spread, making an angle $\theta_d \propto (ka)^{-1}$ with the central axis because of diffraction effects (as can be seen from the result below in Eq. (3.7.5)), and so self-focusing occurs when $\theta_d < \theta_c$. Solving this, we find as the condition for self-focusing $\varepsilon^2 |\varphi_0|^2 > (n_0/n_2)^2 (ka)^{-2}$ (Akhmanov et al. (1966)).

One-dimensional self-focusing (for instance, focusing due to a slab which is independent of z) is governed by the one-dimensional nonlinear Schrödinger equation, and hence the results of the previous section are directly applicable. In this case, an envelope soliton is formed and the initial value problem which is involved will be solved exactly in Section 4.5 of the next chapter.

In the cylindrical case, however, it is difficult to obtain exact solutions of Eq. (3.7.3). Moreover, in general, a solution does not exist, and disruption of the focused beam occurs (Zakharov (1972)). For simplicity, we consider here a nonlinear Schrödinger equation of the form

$$i\varphi_{,t} + \nabla^2 \varphi + |\varphi|^2 \varphi = 0, \tag{3.7.3'}$$

with the boundary conditions $\varphi \to 0$ as $r \to \infty$, and $\partial \varphi / \partial r = 0$ at $r = 0$. It is easily seen that the following conserved quantities exist, which correspond to the conservation of density and energy:

$$I_1 = \int |\varphi|^2 \, dv, \qquad I_2 = \int \{|\nabla \varphi|^2 - \tfrac{1}{2} |\varphi|^4\} \, dv.$$

Introducing A through the integral

$$A = \int r^2 |\varphi|^2 \, dv,$$

and assuming cylindrical symmetry ($dv = r \, dr \, d\theta$), we have by virtue of the boundary conditions

$$\frac{d^2 A}{dt^2} = 8 I_2, \tag{3.7.5}$$

so that $A = 4 I_2 t^2 + c_1 t + c_2$ (where c_1 and c_2 are constants). In the case of no modulation ($\varphi = $ constant), we have $I_2 = -\infty$, and generally if I_2 is negative, A inevitably at some time becomes negative. However, by definition, A cannot be negative, and hence, in the case of cylindrical symmetry, the solution of the nonlinear Schrödinger equation (3.7.3′) can only exist for a finite time.

Let us consider the problem in terms of the distance ξ instead of the time t. That is, for a given φ at $\xi = 0$, we seek a solution for $\xi > 0$. Then we see that a solution can exist only for a finite distance (until the disruption of the light beam). The same effect also occurs in the three-dimensional case with spherical symmetry (Zakharov (1972)).

In this section we found that the nonlinear Schrödinger equation describes the self-focusing effect for some simple examples. It can be shown, however, that Eq. (3.7.3) can also be derived from a fairly general nonlinear equation under the assumptions Eq. (3.7.2). For example, the self-focusing of various plasma waves have been studied in this way (Washimi (1973)).

3.8 Three-wave interaction

As has already been mentioned, the self-modulation of a wave, which is described by the nonlinear Schrödinger equation is to be considered by means of the approximation involving the cube of the amplitude. When, however, several waves of different modes coexist, phenomena appear which need to be considered by means of an approximation involving the square of the amplitude. By way of example, we mention the parametric effect of light, in which the nonlinear part of the refractive index varies in proportion to the amplitude itself, as is often encountered in nonlinear optics, and also the process of parametric amplification used in an electric circuit. As a simple and very familiar example, we can also consider the case of a child on a swing.

In what follows, we shall discuss phenomena by using what is called the method of time averaging, in which the equations under consideration are averaged over rapid oscillations in order that they may be reduced to the equations for slowly varying quantities.

A child on a swing amplifies its swaying motion by lowering the centre of gravity of its body as the swing descends and raising it as it ascends. The period (frequency) of the up and down motion of the child is half of (twice) the period (characteristic frequency) say ω_0 of the swing. Namely, the effective length of the swing is forced to vary with twice the characteristic frequency of the swing. As a result, the energy of the forced excited oscillation with frequency $2\omega_0$ is transferred to the eigenmode with the characteristic frequency ω_0. This example is a case in which two eigenmodes are degenerate. However, when the frequency of the excited oscillation is equal to the sum of the frequencies of two independent modes, coupling occurs between the modes and the energy of the excited oscillation is transferred to the two modes. This is called parametric excitation. The same also holds for the cases when the excited oscillation is one of the independent eigenmodes of a system. Consequently, when the resonance condition (or, as it is often called, the matching condition)

$$\omega_0 = \omega_1 + \omega_2 \tag{3.8.1}$$

is satisfied, a strong interaction occurs between independent modes.

As such an example, we consider three interacting harmonic oscillators. Let the displacement and the momentum of each oscillator be x_j and p_j ($j = 0, 1, 2$), respectively, and let the Hamiltonian be given by

$$\mathcal{H} = \sum_{j=0}^{2} \left(\frac{p_j^2}{2} + \frac{\omega_j^2 x_j^2}{2} \right) + \beta x_0 x_1 x_2, \tag{3.8.2}$$

where ω_j is the characteristic frequency of each oscillator and β is a coupling constant between the three oscillators. Accordingly, the equations of motion for the three oscillators take the form

$$\frac{d^2 x_0}{dt^2} + \omega_0^2 x_0 = -\beta x_1 x_2, \tag{3.8.3a}$$

$$\frac{d^2 x_1}{dt^2} + \omega_1^2 x_1 = -\beta x_0 x_2, \tag{3.8.3b}$$

$$\frac{d^2 x_2}{dt^2} + \omega_2^2 x_2 = -\beta x_0 x_1. \tag{3.8.3c}$$

Without any interaction, each oscillator x_j vibrates with the characteristic frequency ω_j, and the amplitude is determined by the initial condition, and does not change. When the displacement x_j is small, the time variation of the displacement x_j due to the nonlinear interaction of the order $O(x_j^2)$ is slow compared to that associated with the characteristic frequency. Consequently, the displacement x_j can be expressed as the product of a slowly varying amplitude and a factor which oscillates with the frequency ω_j:

$$x_j(t) = C_j(t) e^{i\omega_j t} + C_j^*(t) e^{-i\omega_j t}. \tag{3.8.4a}$$

Substituting the solution (3.8.4) into Eq. (3.8.3), we get

$$\frac{d^2 C_0}{dt^2} + 2i\omega_0 \frac{dC_0}{dt} = -e^{-2i\omega_0 t}\left(\frac{d^2 C_0^*}{dt^2} - 2i\omega_0 \frac{dC_0^*}{dt}\right) - \beta C_1 C_2 e^{i(\omega_1 + \omega_2 - \omega_0)t}$$
$$- \beta C_1^* C_2 e^{i(-\omega_1 + \omega_2 - \omega_0)t} - \beta C_1 C_2^* e^{i(\omega_1 - \omega_2 - \omega_0)t}$$
$$- \beta C_1^* C_2^* e^{-i(\omega_1 + \omega_2 + \omega_0)t}, \tag{3.8.4b}$$

together with similar equations for C_1 and C_2.

Since the amplitude varies slowly with time, it can be considered constant during a time interval of the order ω_j^{-1}. Therefore, averaging Eq. (3.8.4b) over a time scale of the order ω_j^{-1}, in general, all the terms on the right-hand side will vanish, since they contain rapidly oscillating parts. That is, we have the approximation C_j = constant. However, when, for example, the resonance condition (3.8.1) is satisfied, the rapidly oscillating part of the second term on the right-hand side of Eq. (3.8.4b) disappears, and consequently this term does not vanish when averaging with respect to the time. In other words, such a term causes a slow time variation of amplitude.

Thus, if we assume the resonance condition (3.8.1) to be satisfied, we find by means of time-averaging that

$$\frac{dC_0}{dt} = i\left(\frac{\beta}{2\omega_0}\right) C_1 C_2, \tag{3.8.5a}$$

$$\frac{dC_1}{dt} = i\left(\frac{\beta}{2\omega_1}\right)C_0 C_2^*, \tag{3.8.5b}$$

$$\frac{dC_2}{dt} = i\left(\frac{\beta}{2\omega_2}\right)C_0 C_1^*. \tag{3.8.5c}$$

Here we have neglected the terms $d^2 C_j/dt^2$, since the amplitudes only vary slowly. Thus, when a resonance condition is satisfied, interaction takes place between the eigenmodes and the amplitudes vary with time.

As an example of possible initial conditions, let us consider the case in which the amplitude of one oscillation is large in comparison with those of the other two oscillations. For instance, suppose that $|C_0| \gg |C_1|, |C_2|$, then Eqns (3.8.5) can be linearized with respect to C_1, C_2 to yield

$$\frac{dC_0}{dt} = 0, \tag{3.8.6a}$$

$$\frac{dC_1}{dt} = i\left(\frac{\beta}{2\omega_1}\right)C_0 C_2^*, \tag{3.8.6b}$$

$$\frac{dC_2}{dt} = i\left(\frac{\beta}{2\omega_2}\right)C_0 C_1^*, \tag{3.8.6c}$$

so that

$$C_j(t) = C_j(0) e^{\gamma t} \quad (j=1, 2),$$
$$\gamma^2 = \frac{|\beta|^2 |C_0|^2}{4\omega_1 \omega_2}. \tag{3.8.7}$$

Consequently, we find the amplitudes C_1 and C_2 grow with time provided $\omega_1 \omega_2 > 0$. From the resonance condition (3.8.1), in order that the condition $\omega_1 \omega_2 > 0$ holds, we must satisfy the condition

$$|\omega_0| > |\omega_1|, |\omega_2|, \tag{3.8.8}$$

so that parametric excitation occurs only when the frequency of the excited oscillation is higher than the frequencies of the other oscillations.

We now see that, as can be deduced from the motion of a swing, a parametric excitation can take place even if the frequency of the excited oscillation shifts a little from the matching condition (3.8.1). To develop this idea further, let us now consider the case in which there is a shift in the matching condition

$$\omega_0 - \omega_1 - \omega_2 = \Delta, \tag{3.8.9}$$

where the frequency of the shift Δ is assumed to be suitably small compared to the characteristic frequencies; thus we assume $|\Delta| \ll |\omega_j|$.

Extending Eqns (3.8.3b, c) to include a weak dissipation we have

$$\frac{d^2x_1}{dt^2} + 2\nu_1 \frac{dx_1}{dt} + \omega_1^2 x_1 = -\beta x_0 x_2, \tag{3.8.10a}$$

$$\frac{d^2x_2}{dt^2} + 2\nu_2 \frac{dx_2}{dt} + \omega_2^2 x_2 = -\beta x_0 x_1, \tag{3.8.10b}$$

where ν_j denotes the damping rate of the jth mode and is also assumed to be sufficiently small compared to ω_j, so that $0 \leq \nu_j \leq |\omega_j|$.

Now, averaging Eq. (3.7.10) over the time ω_j^{-1} and noticing that the above assumption enables us to neglect the term $\nu_j \, dC_j/dt$ (compared to $\omega_j \, dC_j/dt$), we obtain the following equations which correspond to Eqns (3.8.6):

$$\left(\nu_1 C_1 + \frac{dC_1}{dt}\right) = i\left(\frac{\beta}{2\omega_1}\right) C_0 C_2^* e^{i\Delta t}, \tag{3.8.11a}$$

$$\left(\nu_2 C_2 + \frac{dC_2}{dt}\right) = i\left(\frac{\beta}{2\omega_2}\right) C_0 C_1^* e^{i\Delta t}. \tag{3.8.11b}$$

Suppose that C_0 is constant, then it is readily seen that these equations yield the special solution $C_1 \propto e^{i\omega t}$, $C_2 \propto e^{-i(\omega-\Delta)t - \nu_2 t}$ provided ω satisfies the secular equation

$$(\omega - i\nu_1)(\omega - \Delta - i\nu_2) + \frac{\beta^2 |C_0|^2}{4\omega_1 \omega_2} = 0, \tag{3.8.12}$$

while for another independent solution, $C_2 \propto e^{i\omega t}$, ν_1 and ν_2 in the above secular equation must be interchanged. Here, we shall write the solution ω as

$$\omega = \omega_r - i\gamma. \tag{3.8.13}$$

As will be shown in Eq. (3.8.14), γ is the same for the two independent solutions. Then ω_r corresponds to the frequency change of the excited mode. That is, the characteristic frequency of the x_1 mode changes to $\omega_1 + \omega_r$ on account of the interaction, and also the characteristic frequency of the x_2 mode changes to $\omega_2 + (\Delta - \omega_r)$. Moreover, when $\gamma > 0$, the x_1 and x_2 modes grow in time with the same growth rate γ. Inserting the solution (3.8.13) into Eq. (3.8.12), separating the real and imaginary parts, and eliminating ω_r from the equations so obtained, we find

$$(\gamma + \nu_1)(\gamma + \nu_2)\left[1 + \frac{\Delta^2}{(2\gamma + \nu_1 + \nu_2)^2}\right] = \frac{\beta^2 |C_0|^2}{4\omega_1 \omega_2}. \tag{3.8.14}$$

Since ν_1 and ν_2 are positive, we must also have $\omega_1 \omega_2 > 0$ in order that $\gamma > 0$. Furthermore, since, for $\gamma > 0$, the left-hand side is greater than

$\nu_1 \nu_2$, for the instability to appear $|C_0|^2$ must be at least greater than $4\omega_1\omega_2\nu_1\nu_2/\beta^2$. Thus, in the presence of dissipation, there is a threshold value for the amplitude of the excited oscillation if parametric excitation is to occur. Putting $\gamma = 0$ in Eq. (3.8.14), we obtain the threshold value for the amplitude

$$|C_0|^2 = C_m^2 \left[1 + \frac{\Delta^2}{(\nu_1 + \nu_2)^2} \right], \tag{3.8.15a}$$

$$C_m^2 = \frac{4\omega_1\omega_2\nu_1\nu_2}{\beta^2}. \tag{3.8.15b}$$

Accordingly, with $\nu_1 = 0$ or with $\nu_2 = 0$, parametric excitation occurs even for an excited oscillation of infinitesimal amplitude. Moreover, when $\Delta = 0$, so that the matching condition is satisfied, the threshold value attains a minimum. When the amplitude of the excited oscillation is equal to the threshold value, the frequencies of the respective modes reduce to

$$\omega_1 + \frac{\nu_1}{\nu_1 + \nu_2}\Delta, \qquad \omega_2 + \frac{\nu_2}{\nu_1 + \nu_2}\Delta. \tag{3.8.16}$$

In the case that the amplitude of the excited oscillation is large compared to the threshold value, ν_1 and ν_2 can be neglected in comparison with γ, and hence the growth rate and the frequency change become, respectively,

$$\gamma = \frac{1}{2}\left(\frac{\beta^2 |C_0|^2}{\omega_1 \omega_2} - \Delta^2\right)^{1/2}, \tag{3.8.17a}$$

$$\omega_r = \frac{\Delta}{2}, \tag{3.8.17b}$$

so that the frequency change is then proportional to Δ (the frequency shift for the mismatch).

Next, let us seek the equations for the mode coupling in a dispersive medium. In the presence of dispersion, the wave numbers also have to satisfy the resonance condition

$$k_0 = k_1 + k_2. \tag{3.8.18}$$

For instance, let us consider the propagation of electromagnetic waves in a nonlinear dispersive medium. Then Maxwell's equations reduce to

$$\nabla^2 \boldsymbol{E} - \left(\frac{1}{c^2}\right)(\boldsymbol{E} + 4\pi \boldsymbol{P})_{,tt} = \left(\frac{4\pi}{c^2}\right)\boldsymbol{P}_{,tt}^{(\mathrm{NL})}, \tag{3.8.19a}$$

Here, $\boldsymbol{P}$ and $\boldsymbol{P}^{(\mathrm{NL})}$ are the linear and the nonlinear polarization vectors of the medium, respectively, and generally they are given by nonlocal expressions as follows:

$$\boldsymbol{P}(\boldsymbol{r}, t) = \int_{-\infty}^{\infty} d^3\boldsymbol{r}' \, dt' \chi^{(1)}(\boldsymbol{r}-\boldsymbol{r}', t-t')\boldsymbol{E}(\boldsymbol{r}', t'), \tag{3.8.19b}$$

$$\begin{aligned}\boldsymbol{P}^{(\text{NL})}(\boldsymbol{r}, t) = & \int_{-\infty}^{\infty} d^3\boldsymbol{r}' \, dt' \, d^3\boldsymbol{r}'' \, dt'' \chi^{(2)}(\boldsymbol{r}-\boldsymbol{r}', t-t'; \boldsymbol{r}-\boldsymbol{r}'', \\ & \times t-t'') : \boldsymbol{E}(\boldsymbol{r}', t')\boldsymbol{E}(\boldsymbol{r}'', t'') \\ & + \int_{-\infty}^{\infty} d^3\boldsymbol{r}' \, dt' \, d^3\boldsymbol{r}'' \, dt'' \, d\boldsymbol{r}''' \, dt''' \chi^{(3)} \\ & \times (\boldsymbol{r}-\boldsymbol{r}', t-t'; \boldsymbol{r}-\boldsymbol{r}'', t-t''; \boldsymbol{r}-\boldsymbol{r}''', t-t''') \vdots \boldsymbol{E}(\boldsymbol{r}', t') \\ & \times \boldsymbol{E}(\boldsymbol{r}'', t'')\boldsymbol{E}(\boldsymbol{r}''', t''') + \cdots, \end{aligned} \tag{3.8.19c}$$

where $\chi^{(1)}$, $\chi^{(2)}$, and $\chi^{(3)}$ are the polarization tensors which are determined by the properties of the medium.

In the previous section, we studied the self-focusing of a wave due to a third-order nonlinear effect involving $\chi^{(3)}$. Here we shall investigate the three-wave interaction due to a second-order nonlinear effect involving $\chi^{(2)}$. In addition, in the previous section, we assumed the refractive index to be constant and independent of the frequency, etc., which corresponds to the case in which the polarization coefficient in Eq. (3.8.19b) is assumed to be such that $\chi(r, t) \propto \delta(r)\delta(t)$.

Now, for simplicity, we assume the polarization coefficient χ to be a scalar, and we also suppose that all the electromagnetic modes are propagated along the x-axis and that their amplitudes vary only in the x-direction. Furthermore, we shall suppose that the electromagnetic waves are linearly polarized with the electric vectors aligned, so that vector symbols such as $\boldsymbol{E}$, $\boldsymbol{P}$ may be suppressed and replaced by scalars. Let us suppose that the electric field $E(x, t)$ consists of three quasi-plane wave modes (Eqns (3.5.4)) as follows:

$$E(x, t) = \sum_{j=0}^{2} E_j(x, t), \tag{3.8.20a}$$

$$\begin{aligned}E_j(x, t) &= e^{i(k_j x - \omega_j t)} \int d\delta k E_j'(k_j + \delta k) e^{i\delta k(x - \omega_j' t)} + \text{c.c.} \\ &= \varphi_j(x, t) e^{i(k_j x - \omega_j t)} + \text{c.c.} \quad (j = 0, 1, 2), \end{aligned} \tag{3.8.20b}$$

$$\omega_j' = \frac{\partial \omega_j}{\partial k_j} = v_{gj},$$

where c.c. again denotes the complex conjugate of the preceding quantities, so that in what follows the terms of the complex conjugate expression will be omitted unless otherwise stated.

3.8 THREE-WAVE INTERACTION

The variation of the amplitude $\varphi_j(x, t)$ caused by the nonlinear interaction will be assumed to be suitably slow in comparison to those due to k_j^{-1} and ω_j^{-1}. Without the nonlinear polarization $P^{(\mathrm{NL})}$, we have $\varphi_j(x, t) = a_j =$ constant, and hence substituting Eq. (3.8.20) into Eq. (3.8.19), and making use of the Fourier transform, we obtain

$$\sum_{j=0}^{2}\left[k^2 - \frac{\omega^2}{c^2}(1 + 4\pi\chi_{k_j,\omega_j})\right] a_j \delta(k - k_j)\delta(\omega - \omega_j) = 0. \quad (3.8.21a)$$

Consequently, (k_j, ω_j) must satisfy the linear dispersion relation

$$D(k_j, \omega_j) = 1 + 4\pi\chi_{k_j,\omega_j} - \frac{k_j^2 c^2}{\omega_j^2} = 0, \quad (3.8.21b)$$

where $\chi_{k,\omega}$ is the Fourier component of the linear polarization coefficient $\chi(x, t)$, so that

$$\chi(x, t) = (2\pi)^{-2} \iint_{-\infty}^{\infty} dk\, d\omega\, \chi_{k,\omega} e^{i(kx - \omega t)}, \quad (3.8.22a)$$

$$\chi_{k,\omega} = \chi^*_{-k,-\omega}. \quad (3.8.22b)$$

Inserting Eq. (3.8.22a) into the linear polarization expression (3.8.19b), we find

$$P(x, t) = (2\pi)^{-2} \int_{-\infty}^{\infty} dk\, d\omega\, \chi_{k,\omega} E_{k,\omega} e^{i(kx - \omega t)}, \quad (3.8.23)$$

where $E_{k,\omega}$ is the Fourier component of $E(x, t)$, and using Eq. (3.8.20) this is found to be given by

$$E_{k,\omega} = \sum_{j=0}^{2} (2\pi)^2 \int d\delta k\, E'_j(k_j + \delta k)\, \delta(k - k_j - \delta k)\, \delta(\omega - \omega_j - \delta k v_{gj}). \quad (3.8.24)$$

Using this in Eq. (3.8.23) then yields

$$P(x, t) = \sum_{j=0}^{2} e^{i(k_j x - \omega_j t)} \int d\delta k\, \chi_{k_j + \delta k, \omega_j + \delta k v_{gj}} E'_j(k_j + \delta k) e^{i\delta k(x - v_{gj} t)}. \quad (3.8.25)$$

Now, expanding $\chi_{k_j + \delta k, \omega_j + \delta k v_{gj}}$ about (k_j, ω_j), and introducing the following notations:

$$\chi_{k_j} \equiv \chi_{k_j, \omega_j}, \quad \frac{\partial \chi}{\partial k_j} \equiv \frac{\partial \chi_{k,\omega}}{\partial k}\bigg|_{\substack{k = k_j \\ \omega = \omega_j}}, \quad \frac{\partial \chi}{\partial \omega_j} \equiv \frac{\partial \chi_{k,\omega}}{\partial \omega}\bigg|_{\substack{k = k_j \\ \omega = \omega_j}}, \quad (3.8.26)$$

expression (3.8.25) becomes

$$P(x,t) = \sum_{j=0}^{2} e^{i(k_j x - \omega_j t)} \left[\chi_{k_j} \varphi_j - i\left(\frac{\partial \chi}{\partial k_j}\right) \int d\delta k E'_j(k_j + \delta k) \frac{\partial}{\partial x} e^{i\delta k(x - v_g t)} \right.$$

$$\left. + i\left(\frac{\partial \chi}{\partial \omega_j}\right) \int d\delta k E'_j(k_j + \delta k) \frac{\partial}{\partial t} e^{i\delta k(x - v_g t)} \right]$$

$$= \sum_{j=0}^{2} \left[\chi_{k_j} \varphi_j - i\left(\frac{\partial \chi}{\partial k_j}\right) \frac{\partial \varphi_j}{\partial x} + i\left(\frac{\partial \chi}{\partial \omega_j}\right) \frac{\partial \varphi_j}{\partial t} \right]$$

$$\times e^{i(k_j x - \omega_j t)} + \text{c.c.} \qquad (3.8.27)$$

Let us find the nonlinear polarization $P^{(\text{NL})}$ by neglecting the nonlinear terms of the third order or more. Substituting

$$\chi^{(2)}(x', t'; x'', t'') = \frac{1}{(2\pi)^4} \int_{-\infty}^{\infty} dk' \, d\omega' \, dk'' \, d\omega'' \chi^{(2)}_{k',\omega';k'',\omega''}$$

$$\times e^{i(k'x' - \omega't') + i(k''x'' - \omega''t'')} \qquad (3.8.28)$$

into Eq. (3.8.19c) for $P^{(\text{NL})}$, we have

$$P^{(\text{NL})}(x, t) = \frac{1}{(2\pi)^4} \int_{-\infty}^{\infty} dk' \, d\omega' \, dk'' \, d\omega'' \chi^{(2)}_{k',\omega';k'',\omega''} E_{k',\omega'} E_{k'',\omega''}$$

$$\times e^{i((k'+k'')x - (\omega'+\omega'')t)}. \qquad (3.8.29)$$

Then, substituting $E_{k,\omega}$ given by Eq. (3.8.24) into Eq. (3.8.29), the nonlinear polarization $P^{(\text{NL})}(x, t)$ can be found by a procedure analogous to the one used when deriving the linear polarization $P(x, t)$. As we have already seen in the example of the harmonic oscillator, the time variation of the amplitude associated with the nonlinear interaction is $\partial \varphi_j / \partial t \sim \varepsilon \omega_j \varphi_j$, where ε characterizes the magnitude. Therefore, if we assume a sufficiently small amplitude, the terms containing $\partial \varphi_j / \partial t$, $\partial \varphi_j / \partial x$, which appear in the nonlinear polarization, can be neglected. As a result, the nonlinear polarization is given by

$$P^{(\text{NL})}(x, t) = \sum_{l,m=0}^{2} [\chi_{k_l, k_m} \varphi_l \varphi_m e^{i((k_l + k_m)x - (\omega_l + \omega_m)t)}$$

$$+ \chi_{k_l, -k_m} \varphi_l \varphi_m^* e^{i((k_l - k_m)x - (\omega_l - \omega_m)t)}]$$

$$+ \text{c.c.} \qquad (3.8.30)$$

Substituting the linear polarization $P(x, t)$ and the nonlinear one $P^{(\text{NL})}(x, t)$, thus obtained, into the wave equation (3.8.19a) and neglecting

3.8 THREE-WAVE INTERACTION

the terms containing $\partial^2\varphi_j/\partial t^2$, $\partial^2\varphi_j/\partial x^2$, $(\partial\varphi_j/\partial t)\varphi_j$, etc., gives

$$\sum_{j=0}^{2}(i\omega_j)\left[\left\{2(1+4\pi\chi_{k_j})+4\pi\omega_j\frac{\partial\chi}{\partial\omega_j}\right\}\frac{\partial\varphi_j}{\partial t}\right.$$
$$\left.+\left(\frac{2k_jc^2}{\omega_j}-4\pi\omega_j\frac{\partial\chi}{\partial k_j}\right)\frac{\partial\varphi_j}{\partial x}\right]e^{i(k_jx-\omega_jt)}+\text{c.c.}$$
$$=-4\pi\sum_{l,m=0}^{2}\left[(\omega_l+\omega_m)^2\chi_{k_l,k_m}\varphi_l\varphi_m e^{i((k_l+k_m)x-(\omega_l+\omega_m)t)}\right.$$
$$\left.+(\omega_l-\omega_m)^2\chi_{k_l,-k_m}\varphi_l\varphi_m^* e^{i((k_l-k_m)x-(\omega_l-\omega_m)t)}\right]+\text{c.c.} \quad (3.8.31)$$

The equations for mode coupling are obtained by multiplying both sides of Eq. (3.8.31) by $\exp\{-i(k_jx-\omega_jt)\}$ ($j=0,1,2$) and averaging over the time period ω_j^{-1}. Using the dispersion relation (3.8.21), the group velocity v_{gj} becomes

$$v_{gj}\equiv\frac{\partial\omega_j}{\partial k_j}=\frac{2k_jc^2/\omega_j-4\pi\omega_j(\partial\chi/\partial k_j)}{2(1+4\pi\chi_{k_j})+4\pi\omega_j(\partial\chi/\partial\omega_j)} \quad (3.8.32)$$

and, similarly, we have

$$\frac{\partial}{\partial\omega_j}[\omega_jD(k_j,\omega_j)]=2(1+4\pi\chi_{k_j})+4\pi\omega_j\left(\frac{\partial\chi}{\partial\omega_j}\right). \quad (3.8.33)$$

With the aid of these relations, the equations for mode coupling finally reduce to

$$\frac{\partial\varphi_0}{\partial t}+v_{g0}\frac{\partial\varphi_0}{\partial x}=i8\pi\omega_0\left(\frac{\partial(\omega D)}{\partial\omega_0}\right)^{-1}\chi_{k_0,k_1,k_2}\varphi_1\varphi_2, \quad (3.8.34a)$$

$$\frac{\partial\varphi_1}{\partial t}+v_{g1}\frac{\partial\varphi_1}{\partial x}=i8\pi\omega_1\left(\frac{\partial(\omega D)}{\partial\omega_1}\right)^{-1}\chi_{k_1,k_0,-k_2}\varphi_0\varphi_2^*, \quad (3.8.34b)$$

$$\frac{\partial\varphi_2}{\partial t}+v_{g2}\frac{\partial\varphi_2}{\partial x}=i8\pi\omega_2\left(\frac{\partial(\omega D)}{\partial\omega_2}\right)^{-1}\chi_{k_2,k_0,-k_1}\varphi_0\varphi_1^*, \quad (3.8.34c)$$

where

$$\frac{\partial(\omega D)}{\partial\omega_j}\equiv\frac{\partial}{\partial\omega_j}[\omega_jD(k_j,\omega_j)], \quad (3.8.35)$$

$$\chi_{k_j,k_l,k_m}\equiv\chi_{k_l,k_m}.$$

Although χ_{k_j,k_l,k_m} is determined by the medium, for the following argument we shall assume

$$\chi^{(\text{NL})}\equiv\chi_{k_j,k_l,k_m}=\chi_{-k_l,-k_j,k_m}=\chi_{-k_m,k_l,-k_j},$$
$$\chi^*_{k_j,k_l,k_m}=\chi_{-k_j,-k_l,-k_m}=\chi_{k_j,k_l,k_m}. \quad (3.8.36)$$

As is evident from the coupling equations, without the nonlinear interaction the amplitude $\varphi_j(x, t)$ of each mode is constant along the corresponding characteristic $\xi_j = x - v_{gj}t$. First, let us suppose that the spatial variation of the amplitude $\varphi_j(x, t)$ can be neglected, so that

$$\frac{d\varphi_0}{dt} = i8\pi\omega_0 \left(\frac{\partial(\omega D)}{\partial \omega_0}\right)^{-1} \chi^{(NL)} \varphi_1 \varphi_2, \tag{3.8.37a}$$

$$\frac{d\varphi_1}{dt} = i8\pi\omega_1 \left(\frac{\partial(\omega D)}{\partial \omega_1}\right)^{-1} \chi^{(NL)} \varphi_0 \varphi_2^*, \tag{3.8.37b}$$

$$\frac{d\varphi_2}{dt} = i8\pi\omega_2 \left(\frac{\partial(\omega D)}{\partial \omega_2}\right)^{-1} \chi^{(NL)} \varphi_0 \varphi_1^*. \tag{3.8.37c}$$

Then we find the following conservation laws. The energy density of wave in a dispersive medium is given by

$$W_{k_j} \equiv \frac{1}{8\pi} \frac{\partial(\omega D)}{\partial \omega_j} |\varphi_j|^2 \tag{3.8.38}$$

(Landau and Lifshitz (1960)).

Multiply Eqns (3.8.37) by the corresponding terms $(\frac{1}{8}\pi)[\partial(\omega D)/\partial \omega_j]\varphi_j^*$, respectively, and add the three resulting equations. Then the conservation law of energy is obtained from the resonance condition (3.8.1) in the form

$$\frac{d}{dt} \sum_{j=0}^{2} W_{k_j} = i(\omega_0 - \omega_1 - \omega_2) \chi^{(NL)} \varphi_0^* \varphi_1 \varphi_2 = 0. \tag{3.8.39}$$

Also, from the momentum density of the wave

$$\boldsymbol{P}_{k_j} = \frac{1}{8\pi} \frac{\boldsymbol{k}_j}{\omega_j} \frac{\partial(\omega D)}{\partial \omega_j} |\varphi_j|^2, \tag{3.8.40}$$

and from the resonance condition (3.8.18), it follows similarly that the law of conservation of momentum holds so that

$$\frac{d}{dt} \sum_{j=0}^{2} \boldsymbol{P}_{k_j} = i(\boldsymbol{k}_0 - \boldsymbol{k}_1 - \boldsymbol{k}_2) \chi^{(NL)} \varphi_0^* \varphi_1 \varphi_2 = 0. \tag{3.8.41}$$

From these, there follow the resonance conditions for the frequency and the wave number which are called the conservation laws of energy and momentum, respectively.

Moreover, introducing the complex amplitude by means of the equation

$$C_j = \varphi_j \left|\frac{1}{8\pi\omega_j} \frac{\partial(\omega D)}{\partial \omega_j}\right|^{1/2}, \tag{3.8.42}$$

3.8 THREE-WAVE INTERACTION

we find that

$$|C_j|^2 = \frac{W_{k_j}}{\omega_j} = \hbar n_j \qquad (3.8.43)$$

corresponds to the action of the jth wave, and n_j is the number density of the wave quanta. We hereafter set $\hbar = 1$. Expressing Eqns (3.8.37) in terms of C_j, we have

$$\frac{dC_0}{dt} = i\beta C_1 C_2, \qquad (3.8.44a)$$

$$\frac{dC_1}{dt} = i\beta C_0 C_2^*, \qquad (3.8.44b)$$

$$\frac{dC_2}{dt} = i\beta C_0 C_1^*, \qquad (3.8.44c)$$

where

$$\beta = \chi^{(NL)} \left| \prod_{j=0}^{2} \left(\frac{1}{8\pi\omega_j}\right) \frac{\partial(\omega D)}{\partial \omega_j} \right|^{-1/2}. \qquad (3.8.44d)$$

Letting the complex amplitude C_j be $C_j(t) = |C_j(t)| \exp[i\theta_j(t)]$, the real and the imaginary parts of Eqns (3.8.44) become, respectively,

$$\frac{d|C_0|}{dt} = \beta |C_1| |C_2| \sin \theta, \qquad (3.8.45a)$$

$$\frac{d|C_1|}{dt} = -\beta |C_0| |C_2| \sin \theta, \qquad (3.8.45b)$$

$$\frac{d|C_2|}{dt} = -\beta |C_0| |C_1| \sin \theta, \qquad (3.8.45c)$$

$$\frac{d\theta}{dt} = -\beta \left(\frac{|C_0||C_2|}{|C_1|} + \frac{|C_0||C_1|}{|C_2|} - \frac{|C_1||C_2|}{|C_0|} \right) \cos \theta$$

$$= \cot \theta \frac{d}{dt} \log(|C_0| |C_1| |C_2|), \qquad (3.8.45d)$$

where $\theta = \theta_0 - \theta_1 - \theta_2$.

The following conservation laws may be obtained from (three pairs of) Eqns (3.8.45a to c):

$$m_1 \equiv |C_0|^2 + |C_1|^2 = \frac{W_{k_0}}{\omega_0} + \frac{W_{k_1}}{\omega_1} = n_0 + n_1 = \text{const.}, \qquad (3.8.46a)$$

$$m_2 \equiv |C_0|^2 + |C_2|^2 = \frac{W_{k_0}}{\omega_0} + \frac{W_{k_2}}{\omega_2} = n_0 + n_2 = \text{const.}, \qquad (3.8.46b)$$

$$m_3 \equiv |C_1|^2 - |C_2|^2 = \frac{W_{k_1}}{\omega_1} - \frac{W_{k_2}}{\omega_2} = n_1 - n_2 = \text{const.}, \qquad (3.8.46c)$$

which are called the Manley–Rowe relations. These conservation laws can be interpreted in terms of a quantum mechanics analogy as follows; the annihilation of a single ω_0 quantum creates two quanta, ω_1 and ω_2, and so $\Delta n_0 = -1$, $\Delta n_1 = \Delta n_2 = 1$. This process is similar to the process in which one particle decays into two particles. In this sense, the excitation of φ_1 and φ_2 by φ_0 is often referred to as the decay-type instability.

We here derived the conservation laws for the case when the coupling coefficients of each mode are identical. For the case when there are different coupling coefficients, so that β_j ($j = 0, 1, 2$) are different, we can also derive Eqns (3.8.44), where now $\beta = |\beta_0\beta_1\beta_2|^{1/2}$, by means of the transformation $C'_j = C_j/|\beta_j|^{1/2}$, provided β_j is either real or purely imaginary. Thus the conservation laws (3.8.46) hold quite generally (cf. Eqns (3.8.54) to (3.8.56)).

In the example of the harmonic oscillator, we have assumed that the amplitude C_0 of excited oscillation is large compared to C_1 and C_2, so that the system may be linearized. As is evident from the coupled equations, however, the change of C_0 cannot continue to be neglected as the growing amplitudes C_1 and C_2 become large. The general solution of the nonlinear system of equations (3.8.44) can be expressed in terms of an elliptic function. Integrating Eq. (3.8.45d) gives the following conservation law:

$$m_0 \equiv |C_0| |C_1| |C_2| \cos \theta = \text{const.} \quad (3.8.46\text{d})$$

By means of the conservation laws (3.8.46), Eq. (3.8.45a) reduces to

$$\frac{dn_0}{dt} = 2\beta[n_0(n_0 - m_1)(n_0 - m_2) - m_0^2]^{1/2}. \quad (3.8.47)$$

Let the roots of $n_0(n_0 - m_1)(n_0 - m_2) - m_0^2 = 0$ be given by $n_a \geq n_b \geq n_c \geq 0$, then $n_0(t)$ becomes

$$n_0(t) = n_c + (n_b - n_c) \operatorname{sn}^2 [\beta(n_a - n_c)^{1/2}(t - t_0), m],$$

$$m \equiv \left(\frac{n_b - n_c}{n_a - n_c}\right)^{1/2}, \qquad n_0(t_0) \equiv n_c. \quad (3.8.48)$$

Suppose, for instance, that we choose $n_0(0) \gg n_1(0) > 0$ and $n_2(0) = 0$ at $t = 0$. Then, since we may assume $m_0 = 0$ without loss of generality, we have

$$n_a = m_1 \equiv n_0(0) + n_1(0) > n_b = m_2 \equiv n_0(0) > n_c = 0,$$

and

$$m \equiv [n_0(0)/\{n_0(0) + n_1(0)\}]^{1/2} \leq 1.$$

Thus we find

$$n_0(t) = n_0(0) \, \text{sn}^2 \, [\beta(t-t_0)n_a^{1/2}, m], \quad (3.8.49\text{a})$$
$$n_1(t) = n_1(0) + n_0(0)\{1 - \text{sn}^2 \, [\beta(t-t_0)n_a^{1/2}, m]\}, \quad (3.8.49\text{b})$$
$$n_2(t) = n_0(0)\{1 - \text{sn}^2 \, [\beta(t-t_0)n_a^{1/2}, m]\}. \quad (3.8.49\text{c})$$

The time variation of these solutions is shown in Fig. 3.3(a). Since

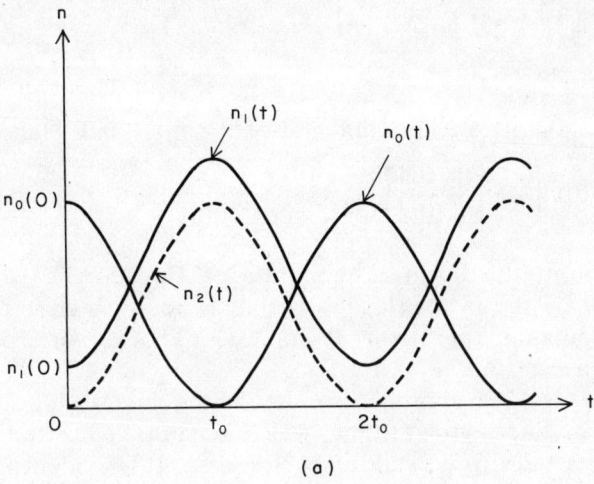

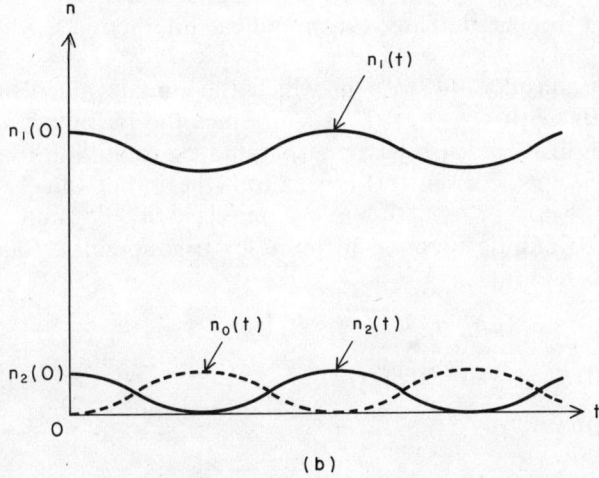

Fig. 3.3 The time evolution of n (Sagdeev and Galeev (1969)).

$n_2(0) = 0$, we have

$$\operatorname{sn}^2[\beta t_0 n_a^{1/2}, m] = 1, \qquad (3.8.50)$$

so that $\beta t_0 n_a^{1/2}$ is a quarter of the period of the elliptic function sn when

$$\beta t_0 n_a^{1/2} = K(m) \simeq -\left[\left(\frac{m'}{2}\right)^2 + O(m'^4)\right] + \log\left(\frac{1}{m'}\right) \\ \times \left[1 + \left(\frac{m'}{2}\right)^2 + O(m'^4)\right], \qquad (3.8.51)$$

$$m' = (1 - m^2)^{1/2},$$

where K is the complete elliptic integral of the first kind. Hence,

$$t_0^{-1} \sim \beta |C_0(0)| \left[\log\left|\frac{C_0(0)}{C_1(0)}\right|\right]^{-1}, \qquad (3.8.52)$$

which differs from the linear growth rate of Eq. (3.8.7), which was $\alpha = \beta |C_0|$, only by the factor $[\log|C_0/C_1|]^{-1}$. Here, as is clear from Eq. (3.8.6), the coupling coefficient β in Eq. (3.8.44) corresponds to $|\beta|/(4\omega_1\omega_2)^{1/2}$ in Eq. (3.8.7).

Consequently, as $C_1(0)$ approaches $C_0(0)$, so the time scale Eq. (3.8.52) deviates from the linear growth time, which is natural since the assumption of linearity ceases to be valid in such a case. It is evident from Fig. 3.3(a) that the initial energy of n_0 is transferred to n_1 and n_2, and that these modes grow (parametric excitation or decay-type instability). We have $n_0 = 0$ at $t = t_0$, and after that the energy is transferred back to n_0 and this cycle is repeated. Thus, this nonlinear interaction is a reversible process.

Next, let us consider the case in which the initial values satisfy the conditions $n_1(0) \gg n_2(0)$, $n_0(0) = 0$, $m_0 = 0$. Since the frequency ω_1 of the excited oscillation n_1 is $|\omega_1| < |\omega_0|$, the parametric excitation (decay-type instability) does not occur in this case. From the initial values, we have $n_a = m_1 \equiv n_1(0) \gg n_b = m_2 \equiv n_2(0) > n_c = 0$, $m \equiv [n_2^{(0)}/n_1^{(0)}]^{1/2}$, and so we can approximate the elliptic function in terms of trigonometric functions to find

$$n_0(t) \approx n_2(0) \sin^2(\beta n_a^{1/2} t), \qquad (3.8.53a)$$

$$n_1(t) \approx n_1(0) - n_2(0) \sin^2(\beta n_a^{1/2} t), \qquad (3.8.53b)$$

$$n_2(t) \approx n_2(0) \cos^2(\beta n_a^{1/2} t) \qquad (3.8.53c)$$

(cf. Fig. 3.3(b)).

As is obvious from the figure, the amplitudes of small oscillations are determined by $n_2(0)$, and the energy of n_1 is never transferred either to n_2 or to n_0.

As our next step, let us take into account the spatial variation of the amplitude (Nozaki and Taniuti (1973)). In this case, a soliton solution exists, as will be shown below. For simplicity, we shall assume the coupling equations (3.8.34) to be in form

$$\frac{\partial \varphi_0}{\partial t} + v_{g0}\frac{\partial \varphi_0}{\partial x} = \beta_0 \varphi_1 \varphi_2, \tag{3.8.54a}$$

$$\frac{\partial \varphi_1}{\partial t} + v_{g1}\frac{\partial \varphi_1}{\partial x} = -\beta_1 \varphi_0 \varphi_2^*, \tag{3.8.54b}$$

$$\frac{\partial \varphi_2}{\partial t} + v_{g2}\frac{\partial \varphi_2}{\partial x} = -\beta_2 \varphi_0 \varphi_1^*. \tag{3.8.54c}$$

In the following we shall suppose that $\beta_j > 0$ and φ_j are real. In order to find the soliton solution propagated with a constant speed λ, let φ_j be a function of $\xi = x - \lambda t$ alone. Then, with the aid of the transformation $\varphi_j' = \varphi_j/|\tilde{\beta}_j|^{1/2}$, the coupling coefficients of each mode become $\tilde{\beta} = |\tilde{\beta}_0 \tilde{\beta}_1 \tilde{\beta}_2|^{1/2}$, where $\tilde{\beta}_j = \beta_j/(v_{gj} - \lambda)$. For example, in the case when $\lambda > \max(v_{gj})$ or $\lambda < \min(v_{gj})$, Eqns (3.8.54) reduce to

$$\frac{d\varphi_0'}{d\xi} = \mp \tilde{\beta} \varphi_1' \varphi_2', \tag{3.8.55a}$$

$$\frac{d\varphi_1'}{d\xi} = \pm \tilde{\beta} \varphi_0' \varphi_2', \tag{3.8.55b}$$

$$\frac{d\varphi_2'}{d\xi} = \pm \tilde{\beta} \varphi_0' \varphi_1'. \tag{3.8.55c}$$

The signs on the right-hand side correspond to $\lambda > \max(v_{gj})$ and $\lambda < \min(v_{gj})$, respectively.

Therefore, putting $\varphi_j'^2 = n_j$, we find that the following conservation laws hold:

$$m_1 = n_0 + n_1 = \text{const.}, \tag{3.8.56a}$$
$$m_2 = n_0 + n_2 = \text{const.}, \tag{3.8.56b}$$
$$m_3 = n_1 - n_2 = \text{const.}, \tag{3.8.56c}$$

and setting $m_1 \geq m_2$, $n_0(\xi_0) = 0$, $m \equiv (m_2/m_1)^{1/2}$, the general solution is given by

$$n_0(\xi) = m_2 \operatorname{sn}^2 [\mp \tilde{\beta} m_1^{1/2}(\xi - \xi_0), m]. \tag{3.8.57}$$

In particular, putting $m = 1$, so that $\bar{n}_0 \equiv m_1 = m_2$, the φ_j becomes, respectively,

$$\varphi_0 = \alpha_0 \tanh \gamma(x - \lambda t), \qquad (3.8.58a)$$

$$\varphi_1 = \alpha_1 \operatorname{sech} \gamma(x - \lambda t), \qquad (3.8.58b)$$

$$\varphi_2 = \alpha_2 \operatorname{sech} \gamma(x - \lambda t), \qquad (3.8.58c)$$

where

$$(v_{g0} - \lambda)\frac{\alpha_0^2}{\beta_0} = (v_{g1} - \lambda)\frac{\alpha_1^2}{\beta_1} = (v_{g2} - \lambda)\frac{\alpha_2^2}{\beta_2} = \gamma^2 \frac{(v_{g0} - \lambda)(v_{g1} - \lambda)(v_{g2} - \lambda)}{\beta_0 \beta_1 \beta_2}$$
$$= \mp \bar{n}_0. \qquad (3.8.59)$$

Since β_j, $n_0 > 0$, we must have $\lambda > \max(v_{gj})$ and $\lambda < \min(v_{gj})$, corresponding to the $-$ and $+$ signs in the last expression, respectively. The solution depends on the amplitudes α_j ($j = 0, 1, 2$), the pulse width γ, and the propagation speed λ, but it is readily seen from Eq. (3.8.59) that two of them can be taken arbitrarily as parameters. For instance, if the amplitudes α_0, α_1 are given, then the amplitude α_2, the pulse width γ, and the propagation speed λ are determined from Eqns (3.8.59). As may be seen clearly from the Fig. 3.4, in front of the pulse, the energy of φ_0 is transferred to φ_1 and φ_2, while behind the pulse the energy is transferred back to φ_0.

From Eqns (3.8.59), we have

$$(v_{g1} - \lambda)(v_{g2} - \lambda)\gamma^2 = \alpha_0^2 \beta_1 \beta_2 \equiv \gamma_d^2, \qquad (3.8.60)$$

and hence when $|\lambda|$ is large enough compared to v_{g1} and v_{g2}, the duration time $(\lambda\gamma)^{-1}$ of the pulse is of the same order of magnitude as the growth time γ_d^{-1} of the decay-type instability (Problem 3.8.2). We have examined here only the case when $\lambda > \max(v_{gj})$ or $\lambda < \min(v_{gj})$, while we have $\varphi_1 \propto \tanh \gamma\xi$ for $v_{g2} < \lambda < \min(v_{g0}, v_{g1})$ or for $v_{g2} > \lambda > \max(v_{g0}, v_{g1})$, and

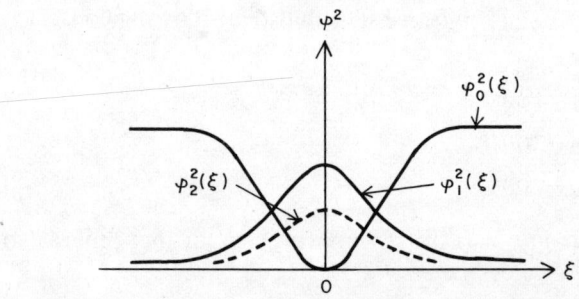

Fig. 3.4

3.8 THREE-WAVE INTERACTION

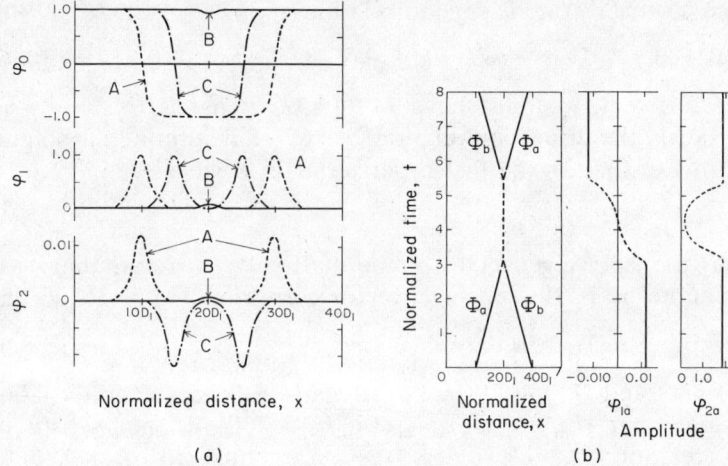

Fig. 3.5 Collisions of two solitons. (a) Wave profile of three waves at each instance (A: $t = 0$, B: $t = 400$, C: $t = 700$). (b) Trajectories of two solitons, the right two figures show the time evolution of the amplitudes φ_1, φ_2 (Nishihara (1975)).

also $\varphi_2 \propto \tanh \gamma \xi$ for $v_{g1} < \lambda < \min(v_{g0}, v_{g2})$ or for $v_{g1} > \lambda > \max(v_{g0}, v_{g2})$ (Ohsawa and Nozaki (1974)). Moreover, by means of numerical calculations, we can be certain that the solitary wave solutions (3.8.58) represent solitons which behave like particles (cf. Fig. 3.5), and also that the solitons propagate steadily, even when the phases vary in space due to inhomogeneity (Nozaki (1974b); Nishihara (1975)).

We have seen that in the presence of three waves satisfying the resonance conditions (3.8.1) and (3.8.18), the wave equation (3.8.19) for electromagnetic waves in a nonlinear dispersive medium can be reduced to the nonlinear coupling equation (3.8.34) by means of the averaging procedure. It can be easily shown by means of the reductive perturbation method that a general partial differential equation (3.6.3) can also be reduced to an analogous set of nonlinear coupling equations. (Of course, the method can also be applied to integro-differential equations such as (3.8.19) (Taniuti and Yajima (1973)). On the assumption that the variation of the amplitude caused by the resonant interaction is proportional to the amplitude, we expand the solution U of Eq. (3.6.3) in powers of a parameter ε as in Eq. (3.6.9a), to find

$$U^{(1)} = \sum_{j=\pm 0}^{\pm 2} U_j^{(1)}(\sigma, \tau) e^{i(k_j x - \omega_j t)}, \qquad (3.8.61a)$$

$$\sigma \equiv \varepsilon x, \qquad \tau \equiv \varepsilon t, \qquad (3.8.61b)$$

where (k_j, ω_j) satisfies the dispersion relation

$$\det |W_j| = 0, \qquad W_j \equiv -i\omega_j I + ik_j A_0 + \nabla_u B_0 \qquad (j = 0, 1, 2). \qquad (3.8.62)$$

Since $U^{(1)}$ is real, we must have $U^{(1)*}_{-j} = U^{(1)}_j$; that is, $k_{-j} = -k_j$ and $\omega_{-j} = -\omega_j$ satisfy the dispersion relation (3.8.62). Substituting the solution (3.6.9a) into Eq. (3.6.3), the first-order terms in ε yield

$$U^{(1)}_j = \varphi_j(\sigma, \tau) R_j, \qquad (3.8.63)$$

where R_j is the right eigenvector of the matrix W_j corresponding to λ_j. Here we denote by L_j the left eigenvector corresponding to R_j, so that

$$W_j R_j = 0, \qquad L_j W_j = 0. \qquad (3.8.64)$$

When (k_j, ω_j) and (k_{-j}, ω_{-j}) $(j = 0, 1, 2)$ satisfy the dispersion relation, let us assume that $(k_m + k_l, \omega_m + \omega_l)$ $(m, l = 0, 1, 2)$ do not satisfy the dispersion relation (3.8.62) unless they are equal to (k_j, ω_j) or to (k_{-j}, ω_{-j}), respectively. Then, from Eqns (3.8.61), $U^{(2)}$ becomes

$$U^{(2)} = \sum_{l,m=\pm 0}^{\pm 2} U^{(2)}_{l,m}(\sigma, \tau) e^{i((k_l+k_m)x - (\omega_l+\omega_m)t)}. \qquad (3.8.65)$$

Inserting Eq. (3.8.65) into Eq. (3.6.3), the second order terms in ε give

$$W_{l+m} U^{(2)}_{l,m} + \left(\frac{\partial U^{(1)}_j}{\partial \tau} + A_0 \frac{\partial U^{(1)}_j}{\partial \sigma} \right) \delta(k_l + k_m - k_j) \, \delta(\omega_l + \omega_m - \omega_j)$$

$$+ \sum_{l',m'=\pm 0}^{\pm 2} [ik_{l'}(U^{(1)}_{m'} \cdot \nabla A_0) U^{(1)}_{l'} + \tfrac{1}{2} \nabla\nabla B_0 : U^{(1)}_{l'} U^{(1)}_{m'}]$$

$$\times \delta(k_{l'} + k_{m'} - k_l - k_m) \, \delta(\omega_{l'} + \omega_{m'} - \omega_l - \omega_m) = 0, \qquad (3.8.66)$$

in which δ-functions in the second term show the resonance conditions. When the resonance conditions $k_j = k_l + k_m$ and $\omega_j = \omega_l + \omega_m$, are satisfied, then by hypothesis we have $\det W_{l+m} = 0$, and hence the compatibility condition must hold.

Consequently, multiplying Eq. (3.8.66) by L_j from the left and using the result $L_j(\partial W_j/\partial k_j) R_j = 0$, we find

$$\frac{\partial \varphi_j}{\partial \tau} + v_{gj} \frac{\partial \varphi_j}{\partial \sigma} = \frac{1}{2} \sum_{l,m=\pm 0}^{\pm 2} \beta_j \varphi_l \varphi_m \, \delta(k_l + k_m - k_j) \, \delta(\omega_l + \omega_m - \omega_j), \qquad (3.8.67a)$$

where the coupling coefficient β_j is

$$\beta_j \equiv -\frac{L_j}{L_j \cdot R_j} [ik_l(R_m \cdot \nabla A_0) R_l + ik_m(R_l \cdot \nabla A_0) R_m + \nabla\nabla B_0 : R_l R_m].$$

$$(3.8.67b)$$

Assuming the resonance conditions (3.8.1) and (3.8.18), Eqns (3.8.67) become

$$\frac{\partial \varphi_0}{\partial \tau} + v_{g0} \frac{\partial \varphi_0}{\partial \sigma} = \tfrac{1}{2}\beta_0 \varphi_1 \varphi_2, \qquad (3.8.68a)$$

$$\frac{\partial \varphi_1}{\partial \tau} + v_{g1} \frac{\partial \varphi_1}{\partial \sigma} = \tfrac{1}{2}\beta_1 \varphi_0 \varphi_2^*, \qquad (3.8.68b)$$

$$\frac{\partial \varphi_2}{\partial \tau} + v_{g2} \frac{\partial \varphi_2}{\partial \sigma} = \tfrac{1}{2}\beta_2 \varphi_0 \varphi_1^*, \qquad (3.8.68c)$$

In Eq. (3.8.34), the coupling coefficient was proportional to $[\partial D/\partial \omega_j]^{-1}$, so let us show here that the coupling coefficient given by Eq. (3.8.67b) also has this property. Denoting an element of the determinant det W by W_{mn}, and its cofactor by Δ_{mn}, we have det $W = \sum_{m,n} W_{mn} \Delta_{mn} = 0$, while from Eqns (3.8.64) we have $\sum_n w_{mn} r_n = 0$, where r_n denotes the nth component of R_j, and consequently $\Delta_{mn} \propto r_n$. The same relationship holds for L_j, and hence, letting a component of L_j be l_m, we find

$$\Delta_{mn} = l_m r_n. \qquad (3.8.69)$$

Therefore, using $(\partial/\partial w) \det W = -i \sum_n (\partial/\partial w_{nn}) \det W = i \sum_n \Delta_{nn}$, we have

$$L_j R_j = \sum_n \Delta_{nn} = i \frac{\partial}{\partial \omega} \det W \bigg|_{\substack{\omega = \omega_j \\ k = k_j}}, \qquad (3.8.70)$$

Putting $x_j = \varepsilon C_j(\tau) e^{i\omega_j t}$ and letting $\tau = \varepsilon t$ in Eq. (3.8.3), from the second-order terms in ε we obtain the coupling equations (3.8.5) for C_j.

Problem 3.8.1

Find the conservation laws (Manley–Rowe relations) for the harmonic oscillators given by Eqns (3.8.3).
Hint The energy of each oscillator is $\omega_i^2 x_i^2 / 2$.

Problem 3.8.2

In Eqns (3.8.55), neglect the spatial variation of φ and assume that $|\varphi_0| \gg |\varphi_1|, |\varphi_2|$. Then show that the growth rates of φ_1, φ_2 reduce to the γ_d given in Eq. (3.8.60).

3.9 Self-induced transparency and the Sine–Gordon equation

So far, using the example of nonlinear optics, we have explained interesting phenomena such as self-focusing and the three-wave interaction. In the present section we shall further discuss self-induced transparency

(SIT), in which a medium transmits a light pulse at the resonant frequency without resonance absorption (for example, McCall and Hahn (1969)).

First of all we consider an electromagnetic wave incident on a system of two-level atoms. That is, an ensemble of atoms with the ground state of energy E_1 and the excited state E_2. When the frequency of the electromagnetic wave equals the transition frequency $\omega_0 = (E_2 - E_1)/\hbar$, the transition from the ground state to the excited one takes place as a result of the resonance absorption of the electromagnetic wave. In the present system composed of a large number of atoms, the atoms excited to the upper level lose energy through collisions with atoms, etc. When, however, the electromagnetic wave is incident as a pulse, and its pulse width is short enough compared to the relaxation time, the relaxation of the excited states like those due to the atomic collisions is not significant and the population of the excited levels induced by the pulse and the resulting high-frequency polarization of the system are kept in a definite phase relation with the incident electromagnetic wave.

Accordingly, when a sufficiently intense electromagnetic wave is incident on a system of two-level atoms in the form of a short pulse, then in the leading part of the pulse the number of atoms in the excited state exceeds that in the ground state on account of the absorption of the electromagnetic wave. However, in the rear of the pulse, the electromagnetic wave is again emitted by virtue of induced emission, so that the system transmits the pulse without absorption. We call this self-induced transparency.

When certain conditions for this process are satisfied, the pulse propagates steadily without attenuation and with a velocity considerably slower than the phase velocity of the linearized system. As will be shown later, the pulse propagates as a soliton. In particular, in a given approximation, this phenomenon is described by a simple nonlinear equation referred to as the Sine–Gordon equation, which is completely integrable (Section 4.6 and Section 4.7).

In the SIT, we may regard the electromagnetic wave as a modulated circularly polarized plane wave propagating along the x axis:

$$\mathbf{E}(x, t) = \mathcal{E}(x, t)(\mathbf{e}_y \cos \zeta + \mathbf{e}_z \sin \zeta)$$
$$\zeta = kx - \omega t + \varphi(x), \quad (3.9.1)$$

where $k \equiv n\omega/c$, and the refractive index n is determined by the linear dispersion relation of the medium. Here, the variations of the amplitude $\mathcal{E}(x, t)$ and the phase $\varphi(x)$ are slow enough in comparison with that of the carrier wave, so that

$$\left|\frac{\partial \mathcal{E}}{\partial t}\right| \ll |\omega \mathcal{E}|, \qquad \left|\frac{\partial \mathcal{E}}{\partial x}\right| \ll |k\mathcal{E}|, \quad (3.9.2)$$

and a similar inequality holds for the phase φ.

3.9 SELF-INDUCED TRANSPARENCY

The electric field $\boldsymbol{E}(x, t)$ is governed by the wave equation obtained from Maxwell's equations:

$$\frac{\partial^2 \boldsymbol{E}}{\partial x^2} - \frac{n^2}{c^2}\frac{\partial^2 \boldsymbol{E}}{\partial t^2} = \frac{4\pi}{c^2}\frac{\partial^2 \boldsymbol{P}}{\partial t^2}, \tag{3.9.3}$$

where $\boldsymbol{P}(x, t)$ is the polarization of the medium (a system of two-level atoms) caused by the electromagnetic wave. If the equation for $\boldsymbol{P}$ is given, we can find the basic equations of this system by combining that equation with Eq. (3.9.3).

Now let us suppose that the medium is made up of N two-level atoms, and that the transition frequencies ω_0 of individual atoms are not all the same, but are distributed about the frequency ω of the incident wave according to the spectral density function $g(\Delta\omega)$, which is normalized by the requirement that

$$\int_{-\infty}^{\infty} g(\Delta\omega)\, d\Delta\omega = 1, \qquad \Delta\omega = \omega_0 - \omega. \tag{3.9.4}$$

Consequently, the polarization $\boldsymbol{P}(x, t)$ of the medium becomes

$$\boldsymbol{P}(x, t) = N \int_{-\infty}^{\infty} g(\Delta\omega)\boldsymbol{p}(\Delta\omega, x, t)\, d\Delta\omega, \tag{3.9.5}$$

where $\boldsymbol{p}(\Delta\omega, x, t)$ is the polarization of a two-level atom.

The same assumptions as those of Eq. (3.9.2) may be considered valid for the polarization $\boldsymbol{p}(\Delta\omega, x, t)$ of a two-level atom induced by the electromagnetic wave. Therefore, when the polarization $\boldsymbol{p}(\Delta\omega, x, t)$ comprises two parts, one is in phase with the carrier wave of the electric field $\boldsymbol{E}(x, t)$ and the other is out of phase; namely, it may be written

$$\boldsymbol{p}(\Delta\omega, x, t) = u(\Delta\omega, x, t)(\boldsymbol{e}_y \cos \zeta + \boldsymbol{e}_z \sin \zeta)$$
$$+ v(\Delta\omega, x, t)(\boldsymbol{e}_y \sin \zeta - \boldsymbol{e}_z \cos \zeta), \tag{3.9.6}$$

where u and v correspond to the modulational parts of the electric field and become slowly varying functions of x and t.

Thus the problem reduces to finding the equations for u and v, which may be obtained from the Schrodinger equation for a two-level atom in the incident electric field. In the electric dipole approximation, neglecting the relaxation effect, it is possible to prove that the following equations hold (McCall and Hahn (1969)):

$$\frac{\partial u}{\partial t} = v\Delta\omega, \tag{3.9.7a}$$

$$\frac{\partial v}{\partial t} = -u\Delta\omega - \left(\frac{\kappa^2}{\omega}\right)\mathscr{E}W, \tag{3.9.7b}$$

$$\frac{\partial W}{\partial t} = v\mathscr{E}\omega, \tag{3.9.7c}$$

where $k = 2p/\hbar$, p is the magnitude of the electric dipole moment, and W is the energy of the two-level atom. That is, $W = -(\frac{1}{2})\hbar\omega_0$ in the ground state and $W = (\frac{1}{2})\hbar\omega_0$ in the excited state. Usually, u and v are called the dispersive and the absorptive component of the electric dipole, respectively. We suppose, moreover, that all the two-level atoms are in the ground state until the dipole transitions occur. (Eqns (3.9.7) take the same form as the Bloch equations for a nuclear induced magnetic moment (Bloch (1946)).)

Thus, Eqns (3.9.7) for u, v, and ω and Eq. (3.9.3) for E comprise the basic equations. However, since u, v, and ω are slowly varying functions (corresponding to the modulation of p), we must derive equations for the modulational functions, $\mathscr{E}$ and φ from Eq. (3.9.3). Inserting Eqns (3.9.1), (3.9.5), and (3.9.6) into Eq. (3.9.3), and using the assumption expressed by the inequalities (3.9.2), we arrive at the following equations for the slowly varying parts:

$$\frac{\partial \mathscr{E}}{\partial x} + \frac{n}{c}\frac{\partial \mathscr{E}}{\partial t} = -\frac{2\pi\omega N}{nc}\int_{-\infty}^{\infty} vg \, d\Delta\omega, \tag{3.9.8a}$$

$$\mathscr{E}\frac{\partial \varphi}{\partial x} = \frac{2\pi\omega N}{nc}\int_{-\infty}^{\infty} ug \, d\Delta\omega. \tag{3.9.8b}$$

Thus the self-induced transparency phenomenon is described by Eqns (3.9.7) and (3.9.8) for u, v, W, $\mathscr{E}$, and φ.

In particular, when all the transition frequencies of N two-level atoms are the same, so that $g(\Delta\omega) = \delta(\Delta\omega)$, and the frequency ω of the electromagnetic wave is equal to that frequency ($\Delta\omega = 0$), Eqns (3.9.7) and (3.9.8) can be reduced to the Sine–Gordon equation in the following way. Since Eqns (3.9.7a) and (3.9.8b) enable us to set $u \equiv 0$ and $\varphi \equiv 0$, Eqns (3.9.7) and (3.9.8) take the form

$$\frac{\partial v}{\partial t} = -\left(\frac{\kappa^2}{\omega_0}\right)\mathscr{E}W, \tag{3.9.9a}$$

$$\frac{\partial W}{\partial t} = v\mathscr{E}\omega_0, \tag{3.9.9b}$$

$$\frac{\partial \mathscr{E}}{\partial x} + \frac{n}{c}\frac{\partial \mathscr{E}}{\partial t} = -\frac{2\pi\omega_0 N}{nc} v. \tag{3.9.9c}$$

Introducing the function $\phi(x, t)$ through

$$\phi(x, t) = \kappa \int_{-\infty}^{t} \mathscr{E}(x, t') \, dt', \tag{3.9.10}$$

from Eqns (3.9.9a, b), $W(x, t)$ and $v(x, t)$ become

$$W(x, t) = W_0 \cos \phi(x, t), \qquad W_0 = -\frac{\hbar \omega_0}{2}, \qquad (3.9.11a)$$

$$v(x, t) = p \sin \phi(x, t). \qquad (3.9.11b)$$

Here we have used the boundary condition that as $t \to -\infty$, all the two-level atoms are in the ground state (that is, $W(x, -\infty) = -\hbar\omega_0/2$). From Eqns (3.9.10) and (3.9.11b), Eq. (3.9.9c) reduces to

$$\phi_{,xt} + \frac{1}{c'} \phi_{,tt} = -\gamma^2 \sin \phi, \qquad c' = \frac{c}{n}, \qquad \gamma^2 = \frac{\pi N \omega_0 \hbar}{nc}. \qquad (3.9.12)$$

By a suitable transformation of variables, Eq. (3.9.12) can be reduced to the Sine–Gordon equation

$$\phi_{,xx} - \phi_{,tt} = \sin \phi. \qquad (3.9.13)$$

Rewriting it in terms of the characteristic coordinates

$$\xi = \pm \frac{x-t}{2}, \qquad \eta = \frac{x+t}{2}, \qquad (3.9.14)$$

we obtain

$$\phi_{,\xi\eta} = \pm \sin \phi. \qquad (3.9.15)$$

This characteristic form will be useful when discussing the Bäcklund transform in the next chapter. In what follows, we shall refer to the solutions ϕ corresponding to the $+$ and $-$ signs in Eq. (3.9.15) as solutions of the positive and negative type, respectively. (Eqns (3.9.12) can be transformed into Eq. (3.9.15) by means of the transformation $\xi = \mp \gamma x$, $\eta = \gamma \{t - (1/c')x\}$.)

Example 3.9.1 System of pendulums connected by a torsional spring

As shown in Fig. 3.6, let us consider the torsional motion of N pendulums suspended by a horizontal spring. The pendulums rotate in a plane perpendicular to the spring, and the rotational motion of one pendulum is transmitted to adjacent pendulums by means of the spring. The equation of motion for the ith pendulum is

$$I \frac{d^2 \phi_i}{dt^2} = \kappa(\phi_{i+1} - 2\phi_i + \phi_{i-1}) - F \sin \phi,$$

where I is the moment of inertia of the pendulum, κ is the torsion constant of the spring, F is the restoring force due to gravity, and all of

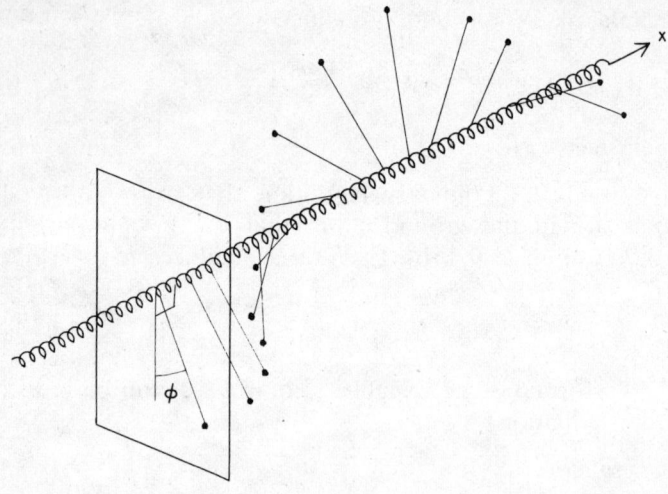

Fig. 3.6

these quantities are constant. Denoting the distance between adjacent pendulums by Δx, the above equation can be written as

$$\frac{1}{\Delta x}\left(\frac{\phi_{i+1}-\phi_i}{\Delta x}-\frac{\phi_i-\phi_{i-1}}{\Delta x}\right)-\frac{I/\Delta x}{\kappa\,\Delta x}\frac{\mathrm{d}^2\phi_i}{\mathrm{d}t^2}=\frac{F/\Delta x}{\kappa\,\Delta x}\sin\phi,$$

and hence, in the limit of a continuous system, the equation reduces to the Sine–Gordon equation (3.9.13). Here the distance and the time are scaled in terms of the units $(\kappa/F)^{1/2}\,\Delta x$ and $(I/F)^{1/2}$, respectively.

The Sine–Gordon equation can also be used to describe various other physical phenomena. (We refer, for instance, to Scott *et al.* (1973).) Also, the equations for the three-wave interactions mentioned in the previous section can be reduced to the Sine–Gordon equation (3.9.13) by means of a transformation analogous to Eqns (3.9.10) and (3.9.11), when the group velocities of two waves are equal (Nozaki (1974a)).

The steady solution of the Sine–Gordon equation (3.9.13) can be obtained easily. Introduce a new wave frame by means of the transformation

$$\zeta = x - ut. \tag{3.9.16}$$

Then Eq. (3.9.13) takes the same form as the equation of motion of a pendulum:

$$\frac{\mathrm{d}^2\phi}{\mathrm{d}\zeta^2}=\frac{1}{1-u^2}\sin\phi. \tag{3.9.17}$$

First of all, let us consider the solutions which approach constant states at infinity. Then, for $|u|<1$ we find

$$\phi(x-ut) = 4\tan^{-1}\left[\exp\left\{\pm\frac{x-ut}{(1-u^2)^{1/2}}\right\}\right] \quad (|u|<1). \qquad (3.9.18a)$$

For the solution with the positive sign in the exponent, we find $\phi \to 0$ as $x \to -\infty$ and $\phi \to 2\pi$ as $x \to \infty$. For the pendulum in the example 3.9.1, this means that the pendulum rotates by 2π when going from $x \to -\infty$ to $x \to \infty$, and the rotation occurs in the direction of a right-handed screw along the x-axis. Also, in the case of the negative sign, it represents the solution corresponding to a rotation in the opposite sense to that of a right-handed screw (i.e., a left-handed screw). Since ϕ approaches asymptotically different constant values as $|x| \to \infty$, it has a shock-like behaviour. As there is no dissipation, however, it is essentially different from a shock. Actually, $\phi = 0$ and $\phi = 2\pi$ correspond to the same physical state for the pendulum. In this sense we might call such a solution a soliton, though, in general, it is referred to as a kink. We call it either a kink or an anti-kink, according as ϕ increases or decreases with the increase of x. Furthermore, for $|u|>1$, we have

$$\phi(x-ut) = 4\tan^{-1}\left[\exp\left\{\pm\frac{x-ut}{(u^2-1)^{1/2}}\right\}\right] + \pi \quad (|u|>1). \qquad (3.9.18b)$$

For the pendulum, such a solution gives $\phi = \pi$ at infinity, and hence represents the unstable equilibrium state.

Instead of Eq. (3.9.13), we may consider a Sine–Gordon equation of negative type,

$$\phi_{,xx} - \phi_{,tt} = -\sin\phi, \qquad (3.9.19)$$

and then we have

$$\frac{d^2\phi}{d\zeta^2} = -\frac{1}{1-u^2}\sin\phi.$$

Since this equation may be derived from Eq. (3.9.13) by putting $\zeta = ux - t$, it is readily seen that, for the negative type of equation, the solutions for $|u|>1$ and $|u|<1$ correspond to Eq. (3.9.18a) ($|u|<1$) and Eq. (3.9.18b) ($|u|>1$), respectively. Since the Sine–Gordon equation is Lorentz invariant, in the relativistic sense, the solutions for $|u|>1$ represent a pulse propagating with a superlight velocity. Linearizing the Sine–Gordon equation of negative type, we obtain a Klein–Gordon equation with an imaginary mass, and hence it is not surprising that such a pulse does not exist. We wish to observe, however, that the equation of positive type, (3.9.13), likewise admits a solution propagating with a superlight velocity. In all of these cases, the solutions extend asymptotically up to $x \to \pm\infty$, so that the velocity of the wavefront is not of significance for these solutions. On the other hand, the Sine–Gordon equation is a semilinear hyperbolic equation. Hence, for a wave propagating into a constant state, the wavefront exists as a boundary between the constant state and the disturbed region ahead of the wave and its speed is always equal to the speed of light ($|u|=1$) (cf. Section 2.3).

As is evident from Eq. (3.9.10), the amplitude $\mathscr{E}(x,t)$ of the electric field is given by the time derivative of $\phi(x,t)$. Therefore, differentiating the steady solution (3.9.18a) for $\phi(x,t)$ with respect to t, it follows that the electric field propagates as a steady pulse; namely, as the soliton

$$\mathscr{E}(x,t) \propto \text{sech}\,(x-ut). \qquad (3.9.20)$$

4

The inverse scattering method for the initial value problem for a nonlinear evolution equation

4.1 The inverse scattering method for the Korteweg–de Vries equation

Zabusky and Kruskal discovered, by means of numerical computation, that solitary waves behave like particles involved in mutual collisions, and called them *solitons*. They also noticed the apparent similarity in form between Burgers' equation and the KdV equation, and as a result various people attempted to transform the KdV equation into a linear equation, in a way similar to that used for the solution of Burgers' equation. Though those attempts were not successful, Gardner *et al.* (1967) found that the KdV equation can be related to the eigenvalue problem of the Schrödinger equation through the inverse scattering method developed for quantum mechanics.

An indication that this important connection existed was the relationship found by Miura (1968) between the modified KdV equation and the KdV equation. Namely, the fact that if v satisfies the modified KdV equation

$$Q(v) = v_{,t} - 6v^2 v_{,x} + v_{,xxx} = 0, \qquad (4.1.1)$$

then the function u given by

$$u = v^2 + v_{,x} \qquad (4.1.2)$$

satisfies the KdV equation

$$P(u) = u_{,t} - 6uu_{,x} + u_{,xxx} = 0. \qquad (4.1.3)$$

In fact, using Eq. (4.1.2), after some manipulation we can show that

$$P(u) = \left(\frac{\partial}{\partial x} + 2v\right) Q(v). \qquad (4.1.4)$$

(Because of the first factor $(\partial/\partial x + 2v)$ on the right-hand side, the inverse statement (if $P(u)=0$ then $Q(v)=0$) is not valid.) By means of Eq. (4.1.4), Kruskal et al. (1970) derived an infinite number of conservation laws for the KdV equation (Section 4.2). We now remark that if u is given, Eq. (4.1.2) becomes a Riccati equation for v, so that the ordinary transformation to linearize the Riccati equation

$$v = \frac{\psi_{,x}}{\psi}, \tag{4.1.5}$$

leads to the equation

$$\psi_{,xx} - u\psi = 0. \tag{4.1.6}$$

Here u is a solution of the KdV equation (4.1.3).

If we now notice that the KdV equation involves a third-order derivative with respect to x, which is one order higher than that of Burgers' equation, then we may deduce that transformation (4.1.6) is a natural extension of the Cole–Hopf transformation (1.2.18), which for $2\mu/\alpha = -1$ may be written in the form

$$\psi_{,x} - u\psi = 0.$$

Substituting Eq. (4.1.6) for u in the KdV equation does not lead to any simplification, or to a useful result; however, if we take note of the Galilean invariance of the KdV equation (4.1.3), which is invariant under the transformation $u \to u - \lambda$, $x \to x + 6\lambda t$, we can at once generalize Eq. (4.1.6) to

$$-\psi_{,xx} + u\psi = \lambda\psi. \tag{4.1.7}$$

This is just the Schrödinger equation for ψ with the potential u. Nevertheless, it is a result which is essentially different from the Schrödinger equation in quantum mechanics, because of the fact that u is the solution of the KdV equation, so that it changes with time. Hence, in Eq. (4.1.7), the time must be considered as a parameter. In other words, it is necessary that at each instant of time Eq. (4.1.7) is valid for $u(x, t)$ at that time.

The eigenvalue λ would thus also seem to be time-dependent. Surprisingly, however, all the eigenvalues λ are time-independent, provided only that u decreases sufficiently rapidly at infinity with respect to x, or that it satisfies a periodic boundary condition. (These boundary conditions, of course, imply that the operator $-\partial^2/\partial x^2 + u$ is self-adjoint. They are physically relevant conditions for certain solutions of the KdV equation.)

A direct proof of this important property (the invariance of the spectrum with time) can be accomplished as follows. Compute $u_{,t}$, $u_{,x}$, $u_{,xxx}$ from Eq. (4.1.7), that is, from $u = \lambda + (\psi_{,xx}/\psi)$, and substitute the

4.1 THE KORTEWEG–DE VRIES EQUATION

results into the KdV equation (4.1.3). It is then readily seen that $u_{,t}\psi^2$ becomes $\lambda_{,t}\psi^2 + (\psi_{,xt}\psi - \psi_{,x}\psi_{,t})_{,x}$. Also, using Eq. (4.1.7) repeatedly, after some manipulation we can show that $(-6uu_{,x} + u_{,xxx})\psi^2$ may be written in the form of a perfect differential $(S_{,x}\psi - S\psi_{,x})_{,x}$ where $S \equiv \psi_{,xxx} - 3(u+\lambda)\psi_{,x}$. We thus obtain the result

$$\lambda_{,t}\psi^2 + [\psi R_{,x} - \psi_{,x}R]_{,x} = 0, \quad (4.1.8)$$
$$R \equiv \psi_{,t} + \psi_{,xxx} - 3(u+\lambda)\psi_{,x}.$$

Hence, by means of the boundary conditions, it follows that

$$\lambda_{,t} \int \psi^2 \, dx = 0,$$

showing that λ is independent of the time t. In what follows we restrict ourselves to the boundary condition that u decays sufficiently rapidly as $|x| \to \infty$. Because of $\lambda_{,t} = 0$, Eq. (4.1.8) reduces to

$$\psi R_{,x} - \psi_{,x} R = \text{const.} \quad (4.1.9)$$

Since the left-hand side of this equation is a Wronskian, R can be expressed as the linear combination of the two functions ψ and ϕ, where ϕ is a solution linearly independent of ψ:

$$R \equiv \psi_{,t} + \psi_{,xxx} - 3(u+\lambda)\psi_{,x} = C\psi + D\phi$$

$$\phi \equiv \psi \int^x \frac{1}{\psi^2} \, dx. \quad (4.1.10)$$

This is the equation which determines the time-evolution of ψ. We remark that the derivation of Eqns (4.1.7) and (4.1.10) can be formulated in a more general mathematical framework, and that it can also be accomplished more simply. This was shown by Lax (1968), and his approach will be given in Section 4.4. Readers who are familiar with quantum mechanics may first read Section 4.4.

It is worth while deducing a certain correspondence between the KdV equation and the Schrödinger equation. When u is infinitesimal in the KdV equation (4.1.3), the dispersion relation (3.3.1) becomes $\omega + k^3 = 0$. Consequently, the phase velocity $\lambda_p = \omega/k$ is equal to $-k^2$, and hence the solution is a plane wave proceeding in the negative x direction. When $|u|$ is large, but $u \to 0$ as $|x| \to \infty$, the linear approximation is also valid as $|x| \to \infty$, because in that case $|u|$ is sufficiently small. For the case of a soliton, u decays exponentially as $|x| \to \infty$, so that k must be purely imaginary. That is, $k = i\kappa_p$ ($\kappa_p \gtreqless 0$ for $x \to \pm\infty$, respectively) and λ_p becomes κ_p^2, so that the soliton proceeds in the positive x direction.

This is, of course, readily seen from the soliton solution

$$u = -2\kappa^2 \text{sech}^2[\kappa(x - 4\kappa^2 t)], \qquad (4.1.11)$$

which becomes proportional to $e^{\pm 2\kappa x}$ as $|x| \to \infty$. Hence putting $2\kappa = \kappa_p$ yields the soliton velocity $4\kappa^2(=\kappa_p^2)$. On the other hand, if $|u|$ is sufficiently small, the Schrödinger equation (4.1.7) reduces to $-\psi_{,xx} \simeq \lambda \psi$ to give $\lambda \simeq k^2$ and $\psi \sim e^{\pm ikx}$. In general, λ is positive and arbitrary for the scattering state (the continuous spectrum), while for the bound state λ becomes negative and discrete, and ψ decays exponentially as $|x| \to \infty$, so that k becomes purely imaginary. In particular, for the one-soliton solution Eq. (4.1.11), the Schrödinger equation (4.1.7) admits one and only one bound state, with the eigenvalue $\lambda = -\kappa^2$ (Landau and Lifschitz (1958)), which implies that the soliton velocity is determined by the eigenvalue.

If $u(x, 0)$ is a potential which admits N bound states, the solution of the KdV equation $u(x, t)$ as $t \to \infty$ is given by N solitons, each proceeding with a velocity equal to four times their respective eigenvalue, and the residual wave train attenuates algebraically with time, which corresponds to the scattered state. This is one of the most important results of the theory which will be developed in this chapter. From this it also follows that in a general dispersive system like the one discussed in Section 3.3, in an asymptotic sense as $t \to \infty$, solutions can be approximated by N solitons.

We now show that when $u(x, 0)$ is given, the following procedure enables $u(x, t)$ to be obtained exactly.

(1) *The direct problem* For the given potential $u(x, 0)$, solve the Schrödinger equation to find the spectrum; that is, find the eigenvalues of the bound states and the scattering data of the continuous spectrum comprising the reflection and transmission coefficients.

(2) *The time evolution of the scattering data* By means of Eq. (4.1.10) and the asymptotic form of ψ as $|x| \to \infty$ obtain the time evolution of the reflection and transmission coefficients for the scattered state and normalization factors for the bound states which will be called collectively the scattering data.

(3) *The inverse problem* By use of the scattering data at each instant of time obtain the potential $u(x, t)$ of the Schrödinger equation at the time t. This process is shown graphically in Fig. 4.1.

By virtue of the assumption that $u(x, t)$ decays rapidly as $|x| \to \infty$ for all t, the Schrödinger equation (4.1.7) admits a finite number of eigenstates of negative energy ($\lambda = -\kappa_n^2$ ($n = 1, 2, \ldots, N$)) and a continuous spectrum of positive energy ($\lambda = k^2$). As $x \to \infty$, the wave function for the eigenstate with $\lambda = -\kappa_n^2$ is given by

$$\psi_n \to c_n(t) e^{-\kappa_n x} \qquad (\kappa_n > 0, x \to \infty). \qquad (4.1.12\text{a})$$

4.1 THE KORTEWEG–DE VRIES EQUATION

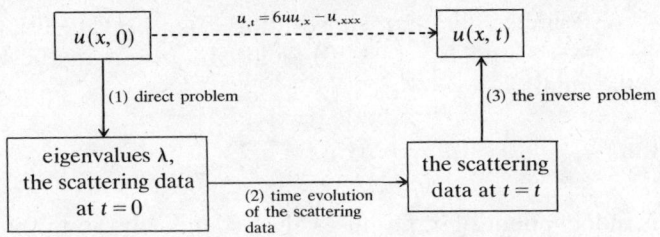

Fig. 4.1

Here ψ_n is normalized by means of the condition

$$\int_{-\infty}^{\infty} \{\psi_n\}^2 \, dx = 1. \tag{4.1.12b}$$

The wave function for the continuous spectrum can be expressed asymptotically as follows:

$$\psi \sim e^{-ikx} + r(k, t)e^{ikx} \quad (x \to \infty) \tag{4.1.13a}$$

$$\psi \sim a_T(k, t)e^{-ikx} \quad (x \to -\infty) \tag{4.1.13b}$$

Since the amplitude of the incident wave as $x \to \infty$ is set equal to unity, $r(k, t)$, $a_T(k, t)$ are, respectively, the reflection and transmission coefficients, and by virtue of the conservation of probability we have

$$|a_T(k, t)|^2 + |r(k, t)|^2 = 1. \tag{4.1.13c}$$

Hence the direct problem (1) comprises finding the scattering data, N, κ_n, $c_n(0)$ $(n = 1, 2, \ldots, N)$, $r(k, 0)$ and $a_T(k, 0) (0 \le k^2 < \infty)$, for the given potential $u(x, 0)$. The inverse problem involves obtaining $u(x, t)$ from this scattering data at an arbitrary time t. Equation (4.1.10) enables us to find $a_T(k, t)$, $r(k, t)$, $c_n(t)$ as follows.

For the discrete eigenstates, $\psi \to e^{-\kappa_n x}$ as $x \to \infty$, while the solution ϕ which is linearly independent of ψ diverges as $x \to \infty$. Consequently, we conclude that $D \equiv 0$. Moreover, multiplying Eq. (4.1.10) by ψ and eliminating u by use of Eq. (4.1.7), we have

$$(\psi^2/2)_{,t} + (\psi\psi_{,xx} - 2\psi_{,x}^2 - 3\lambda\psi^2)_{,x} = C\psi^2.$$

Integrating this equation from $x = -\infty$ to $x = +\infty$ and using the normalization Eq. (4.1.12b) we obtain $C \equiv 0$. Then, substituting Eq. (4.1.12a) for ψ as $x \to \infty$ in Eq. (4.1.10) and noticing that $u \to 0$ as $x \to \infty$, we obtain

$$c_{n,t} - 4\kappa_n^3 c_n = 0 \tag{4.1.14a}$$

leading to

$$c_n(t) = c_n(0)e^{4\kappa_n^3 t}. \tag{4.1.14b}$$

For the continuous spectrum, substituting Eq. (4.1.13b) for the wave function as $x \to -\infty$ into Eq. (4.1.10) yields, by means of the boundary condition, the relation

$$a_{T,t} + 4ik^3 a_T = Ca_T + Da_T \int^x dx/\psi^2.$$

Since a_T is independent of x, D must vanish identically, so that it reduces to

$$a_{T,t} + (4ik^3 - C)a_T = 0. \tag{4.1.15a}$$

Similarly, substituting Eq. (4.1.13a) for ψ as $x \to \infty$ into Eq. (4.1.10), we find from the coefficients of e^{-ikx} and e^{ikx}

$$C = 4ik^3, \qquad r_{,t} - 8ik^3 r = 0, \tag{4.1.15b}$$

and hence that

$$r(k, t) = r(k, 0)e^{8ik^3 t}, \tag{4.1.16a}$$
$$a_T(k, t) = a_T(k, 0). \tag{4.1.16b}$$

The set $\{c_n, \kappa_n, r(k)\}$ gives the scattering data. Equations (4.1.14) and (4.1.16) give the scattering data at an arbitrary time ($t \geq 0$).

We now give a brief introduction to the method of obtaining $u(x, t)$ from the scattering data. This is called the inverse scattering method. For mathematically rigorous discussions of the issues involved we refer to specialized books and papers on the subject.† The basic equation is the Gel'fand–Levitan–Marchenko (G–L–M) equation

$$K(x, y; t) + B(x+y; t) + \int_x^\infty B(y+z; t)K(x, z; t)\, dz = 0 \qquad (y > x),$$
$$\tag{4.1.17}$$

in which $B(x+y)$ is given in terms of the scattering data by

$$B(x+y; t) = (2\pi)^{-1} \int_{-\infty}^\infty r(k, t)e^{ik(x+y)}\, dk + \sum_{n=1}^N c_n^2(t)e^{-\kappa_n(x+y)}.$$
$$\tag{4.1.18}$$

The G–L–M equation (4.1.17) is an integral equation for $K(x, y)$, and the solution $K(x, y)$ is determined uniquely. The required solution $u(x, t)$ to the KdV equation is given by

$$u(x, t) = -2\frac{d}{dx} K(x, x; t). \tag{4.1.19}$$

† For example, Newton (1966); Gel'fand and Levitan (1955), Marchenko (1955), and Kato (1974).

4.1 THE KORTEWEG–DE VRIES EQUATION

A derivation of the G–L–M equation from the Schrödinger equation will be carried out later. We first solve some simple examples of this equation, which will be of help when seeking to understand the framework of the subsequent mathematical discussion. Assume that for an initial condition $u(x, 0)$ Eq. (4.1.17) admits only one bound state ($N = 1$), and that in the scattered state the incident wave does not reflect, but is transmitted perfectly so that $r(k, 0) = 0$, $a_T(k, 0) = 1$. For such a special $u(x, 0)$, solving the ordinary problem (1) determines κ and $c(0)$. Then, from Eqns (4.1.14b), (4.1.18), the G–L–M equation (4.1.17) becomes

$$K(x, y; t) + c^2(0)e^{8\kappa^3 t - \kappa(x+y)} + c^2(0)e^{8\kappa^3 t} \int_x^\infty e^{-\kappa(y+z)} K(x, z; t)\, dz = 0$$

$$(y > x). \qquad (4.1.20a)$$

Differentiating this equation with respect to y yields

$$K(x, y; t)_{,y} = -\kappa K,$$

that is

$$K(x, y; t) = e^{-\kappa y} h(x, t). \qquad (4.1.20b)$$

Introducing Eq. (4.1.20b) into Eq. (4.1.20a), we have

$$K(x, y; t) = -\frac{c^2(0)e^{8\kappa^3 t - \kappa(x+y)}}{1 + c^2(0)(2\kappa)^{-1} e^{8\kappa^3 t - 2\kappa x}}. \qquad (4.1.20c)$$

Therefore, from Eq. (4.1.19), it follows that

$$u(x, t) = -2\kappa^2 \operatorname{sech}^2 [\kappa(x - 4\kappa^2 t) - \delta]$$

$$\delta = 2^{-1} \log\left[\frac{c^2(0)}{2\kappa}\right]. \qquad (4.1.21)$$

Since the phase δ is constant, shifting the origin of x or t reduces Eq. (4.1.21) to the one-soliton solution Eq. (4.1.11). We thus find that a single soliton corresponds to a reflectionless potential which admits only one bound state.

For a reflectionless potential having two bound states, ($N = 2$, $\lambda^{(i)} = -\kappa_i^2$, $c_i^2(0) \equiv m_i$ ($i = 1, 2$) $r(k, 0) = 0$), the G–L–M equation becomes

$$K(x, y; t) + m_1 e^{8\kappa_1^3 t - \kappa_1(x+y)} + m_2 e^{8\kappa_2^3 t - \kappa_2(x+y)}$$

$$+ m_1 e^{8\kappa_1^3 t} \int_x^\infty e^{-\kappa_1(y+z)} K(x, z; t)\, dz \qquad (4.1.22a)$$

$$+ m_2 e^{8\kappa_2^3 t} \int_x^\infty e^{-\kappa_2(y+z)} K(x, z; t)\, dz = 0.$$

Assume for K the form

$$K(x, y; t) = e^{-\kappa_1 y} h_1(x, t) + e^{-\kappa_2 y} h_2(x, t). \tag{4.1.22b}$$

Substituting this expression into Eq. (4.1.22a) and equating the coefficients of $e^{-\kappa_1 y}$, $e^{-\kappa_2 y}$ to zero we obtain, respectively,

$$\left(1 + \frac{m_1}{2\kappa_1} e^{8\kappa_1^3 t - 2\kappa_1 x}\right) h_1 + \frac{m_1}{\kappa_1 + \kappa_2} e^{8\kappa_1^3 t - (\kappa_1 + \kappa_2)x} h_2 = -m_1 e^{8\kappa_1^3 t - \kappa_1 x},$$

$$\frac{m_2}{\kappa_1 + \kappa_2} e^{8\kappa_2^3 t - (\kappa_1 + \kappa_2)x} h_1 + \left(1 + \frac{m_2}{2\kappa_2} e^{8\kappa_2^3 t - 2\kappa_2 x}\right) h_2 = -m_2 e^{8\kappa_2^3 t - \kappa_2 x}. \tag{4.1.22c}$$

The Eqns (4.1.22c) can be solved for $e^{-\kappa_i x} h_i$ ($i = 1, 2$) to give

$$e^{-\kappa_1 x} h_1(x, t) = \frac{1}{D} \det \begin{vmatrix} -m_1 e^{-2\kappa_1 \xi_1} & m_1(\kappa_1 + \kappa_2)^{-1} e^{-2\kappa_1 \xi_1 - \kappa_2 \xi_2 - 4\kappa_2^3 t} \\ -m_2 e^{-\kappa_2 \xi_2 + 4\kappa_2^3 t} & 1 + (m_2/2\kappa_2) e^{-2\kappa_2 \xi_2} \end{vmatrix},$$

$$e^{-\kappa_2 x} h_2(x, t) = \frac{1}{D} \det \begin{vmatrix} 1 + (m_1/2\kappa_1) e^{-2\kappa_1 \xi_1} & -m_1 e^{-\kappa_1 \xi_1 + 4\kappa_1^3 t} \\ m_2(\kappa_1 + \kappa_2)^{-1} e^{-\kappa_1 \xi_1 - 2\kappa_2 \xi_2 - 4\kappa_1^3 t} & -m_2 e^{-2\kappa_2 \xi_2} \end{vmatrix},$$

$$D \equiv \det \begin{vmatrix} 1 + (m_1/2\kappa_1) e^{-2\kappa_1 \xi_1} & m_1(\kappa_1 + \kappa_2)^{-1} e^{-\kappa_1 \xi_1 - \kappa_2 \xi_2 + 4(\kappa_1^3 - \kappa_2^3)t} \\ m_2(\kappa_1 + \kappa_2)^{-1} e^{-\kappa_1 \xi_1 - \kappa_2 \xi_2 - 4(\kappa_1^3 - \kappa_2^3)t} & 1 + (m_2/2\kappa_2) e^{-2\kappa_2 \xi_2} \end{vmatrix},$$

$$\tag{4.1.23a}$$

where $\xi_i \equiv x - 4\kappa_i^2 t$ ($i = 1, 2$). Consequently, $u(x, t)$ is found to be

$$u(x, t) = -2 \frac{d}{dx} [e^{-\kappa_1 x} h_1(x, t) + e^{-\kappa_2 x} h_2(x, t)]. \tag{4.1.23b}$$

To find the asymptotic solution as $x \to \infty$, let t approach infinity keeping $\xi_1 = x - 4\kappa_1^2 t$ constant. Without loss of generality take $\kappa_1 > \kappa_2$, and set $\xi_2 = \xi_1 + 4(\kappa_1^2 - \kappa_2^2)t$, so that $\exp(-\kappa_2 \xi_2) \to 0$ as $t \to \infty$. Hence the approximations

$$e^{-\kappa_1 x} h_1(x, t) \sim -\frac{m_1}{D} e^{-2\kappa_1 \xi_1}$$

$$e^{-\kappa_2 x} h_2(x, t) \sim 0, \tag{4.1.24a}$$

$$D \sim 1 + \frac{m_1}{2\kappa_1} e^{-2\kappa_1 \xi_1},$$

are valid, which may be introduced into Eq. (4.1.23b) to give

$$u(x, t) \sim -2\kappa_1^2 \operatorname{sech}^2 [\kappa_1(x - 4\kappa_1^2 t) + \delta_1],$$

$$\delta_1 = -\tfrac{1}{2} \log \frac{m_1}{2\kappa_1}. \tag{4.1.24b}$$

4.1 THE KORTEWEG–DE VRIES EQUATION

Also, in the limit as $t \to \infty$ while $\xi_2 = x - 4\kappa_2^2 t$ is kept constant, by means of the relations $\xi_1 = \xi_2 - 4(\kappa_1^2 - \kappa_2^2)t$, $\kappa_1 > \kappa_2$ and $e^{-\kappa_1 \xi_1} \gg 1$, we obtain

$$e^{-\kappa_1 x} h_1(x, t) = -\frac{m_1}{D}\left(1 + \frac{m_2(\kappa_1 - \kappa_2)}{2\kappa_2(\kappa_1 + \kappa_2)} e^{-2\kappa_2 \xi_2}\right) e^{-2\kappa_1 \xi_2 + 8\kappa_1(\kappa_1^2 - \kappa_2^2)t},$$

$$e^{-\kappa_2 x} h_2(x, t) \sim \frac{m_1 m_2}{2\kappa_1 D} \frac{\kappa_1 - \kappa_2}{\kappa_1 + \kappa_2} e^{-2\kappa_2 \xi_2 - 2\kappa_1 \xi_2 + 8\kappa_1(\kappa_1^2 - \kappa_2^2)t}, \qquad (4.1.24a')$$

$$D \sim \frac{m_1}{2\kappa_1}\left(1 + \frac{m_2}{2\kappa_2}\frac{(\kappa_1 - \kappa_2)^2}{(\kappa_1 + \kappa_2)^2} e^{-2\kappa_2 \xi_2}\right) e^{-2\kappa_1 \xi_2 + 8\kappa_1(\kappa_1^2 - \kappa_2^2)t}.$$

When these are introduced into Eq. (4.1.23b) they give

$$u(x, t) \sim -2\kappa_2^2 \operatorname{sech}^2[\kappa_2(x - 4\kappa_2^2 t) + \delta_2],$$

$$\delta_2 = -\tfrac{1}{2}\log\left[\frac{m_2}{2\kappa_2}\frac{(\kappa_1 - \kappa_2)^2}{(\kappa_1 + \kappa_2)^2}\right]. \qquad (4.1.24b')$$

Namely, when $u(x, 0)$ has two discrete eigenstates ($\lambda^{(i)} = -\kappa_i^2$), in the limit as $t \to \infty$, the solution comprises two solitons proceeding with velocities $4\kappa_i^2$, corresponding to the eigenvalues $-\kappa_i^2$. The explicit form for a two-soliton wave function is given by the initial condition

$$u(x, 0) = -6 \operatorname{sech}^2 x,$$

which yields the eigenvalues $\kappa_1 = 2$, $\kappa_2 = 1$ and, for $t \geq 0$, the solution

$$u(x, t) = -12[3 + 4\cosh(2x - 8t) + \cosh(4x - 64t)] \qquad (4.1.25)$$
$$\times [3\cosh(x - 28t) + \cosh(3x - 36t)]^{-2}.$$

We now give an outline of the derivation of the G–L–M equation from the Schrödinger equation, which can be extended to operators which are not self-adjoint and which will be used later for the nonlinear Schrödinger equation (Zakharov and Shabat (1972)). A review of the derivations, including other standard methods, is to be found in (Kato (1974)).

Let $\varphi(k, x)$ and $\chi(k, x)$ be the two solutions of the scattered state of the Schrödinger equation (4.1.7), which satisfy the boundary conditions

$$\begin{aligned}\varphi(k, x) &\to e^{-ikx} \quad (x \to -\infty), \\ \chi(k, x) &\to e^{ikx} \quad (x \to \infty),\end{aligned} \qquad (4.1.26a)$$

where k is real and $\lambda = k^2$. It is obvious that $\varphi(k, x)$ represents the transmitted wave at $x = -\infty$ while $\chi(k, x)$ is the reflected wave at $x = +\infty$. By virtue of the definition of $\chi(k, x)$, $\chi(-k, x)$ satisfies the boundary conditions

$$\chi(-k, x) \to e^{-ikx} \quad (x \to \infty), \qquad (4.1.26b)$$

and hence it represents the incident wave as $x \to \infty$ (cf. Fig. 4.2). (The

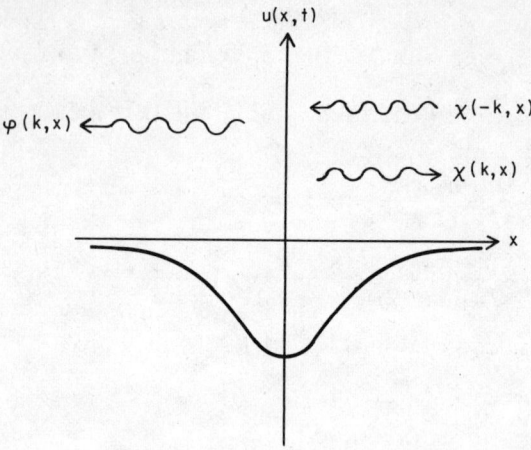

Fig. 4.2

solutions $\varphi(k, x)$ and $\chi(k, x)$ are called the Jost functions.) Since $\chi(-k, x) = \chi^*(k, x)$ also satisfies the Schrödinger equation (4.1.7) as does $\chi(k, x)$, where $*$ denotes the complex conjugate operation, the Wronskian

$$[\chi, \chi^*] \equiv \chi_{,x}\chi^* - \chi\chi^*_{,x}$$

is constant, and can be evaluated at $x = \infty$ to give

$$[\chi, \chi^*] = 2ik.$$

Hence, if $k \neq 0$, χ and χ^* are linearly independent. Therefore φ is given by the following linear combination of χ and χ^*:

$$\varphi(k, x) = a(k)\chi(-k, x) + b(k)\chi(k, x), \qquad (4.1.27)$$

in which $a(k), b(k)$ are the amplitudes of the incident and reflected waves, respectively, and

$$r(k) = \frac{b(k)}{a(k)}. \qquad (4.1.28)$$

Differently from Eq. (4.1.13c), we set the amplitude of the transmitted wave equal to unity, and consequently (from the conservation of probability in quantum mechanics) it follows that

$$|a(k)|^2 = 1 + |b(k)|^2. \qquad (4.1.29)$$

4.1 THE KORTEWEG–DE VRIES EQUATION

Summarizing, we have that for real k

$$\varphi(-k, x) = \varphi^*(k, x), \qquad \chi(-k, x) = \chi^*(k, x) \tag{4.1.30}$$

which are solutions of Eq. (4.1.7) and, moreover, it is readily shown that the Wronskian of $\varphi(k, x)$ and $\chi(k, x)$ becomes constant, which may also be evaluated at $x = \infty$ to give

$$\varphi(k, x)\chi_{,x}(k, x) - \varphi_{,x}(k, x)\chi(k, x) = 2ika(k). \tag{4.1.31}$$

Therefore, if $k \neq 0$, φ and χ are linearly independent solutions.

The bound states can be obtained by extending k to the field of complex numbers. It can be deduced that for Im $k > 0$, the φ and χ which satisfy Eqns (4.1.7) and (4.1.26a) for complex-valued k, are analytic, because from Eq. (4.1.26a), if Im $k > 0$, φ and χ decay exponentially as $x \to -\infty$ and $x \to +\infty$, respectively. When this is valid, by means of Eq. (4.1.31), $a(k)$ is also analytic for Im $k > 0$; moreover, for $|k| \to \infty$, Eq. (4.1.7) may be approximated by $-\psi'' \simeq k^2 \psi$ to give $\varphi \simeq e^{-ikx}$, $\chi \simeq e^{ikx}$, and hence we find (cf. Eq. (4.1.52b))

$$a(k) \to 1 + O(1/k), \qquad |k| \to \infty \ (\text{Im } k \geq 0). \tag{4.1.32}$$

Also, Eqns (4.1.30) and (4.1.31) yield $a(k) = a^*(-k)$ for Im $k = 0$, and consequently for Im $k \geq 0$

$$a(k) = a^*(-k^*). \tag{4.1.33}$$

Then, if for Im $k > 0$, there exists a root k_0 such that $a(k_0) = 0$, k_0 is purely imaginary. In fact, at a root k_0, the Wronskian of φ and χ vanishes, so that φ and χ become linearly dependent and decay exponentially as $x \to -\infty$ and $x \to \infty$, respectively. Thus, at $\lambda = k_0^2$, Eq. (4.1.7) admits a square integrable solution, and consequently λ becomes an eigenvalue for a bound state: that is, it is real and negative and therefore k_0 is purely imaginary. Writing $k_{0n} = i\kappa_n$, so that $\lambda_n = -\kappa_n^2$, we have

$$\varphi_n(i\kappa_n, x) = \tilde{c}_n(i\kappa_n)\chi_n(i\kappa_n, x). \tag{4.1.34}$$

Since $\chi_n \to e^{-\kappa_n x}$ as $x \to \infty$, from Eqns (4.1.12) we have

$$\psi_n = c_n \chi_n$$

and

$$c_n^2 = \left(\int \chi_n^2 \, dx\right)^{-1},$$

where χ_n is real.

Let us now summarize the important properties. The number of eigenvalues $N(\kappa_n^2; n = 1, 2, \ldots, N)$ is finite, which is readily seen from Eq. (4.1.32). (If N is infinite, the zeros of $a(k)$ accumulate at a finite point on the imaginary axis.) Also the eigenstates are not degenerate because of the condition

$$a'_n \equiv \left(\frac{da}{dk}\right)_{k=i\kappa_n} \neq 0 \quad \text{(Problem 4.1.1)}.$$

Furthermore, $\tilde{c}_n$ is given by $\tilde{c}_n = ia'_n c_n^2$ (Problem (4.1.1).

The proof of the analyticity of $\varphi(k, x)$, $\chi(k, x)$ for Im $k \geq 0$, is not given here and we merely note that it is shown by solving the integral equations (cf. Problem 4.1.2)

$$\varphi(k, x) = e^{-ikx} + \int_{-\infty}^{x} \frac{\sin k(x-y)}{k} u(y)\varphi(k, y) \, dy,$$
$$\chi(k, x) = e^{ikx} - \int_{x}^{\infty} \frac{\sin k(x-y)}{k} u(y)\chi(k, y) \, dy, \tag{4.1.35a}$$

which follow immediately from Eqns (4.1.7) and (4.1.26).

The standard derivation of the G–L–M equation is based on the following equation (Problem 4.1.3):

$$\chi(k, x) = e^{ikx} + \int_{x}^{\infty} K(x, y) e^{iky} \, dy, \tag{4.1.35b}$$

in which $K(x, y)$ will become the solution of the G–L–M equation. However, the existence of such a kernel K is not obvious and must be proved (Problem 4.1.3). Therefore, in what follows, we refer to an heuristic argument to elucidate the connection with the soliton solutions. (A rigorous proof is outlined in Problem 4.1.4.)

Rewriting Eq. (4.1.27) as

$$\frac{\varphi(k, x)}{a(k)} = \chi(-k, x) + r(k)\chi(k, x),$$

we find that the function on the left-hand side $\varphi(k, x)/a(k)$ is analytic except at N poles for Im $k > 0$, while $\chi(-k, x)$ is analytic for Im $k < 0$, and for Im $k = 0$ these two functions differ by $r(k)\chi(k, x)$. (Note that $b(k)$ is not necessarily analytic for Im $k > 0$.) Hence the function $\Psi(k, x)$ defined as

$$\Psi(k, x) = \begin{cases} a^{-1}\varphi e^{ikx} & (\text{Im } k > 0) \\ \chi^*(k^*, x) e^{ikx} & (\text{Im } k < 0) \end{cases}, \tag{4.1.36}$$

is analytic for Im $k < 0$ and also for Im $k > 0$ except at N poles, and has a jump across the real axis,

$$\Psi(k_r + i0, x) - \Psi(k_r - i0, x) \equiv \phi(k_r)$$
$$= r(k)\chi(k_r, x) e^{ik_r x}. \tag{4.1.37}$$

4.1 THE KORTEWEG–DE VRIES EQUATION

Hence this function Ψ has singularities given by the scattering data $(c_n, \kappa_n, r(k))$ and this enables us to connect the scattering data with the solutions of the Schrödinger equation $\chi(k, x), \chi(-k, x)$ as follows.

Consider the integral with a contour encircling the upper half of the complex k-plane.

$$\oint \frac{\Psi(k', x)}{k'-k} dk' = \lim_{R\to\infty} \int_0^\pi i\Psi(|k'|=R, x) \, d\theta + \int_{-\infty}^\infty \frac{\Psi(k'+i\delta, x)}{k'-k} dk'. \tag{4.1.38}$$

The first term on the right-hand side is the integral over the upper half circle $k' = Re^{i\theta}$ (R = const.), and by means of the definition of $\Psi(k', x)$ in Eq. (4.1.36) ($\Psi(|k'|\to\infty) = e^{-ik'x}e^{ik'x} = 1$) it becomes $i\pi$. For Im $k < 0$, the integral on the left-hand side of Eq. (4.1.38) is given by the contribution from the residues at the poles $a(k) = 0$; that is, by

$$\oint \frac{\Psi(k', x)}{k'-k} dk' = 2\pi i \sum_n^N \frac{A_n}{i\kappa_n - k}. \tag{4.1.39}$$

Here, in view of Eq. (4.1.34) and Problem (4.1.1), A_n is given by

$$A_n = \tilde{c}_n/a'_n \chi_n(x) e^{-\kappa_n x} = ic_n^2 \chi_n(x) e^{-\kappa_n x}. \tag{4.1.40}$$

Hereafter, ic_n^2 will be denoted by C_n (which is purely imaginary). Hence from Eqns (4.1.38) to (4.1.40), for Im $k < 0$, it follows

$$i\pi + \int_{-\infty}^\infty \frac{\Psi(k'+i\delta, x)}{k'-k} dk' = 2\pi i \sum_n^N \frac{A_n}{i\kappa_n - k}, \tag{4.1.41a}$$

On the other hand, the integral encircling the lower half-plane becomes, in the same way,

$$\oint \frac{\Psi(k', x)}{k'-k} dk' = i\pi - \int_{-\infty}^\infty \frac{\Psi(k'-i\delta, x)}{k'-k} dk'.$$

The integral on the left-hand side is given by the residue at the pole $k' = k$, namely by

$$i\pi - \int_{-\infty}^\infty \frac{\Psi(k'-i\delta, x)}{k'-k} dk' = 2\pi i \chi^*(k^*, x) e^{ikx} \tag{4.1.41b}$$

Adding Eqns (4.1.41a, b) and using Eqns (4.1.40) and (4.1.37), we have, for Im $k < 0$,

$$\chi^*(k^*, x) e^{ikx} = 1 + \frac{1}{2\pi i} \int_{-\infty}^\infty \frac{r(k')\chi(k', x)e^{ik'x}}{k'-k} dk' - \sum_n^N \frac{C_n \chi_n(x) e^{-\kappa_n x}}{i\kappa_n - k}, \tag{4.1.42}$$

Taking the complex conjugate of this equation, and writing k^* for k, yields, for $\text{Im } k > 0$,

$$\chi(k,x)e^{-ikx} = 1 - \frac{1}{2\pi i}\int_{-\infty}^{\infty} \frac{r^*(k')\chi^*(k',x)e^{-ik'x}}{k'-k}dk'$$

$$+ \sum_n^N \frac{C_n^* \chi_n^*(x) e^{-\kappa_n x}}{i\kappa_n + k}. \quad (4.1.43)$$

Putting $k = i\kappa_m$ ($m = 1, 2, \ldots, N$) in this equation we obtain

$$\chi_m(x)e^{\kappa_m x} = 1 + \frac{1}{2\pi i}\int_{-\infty}^{\infty} \frac{r(k')\chi(k',x)e^{ik'x}}{k'+i\kappa_m}dk' - i\sum_n^N \frac{C_n^*\chi_n(x)e^{-\kappa_n x}}{\kappa_n + \kappa_m} \quad (4.1.44\text{a})$$

where, in the derivation, we have used the results that for real k $r^*(k) = r(-k)$, $\chi^*(k,x) = \chi(-k,x)$, and $\chi_n(x)$ is real. In the same way, for $\text{Im } k = 0$, by putting $k = k_r - i\delta$, Eq. (4.1.43) is reduced to

$$\chi(k,x)e^{-ikx} = 1 + \frac{1}{2\pi i}\int_{-\infty}^{\infty}\frac{r(k')\chi(k',x)e^{ik'x}}{k'+k+i\delta}dk' - i\sum_n^N \frac{C_n^*\chi_n(x)e^{-\kappa_n x}}{\kappa_n - ik}. \quad (4.1.44\text{b})$$

Equations (4.1.44a, b) comprise $N+1$ equations which connect the scattering data with $\chi(k,x)$ and χ_m. For example, consider the reflectionless case with one bound state, $r(k) = 0$, $N = 1$. From Eq. (4.1.44a) we have

$$\chi_1(x) = e^{-\kappa_1 x} - \frac{c_1^2 \chi_1}{2\kappa_1} e^{-2\kappa_1 x}, \quad (4.1.45\text{a})$$

or

$$\chi_1(x) = \frac{e^{-\kappa_1 x}}{1 + c_1^2 (2\kappa_1)^{-1} e^{-2\kappa_1 x}}. \quad (4.1.45\text{b})$$

Comparing with Eq. (4.1.20c), we find from Eq. (4.1.45b) that

$$K(x,y,t) = -c_1^2(t)e^{-\kappa_1 y}\chi_1(x). \quad (4.1.46)$$

So, substituting Eq. (4.1.46) for $K(x,y)$ into Eq. (4.1.35b) gives Eq. (4.1.45a) for $k = i\kappa_1$.

The correspondence between K and χ_n illustrated by Eq. (4.1.46) becomes more evident for $N = 2$. Namely, for $N = 2$, $r(k) = 0$, Eq. (4.1.44a) yields

$$\left(1 + \frac{c_1^2(t)}{2\kappa_1}e^{-2\kappa_1 x}\right)\chi_1 + \left(\frac{c_2^2(t)}{\kappa_1+\kappa_2}e^{-(\kappa_1+\kappa_2)x}\right)\chi_2 = e^{-\kappa_1 x}, \quad (4.1.47\text{a})$$

$$\left(\frac{c_1^2(t)}{\kappa_1+\kappa_2}e^{-(\kappa_1+\kappa_2)x}\right)\chi_1 + \left(1 + \frac{c_2^2(t)}{2\kappa_2}e^{-2\kappa_2 x}\right) = e^{-\kappa_2 x}. \quad (4.1.47\text{b})$$

4.1 THE KORTEWEG–DE VRIES EQUATION

Comparing these equations with Eq. (4.1.22b, c) we find for the reflectionless case that

$$K(x, y) = -\sum_{n=1}^{2} c_n^2(t)\chi_n e^{-\kappa_n y}. \tag{4.1.48}$$

The extension to N bound states is obvious. Moreover, noticing that the contribution from the scattering state can be deduced by replacing the summation by an integration over k, we postulate

$$K(x, y) = -\sum_{n=1}^{N} c_n^2 \chi_n(x) e^{-\kappa_n y} - \frac{1}{2\pi} \int_{-\infty}^{\infty} r(k)\chi(k, x)e^{iky}\, dk. \tag{4.1.49}$$

For $y > x$, this kernel $K(x, y)$ satisfies Eq. (4.1.35b). From Eqns (4.1.49), (4.1.43) for Im $k > 0$, the following integral of $K(x, y)$ becomes

$$\int_x^{\infty} K(x, y) e^{iky}\, dy$$

$$= -\sum_{n=1}^{N} \frac{c_n^2 \chi_n(x) e^{(-\kappa_n + ik)x}}{\kappa_n - ik} + \frac{1}{2\pi i} \int_{-\infty}^{\infty} \frac{r(k')\chi(k', x)e^{i(k+k')x}}{k' + k}\, dk',$$

$$\tag{4.1.50}$$

$$= \chi(k, x) - e^{ikx}$$

which is merely Eq. (4.1.35b). Also, it is readily seen that the $K(x, y)$ given above satisfies the G–L–M equation. That is, from the definition of $B(y+z)$ in Eq. (4.1.18) and Eq. (4.1.50), it follows that

$$\int_x^{\infty} B(y+z) K(x, z)\, dz$$

$$= \sum_n c_n^2 e^{-\kappa_n y} \int_x^{\infty} K(x, z) e^{-\kappa_n z}\, dz$$

$$+ \frac{1}{2\pi} \int_{-\infty}^{\infty} dk\, r(k) e^{iky} \int_x^{\infty} K(x, z) e^{ikz}\, dz \tag{4.1.51}$$

$$= \sum_n c_n^2 e^{-\kappa_n y} [\chi_n(x) - e^{-\kappa_n x}] + \frac{1}{2\pi} \int_{-\infty}^{\infty} r(k) e^{iky} [\chi(k, x) - e^{ikx}]\, dk$$

$$= -K(x, y) - B(x+y) \qquad (y > x).$$

The relationship between $u(x, t)$ and $K(x, x; t)$, namely Eq. (4.1.19), is obtained in Problem 4.1.3.

Here we shall derive it from Eqns (4.1.43) and (4.1.35a) by expanding Eq. (4.1.43) in powers of $1/k$ to give

$$\chi(k, x) e^{-ikx} = 1 + \left[\frac{1}{2\pi i} \int_{-\infty}^{\infty} r(k')\chi(k', x) e^{ik'x}\, dk' - i\sum_n^{N} c_n^2 \chi_n(x) e^{-\kappa_n x} \right] \frac{1}{k}$$
$$+ \cdots. \tag{4.1.52a}$$

Solving Eq. (4.1.35a) by iteration yields

$$\chi(k, x)e^{-ikx} = 1 + \left[\frac{i}{2}\int_x^\infty u(y)\,dy\right]\frac{1}{k} + \cdots. \quad (4.1.52b)$$

Comparing these two equations we get

$$\frac{1}{2}\int_x^\infty u(y)\,dy = -\frac{1}{2\pi}\int_{-\infty}^\infty r(k)\chi(k,x)e^{ikx}\,dk - \sum_n^N c_n^2\chi_n(x)e^{-\kappa_n x}. \quad (4.1.53)$$

Differentiating this equation with respect to x and using Eq. (4.1.49) we obtain Eq. (4.1.19). Therefore the N-soliton solution follows from Eqns (4.1.19), (4.1.49) as

$$u(x, t) = 2\frac{d}{dx}\left[\sum_n^N c_n^2 \chi_n(x, t) e^{-\kappa_n x}\right]. \quad (4.1.54)$$

The explicit form of the N-soliton solution can be obtained by solving the G–L–M equation, but may also be found by using Eq. (4.1.44a), which for $r(k') = 0$, reduces to

$$\chi_m(x)e^{\kappa_m x} = 1 - \sum_n^N \frac{c_n^2 \chi_n(x) e^{-\kappa_n x}}{\kappa_n + \kappa_m}.$$

Multiplying this equation by $c_m e^{-\kappa_m x}$ and noticing that

$$\psi_m = c_m \chi_m, \qquad \text{for } m = 1, 2, \ldots, N,$$

we get

$$\psi_m + \sum_n^N D_{mn}\psi_n = c_m e^{-\kappa_m x}, \qquad D_{mn} \equiv \frac{c_m c_n}{\kappa_m + \kappa_n}e^{-(\kappa_m + \kappa_n)x}. \quad (4.1.55a)$$

That is,

$$(I + D)\psi = E, \quad \psi \equiv \{\psi_m\}, \qquad D \equiv \{D_{mn}\},\ E = \{c_m e^{-\kappa_m x}\}. \quad (4.1.55b)$$

Since $\det(I + D)$ is positive definite for all x (Problem 4.1.5), the algebraic equation (4.1.55b) can be solved to give

$$\psi_n(x) = \frac{1}{\Delta}\sum_m^N c_m e^{-\kappa_m x} Q_{mn}, \qquad \Delta \equiv \det(I + D), \quad (4.1.56)$$

where Q_{mn} is the cofactor (m, n) of the matrix $I + D$. Substituting Eq. (4.1.56) for ψ_m into Eq. (4.1.54) we get

$$u(x, t) = -2\frac{d}{dx}\left[\frac{1}{\Delta}\sum_{m,n}^N \frac{dD_{mn}}{dx}Q_{mn}\right]$$

$$= -2\frac{d}{dx}\left(\frac{1}{\Delta}\frac{d\Delta}{dx}\right) = -2\frac{d^2}{dx^2}(\log \Delta). \quad (4.1.57)$$

From Eq. (4.1.57) we can show that the solution $u(x, t)$ decouples asymptotically into N solitons as $|t| \to \infty$ (Wadati and Toda (1972)). This is done by means of a direct extension of the method of solution for

4.1 THE KORTEWEG–DE VRIES EQUATION

$N=2$ based on Eqns (4.1.23), (4.1.24). From Eq. (4.1.14b) $D_{ij} = c_i(0)c_j(0)(\kappa_i+\kappa_j)^{-1}e^{-\kappa_i\xi_i-\kappa_j\xi_j}$, where $\xi_i = x - 4\kappa_i^2 t$. Hence the solution is to be considered in a coordinate frame moving with the constant velocity $\lambda_n = 4\kappa_n^2$ ($n \leq N$), $\xi_n = x - \lambda_n t$. Namely, if ξ_n and t are chosen independent variables, the variable ξ_i becomes $\xi_i = \xi_n - \varepsilon_{in} t$, where $\varepsilon_{in} = 4\kappa_i^2 - 4\kappa_n^2$.

Without loss of generality we may take $\kappa_1 > \kappa_2 > \cdots > \kappa_N$, and then corresponding to the inequalities $1 \leq i < n$, $i = n$, $n < i \leq N$, the relations $\varepsilon_{in} > 0$, $\varepsilon_{in} = 0$, $\varepsilon_{in} < 0$ hold, respectively. Hence, in the limit as $t \to \infty$ keeping ξ_n constant, the elements of the matrix $I + D$ may be approximated by

$$1 + D_{ii} \approx \frac{c_i^2(0)}{2\kappa_i} e^{-2\kappa_i(\xi_n - \varepsilon_{in} t)} \quad (i < n),$$

$$D_{ij} \approx 0 \quad (i, j > n),$$

(4.1.58a)

by means of which $\Delta(x, t)$ becomes

$$\Delta(x, t) = \prod_{i=1}^{n-1} \exp\left[-2\kappa_i(\xi_n - \varepsilon_{in} t)\right]$$

$$\times \begin{vmatrix} \dfrac{c_1^2}{2\kappa_1} & \dfrac{c_1 c_2}{\kappa_1+\kappa_2} & \cdots & \dfrac{c_1 c_n}{\kappa_1+\kappa_n} e^{-\kappa_n \xi_n} & & \\ \dfrac{c_2 c_1}{\kappa_2+\kappa_1} & \dfrac{c_2^2}{2\kappa_2} & \cdots & \dfrac{c_2 c_n}{\kappa_2+\kappa_n} e^{-\kappa_n \xi_n} & 0 & \\ \vdots & & & \vdots & & \\ \dfrac{c_n c_1}{\kappa_n+\kappa_1} e^{-\kappa_n \xi_n} & & \cdots & 1 + \dfrac{c_n^2}{2\kappa_n} e^{-2\kappa_n \xi_n} & & \\ & & & & 1 & 0 \\ & & 0 & & & \ddots \\ & & & & 0 & 1 \end{vmatrix}$$

$$= \prod_{i=1}^{n-1} \exp\left[-2\kappa_i(\xi_n - \varepsilon_{in} t)\right]$$

$$\times \left[\prod_{j=1}^{n-1} c_j^2(0) \begin{vmatrix} (2\kappa_1)^{-1} & (\kappa_1+\kappa_2)^{-1} & \cdots & (\kappa_1+\kappa_{n-1})^{-1} \\ (\kappa_1+\kappa_2)^{-1} & (2\kappa_2)^{-1} & & \vdots \\ \vdots & & \ddots & \\ (\kappa_{n-1}+\kappa_1)^{-1} & \cdots & \cdots & (2\kappa_{n-1})^{-1} \end{vmatrix} \right.$$

$$\left. + \prod_{j=1}^{n} c_j^2(0) e^{-2\kappa_n \xi_n} \begin{vmatrix} (2\kappa_1)^{-1} & (\kappa_1+\kappa_2)^{-1} & \cdots & (\kappa_1+\kappa_n)^{-1} \\ (\kappa_1+\kappa_2)^{-1} & (2\kappa_2)^{-1} & & \vdots \\ \vdots & & \ddots & \\ (\kappa_n+\kappa_1)^{-1} & \cdots & \cdots & (2\kappa_n)^{-1} \end{vmatrix} \right]$$

(4.1.58b)

in which the exponential factors of the rows and columns with $i, j < n-1$ have been taken outside the determinant, and the result $\varepsilon_{in} < 0$ has been used for $i > n$.

Since the determinant with the elements $a_{ij} = (\kappa_i + \kappa_j)^{-1}$ is given by

$$\left[\prod_{i<j}^{n}(\kappa_i - \kappa_j)\right]^2 \left[\prod_{i,j}^{n}(\kappa_i + \kappa_j)\right]^{-1},$$

$\Delta(x, t)$ can be expressed as

$$\Delta(x, t) = \prod_{i=1}^{n-1} e^{-2\kappa_i(\xi_n - \varepsilon_{in} t)} A_n^+ [1 + B_n^+ e^{-2\kappa_n \xi_n}],$$

$$A_n^+ = \prod_{l=1}^{n-1} c_l^2(0) \left[\prod_{i<j}^{n-1}(\kappa_i - \kappa_j)\right]^2 \left[\prod_{i,j}^{n-1}(\kappa_i + \kappa_j)\right]^{-1}, \qquad (4.1.58c)$$

$$B_n^+ = \frac{c_n^2(0)}{2\kappa_n} \left[\prod_{j=1}^{n-1} \frac{\kappa_j - \kappa_n}{\kappa_j + \kappa_n}\right]^2.$$

Introducing this expression into Eq. (4.1.57) enables us to find that in the coordinate frame moving with constant velocity $\lambda_n = 4\kappa_n^2$, as $t \to \infty$, $u(x, t)$ becomes

$$u(x, t) = -2\kappa_n^2 \operatorname{sech}^2(\kappa_n \xi_n + \delta_n^+),$$
$$\delta_n^+ = -\tfrac{1}{2} \log B_n^+. \qquad (4.1.59a)$$

Similarly, as $t \to -\infty$,

$$u(x, t) = -2\kappa_n^2 \operatorname{sech}^2(\kappa_n \xi_n + \delta_n^-),$$
$$\delta_n^- = -\tfrac{1}{2} \log B_n^-, \qquad B_n^- = \frac{c_n^2(0)}{2\kappa_n} \left[\prod_{j=n+1}^{N} \frac{\kappa_n - \kappa_j}{\kappa_n + \kappa_j}\right]^2. \qquad (4.1.59b)$$

Thus, as $|t| \to \infty$, $u(x, t)$ represents N solitons proceeding with the velocities $-4\kappa_n^2$, corresponding to the N discrete eigenvalues $-\kappa_n^2$. Since the velocity of a soliton is proportional to its amplitude, at $t = -\infty$ there exist N solitons with amplitudes arranged in decreasing order of magnitude as x increases. As time evolves overtaking occurs repeatedly amongst these solitons and finally, at $x = +\infty$, N solitons proceed to the right, the amplitudes of which are arranged in increasing order of magnitude as x increases. The form of each soliton, including the amplitude and the velocity at $t = \pm\infty$ does not change, and only the phases change. The phase-difference at $t \to -\infty$ and $t \to +\infty$, $\delta_n = \delta_n^+ - \delta_n^-$, is given by

$$\delta_n = \sum_{j=n+1}^{N} \log \frac{\kappa_n - \kappa_j}{\kappa_n + \kappa_j} - \sum_{j=1}^{n-1} \log \frac{\kappa_j - \kappa_n}{\kappa_j + \kappa_n}. \qquad (4.1.59c)$$

Therefore we find $\sum_{n=1}^{N} \delta n = 0$ (cf. Problem 4.1.6).

The N-soliton solution can be given in another form which is a linear combination of the square of the wave function:

$$u(x, t) = 2 \sum_{m}^{N} c_m(\psi_{m,x} - \kappa_m \psi_m) e^{-\kappa_m x} = -4 \sum_{m}^{N} \kappa_m \psi_m^2. \qquad (4.1.60)$$

This is obtained by differentiating Eq. (4.1.55a), solving an algebraic equation for $\psi_{m,x}$ and using Eq. (4.1.56). It then follows immediately that N solitons at $t = -\infty$ transform to N solitons at $t = +\infty$.

So far we have considered a reflectionless potential for $u(x, 0)$. However, even if $r(k) \neq 0$ in an asymptotic sense, the solution at $x = \infty$ may be considered as consisting of solitons. Namely, it can be shown that the contribution from the scattered state with $r_0(k) \equiv r(k, 0) \neq 0$ attenuates algebraically and exponentially as $t \to \infty$. This can be deduced from the behaviour of $B(x, t)$ as $t \to \infty$. For a proof of this we refer, for example, to Ablowitz et al. (1973a). For simplicity, consider the case without bound states ($c_n = 0$). Then from Eqns (4.1.16a) and (4.1.18), $B(x, t)$ becomes

$$B(x, t) = \frac{1}{2\pi} \int_{-\infty}^{\infty} r_0(k) e^{it(8k^3 + kx/t)} \, dk. \qquad (4.1.61a)$$

As $t \to \infty$, the integrand oscillates rapidly with a small variation of k. Consequently, this integral can be estimated by the contribution from the neighbourhood of the point at which the variation of the phase $it(8k^3 + kx/t)$ is a minimum, so that $d(8k^3 + kx/t)/dk = 0$, that is $k_\pm = \pm(-x/24t)^{1/2}$. (See, for example, Mizohata (1973).) This can be made clearer by rewriting Eq. (4.1.61a) using the transformation $k = k't^{-1/3}$:

$$B(x, t) = \frac{1}{2\pi t^{1/3}} \int_{-\infty}^{\infty} r_0(k't^{-1/3}) \exp\left[i\left(8k'^3 + \frac{k'x}{t^{1/3}}\right)\right] dk' \qquad (4.1.61b)$$

which, for $t \gg 1$ and $x/t^{1/3}$ finite, reduces to

$$B(x, t) \sim \frac{r_0(0)}{2\pi t^{1/3}} \int_{-\infty}^{\infty} \exp\left[i\left(8k'^3 + \frac{k'x}{t^{1/3}}\right)\right] dk'. \qquad (4.1.62)$$

This result shows that in such a region of the (x, t) plane the solution of the G–L–M equation (4.1.17) can be assumed to be of the form

$$K(x, y; t) = \frac{1}{t^{1/3}} K_s(X, Y) + \cdots, \quad X = \frac{x}{t^{1/3}}, \ Y = \frac{y}{t^{1/3}}. \qquad (4.1.63)$$

Introducing Eqns (4.1.62) and (4.1.63) into the G–L–M equation (4.1.17) and transforming the variables x, y, z of K_s to $x/t^{1/3}$, $y/t^{1/3}$, respectively, we can eliminate the parameter t. As a result, $u(x, t)$ is given by

$$u(x, t) \simeq \frac{1}{t^{2/3}} F\left(\frac{x}{t^{1/3}}\right). \qquad (4.1.64)$$

Thus, as $t \to \infty$, in the region $x \sim t^{1/3}$ the solution $u(x, t)$ decays like $t^{-2/3}$. In addition, in the regions $x \gg t^{1/3}$, or, $x \ll -t^{1/3}$, expanding the integral in terms of k in the neighbourhood of $k_{\pm}$, we find that for $x \gg t^{1/3}$, the solution decays exponentially, and for $x \ll -t^{-1/3}$ it decays algebraically (for $-x/t \sim 1$, like $t^{-1/2}$, and for $-x/t \gg 1$, like $t^{-1/2}/(-x/t)^{1/4}$) (Ablowitz and Newell (1973)). As a result, if $u(x, 0)$ admits bound states, the solution comprises solitons only in the asymptotic sense given above as $t \to \infty$. This implies the stability of solitons with respect to small perturbations. Namely, small disturbances imposed on solitons correspond to the effect of these same disturbances on the reflectionless potential which has given rise to the scattered state. However, the contribution from the scattered state, which appears as a wave-train, decays as $t \to \infty$ whilst, in general, the bound states do not change as a result of small disturbances. Hence the solitons do not change, and propagate stably.

Finally, we note that the asymptotic solution Eq. (4.1.64) satisfies the similarity law of the KdV equation $(x \to \varepsilon^{1/2} x, t \to \varepsilon^{3/2} t, u \to \varepsilon u)$.

Problem 4.1.1 (Kato (1974))

Show that $a'_n = -i \int_{-\infty}^{\infty} \varphi_n \chi_n \, dx$, and consequently that $\tilde{c}_n = i a'_n c_n^2$.
Hint Differentiate Eq. (4.1.7) with respect to k to get

$$\frac{d}{dx}[\varphi_{,k}, \chi] = -2k\varphi\chi,$$

$$\frac{d}{dx}[\varphi, \chi_{,k}] = 2k\varphi\chi,$$

then integrate to give

$$\frac{d}{dk}[\chi, \varphi]|_{k=i\kappa_n} = 2i\kappa_n \int_{-\infty}^{\infty} \varphi\chi \, dx.$$

Problem 4.1.2 (Kato (1974))

Show that if $\int_{-\infty}^{\infty} (1+|x|) |u(x)| \, dx < \infty$, the integral equations (4.1.35a) admit continuous solutions $\varphi(k, x)$, $\chi(k, x)$ with the parameter k, which are continuous for Im $k \geq 0$ and analytic for Im $k > 0$ (Kato (1974)).
Hint Solve Eq. (4.1.35a) for χ by iteration:

$$\chi_0(k, x) = e^{ikx}, \qquad \chi_{n+1}(k, x) = -\int_x^{\infty} \frac{\sin k(x-y)}{k} u(y) \chi_n \, dy,$$

if

$$\text{Im } k \geq 0, k \neq 0, \qquad |\chi_n| \leq n!^{-1} |k|^{-n} \sigma(x)^n e^{-\kappa x}$$

$$\sigma(x) = \int_x^{\infty} |u(y)| \, dy;$$

4.1 THE KORTEWEG–DE VRIES EQUATION

for $k=0$, $|\chi_n(x,k)| \leq n!^{-1}\{C(x)\tau(x)\}^n e^{-\kappa x}$, where $C(x)$ is a non-increasing function of x, $\tau(x) \equiv \int_x^\infty (1+|y|)|u(y)|\,dy$, and consequently $\chi(x) = \sum_{n=1}^\infty \chi_n(x,k)$ is uniformly convergent for $\text{Im}\,k \geq 0$, being a continuous function of x, k and analytic for $\text{Im}\,k > 0$. It can be estimated as

$$|\chi(x,k) - e^{ikx}| \leq \frac{C(x)}{1+|k|}\tau(x)e^{-\kappa x}.$$

The proof is parallel to that for $\varphi(k, x)$. (The estimate for $k = 0$ is not straightforward.)

Problem 4.1.3 (Kato (1974))

Show that χ can be expressed as Eq. (4.1.35b) or as

$$\chi(k, x) = e^{ikx}\left[1 + \int_0^\infty \tilde{K}(x, t)e^{2ikt}\,dt\right].$$

Hint Put $t = \frac{1}{2}(y-x)$, $\frac{1}{2}\tilde{K}(x, \frac{1}{2}(y-x)) \equiv K(x, y)$. Substitute the above expression for χ in Eq. (4.1.35a) to give

$$\tilde{K}(x, t) = \int_{x+t}^\infty u(s)\,ds + \int_0^t \left(\int_{x+t-s}^\infty u(\tau)\tilde{K}(\tau, s)\,d\tau\right) ds.$$

Solve by iteration to show the existence of a real-valued continuous solution estimated as

$$|\tilde{K}(x, y)| \leq C(x) \int_{x+y}^\infty |u(t)|\,dt.$$

Then, by the above integral equation, $\tilde{K}$ becomes continuously differentiable. Differentiate once to get

$$u(x) = -2\frac{\partial K(x, x)}{\partial x} \quad \left(= -\frac{\partial \tilde{K}(x, 0)}{\partial x}\right).$$

Show also that

$$\left(\frac{\partial^2}{\partial x^2} - \frac{\partial^2}{\partial y^2}\right) K = u(x)K.$$

Problem 4.1.4 (Kato (1974))

Derive the G–L–M equation using Eqns (4.1.35b) and (4.1.27).
Hint Transform Eq. (4.1.27) to $(1/a(k))\varphi(k,x)e^{ikx} - 1 = \chi^*(k,x)e^{ikx} - 1 + r(k)\chi(k,x)e^{ikx}$. Substitute Eq. (4.1.35b) for χ and χ^* and use the

inverse Fourier transform to get

$$\tilde{K}(x,t) + B_0(x+t) + \int_0^\infty B_0(x+t+s)\tilde{K}(x,s)\,ds$$

$$= \frac{1}{\pi} \int_{-\infty}^\infty \left(\frac{1}{a(k)} \varphi(k,x) e^{ikx} - 1 \right) e^{2ikt}\,dk \qquad (t>0),$$

where

$$B_0(x) = \frac{1}{\pi} \int_{-\infty}^\infty r(k) e^{2ikx}\,dk.$$

Find the integral on the right-hand side by summing the residues. Notice that a further proof is required in order to show mathematically that u is determined by the solution of the G–L–M equation for given scattering data.

Problem 4.1.5

Show that $\det(I+D)$ given by Eq. (4.1.55) is positive for all x. (Gardner et al. (1974).)

Problem 4.1.6

Show that in the interactions of N solitons the collisions are paired, the faster soliton being accelerated while the slower one is decelerated.

4.2 The conservation laws and the canonical form of the KdV equation

In general, the following relation for the field variables I and F,

$$\frac{\partial I}{\partial t} + \frac{\partial F}{\partial x} = 0, \qquad (4.2.1)$$

is called a conservation law. That is, when $F \to 0$ as $|x| \to \infty$, this equation yields the conserved quantity $\int_{-\infty}^\infty I\,dx$, and I is called the conserved density. (In Section 2.4, a system of n conservation laws for n field variables was considered.) The KdV equation may be written in the form of a conservation law for one field variable. However, as will be shown later, there exist an infinite number of the conservation laws for it of the form Eq. (4.2.1) and, consequently, there exist an infinite number of conserved quantities $\int_{-\infty}^\infty I_n\,dx$ $(n = 1, 2, \ldots)$. The corresponding conserved densities† are polynomial functions of u and its derivatives. We

---

† The conserved quantity $\int_{-\infty}^\infty I_n\,dx$ depends on u but takes a numerical value. In general, a quantity which depends on the form of a function f, and takes a numerical value, is called a functional of f. In this sense, the eigenvalues λ of Eq. (4.1.7) can be called functionals of u.

4.2 CONSERVATION LAWS AND CANONICAL FORM

have already remarked that this important property of the KdV equation (the existence of an infinite number of conservation laws) was found also by means of the Miura transformation. Here, as an extension of the inverse scattering method, we refer to the method of Zakharov and Faddeev (1971), by which the precise relationship between the inverse scattering method and the infinite number of conserved quantities will be made clear. In addition, the complete integrability of the KdV equation will be shown.

We first discuss the conservation laws from the viewpoint of the inverse scattering method. Consider again the (left) Jost function $\varphi(k, x)$ for $\operatorname{Im} k > 0$. By means of Eqns (4.1.26a), (4.1.27), as $x \to \pm\infty$ we have, respectively,

$$\varphi(k, x) = a(k)e^{-ikx} = e^{-ikx + \log a(k)} \qquad (x \to +\infty),$$
$$\varphi(k, x) = e^{-ikx} \qquad (x \to -\infty). \qquad (4.2.2)$$

Write φ in the form

$$\varphi = e^{\theta(x)}. \qquad (4.2.3)$$

Then from Eq. (4.2.2) we have

$$\theta(+\infty) - \theta(-\infty) = -[ikx]_{-\infty}^{\infty} + \log a(k). \qquad (4.2.4)$$

Hence, in terms of σ introduced through the definition

$$\sigma \equiv \theta_{,x} + ik, \qquad (4.2.5)$$

Eq. (4.2.4) may be written

$$\int_{-\infty}^{\infty} \sigma \, dx = \log a(k), \qquad (4.2.6)$$

whilst φ is the solution of Eq. (4.1.7) for $\lambda = k^2$ $\operatorname{Im} k > 0$. Hence θ, and consequently σ, depend on u. In fact, introducing Eqns (4.2.3), (4.2.5) into Eq. (4.1.17), we find that σ satisfies the Riccati equation

$$\sigma_{,x} + \sigma^2 - u - 2ik\sigma = 0. \qquad (4.2.7)$$

On the other hand, from Eq. (4.1.16b), $a_{T,t} = 0$ and $a_T = 1/a$. That is, $a(k)_{,t} = 0$, and therefore σ is a conserved density.

Zakharov and Faddeev derived, in the first step of their procedure, the asymptotic expansions of σ and $\log a(k)$ for large $|k|$. They then, in the next step, substituted these expressions into Eq. (4.2.6) and compared both sides, thereby obtaining an infinite number of conserved quantities. In order to expand $\log a(k)$, the explicit form of $a(k)$ expressed in terms of the scattering data is required, which can be obtained as follows. We first notice that on the real axis (Im $k = 0$)

$$|a(k)|^2 = [1 - |r(k)|^2]^{-1}, \tag{4.2.8}$$

which follows from Eqns (4.1.28), (4.1.29).

We now use the following theorem:

Let $f(k)$ be continuous for Im $k \geq 0$ and analytic for Im $k > 0$. Then, if $f(k)$ does not have a zero for Im $k \geq 0$ and $k(f(k)-1)$ is bounded for Im $k \geq 0$,

$$f(k) = \exp\left(\frac{1}{2\pi i}\int_{-\infty}^{\infty}\frac{\log|f(q)|^2}{q-k}dq\right) \quad (\text{Im } k > 0).$$

For a proof, put $F(k) = \log f(k)$, so that $F(k)$ is continuous for Im $k \geq 0$ and analytic for Im $k > 0$, being estimated as $F(k) = 0(k^{-1})$. Then for the integral encircling the upper half-plane

$$\frac{1}{2\pi i}\int_{\partial}\frac{F(q)}{q-k}dq = \begin{cases} F(k) & \text{for Im } k > 0 \\ 0 & \text{for Im } k < 0, \end{cases}$$

whilst the integral on the left-hand side becomes, in either case Im $k \geq 0$,

$$\frac{1}{2\pi i}\int_{-\infty}^{\infty}\frac{F(q)}{q-k}dq.$$

Hence, for Im $k < 0$, taking the complex conjugate yields

$$\frac{1}{2\pi i}\int_{-\infty}^{\infty}\frac{F(q)^*}{q-k}dq = 0.$$

Therefore,

$$F(k) = \frac{1}{2\pi i}\int_{-\infty}^{\infty}\frac{F(q)}{q-k}dq + \frac{1}{2\pi i}\int_{-\infty}^{\infty}\frac{F(q)^*}{q-k}dq$$

$$= \frac{1}{2\pi i}\int_{-\infty}^{\infty}\frac{2\,\text{Re}\,F(q)}{q-k}dq. \qquad\qquad \text{(q.e.d.)}$$

Applying this theorem to the function $f(k)$ given by

$$a(k)\prod_{n=1}^{N}\frac{k+i\kappa_n}{k-i\kappa_n},$$

and taking note of Eq. (4.2.8), we obtain

$$a(k) = \exp\left(\frac{1}{2\pi i}\int_{-\infty}^{\infty}\frac{\log(1-|r(q)|^2)}{k-q}dq\right)\cdot\prod_{n=1}^{N}\frac{k-i\kappa_n}{k+i\kappa_n}. \tag{4.2.9}$$

4.2 CONSERVATION LAWS AND CANONICAL FORM

Consequently,

$$\log a(k) = \frac{1}{2\pi i} \int_{-\infty}^{\infty} \frac{\log(1-|r(q)|^2)}{k-q} dq + \sum_{n=1}^{N} \log(k-i\kappa_n)$$
$$- \sum_{n=1}^{N} \log(k+i\kappa_n). \quad (4.2.10)$$

We now expand $\log a(k)$ in powers of $1/k$ for large k. Since $|r(q)|^2$ is an even function of q, we have

$$\log a(k) = \sum_{m=1}^{\infty} \frac{C_m}{k^m},$$

$$C_{2j} = 0, \quad (4.2.11)$$

$$C_{2j+1} = \frac{1}{2\pi i} \int_{-\infty}^{\infty} k^{2j} \log(1-|r(k)|^2) dk - \frac{2}{2j+1} \sum_{n=1}^{N} (i\kappa_n)^{2j+1}.$$

On the other hand, to get the asymptotic expansion of σ for large k, σ is expanded in powers of $1/ik$, that is, we write

$$\sigma(x, k) = \sum_{m=1}^{\infty} \frac{\sigma_m(u)}{(2ik)^m}. \quad (4.2.12)$$

Introducing this equation into Eq. (4.2.7) yields the recurrence formula

$$\sigma_1(x) = -u(x),$$
$$\sigma_m(x) = \sigma_{m-1,x} + \sum_{l=1}^{m-2} \sigma_{m-l-1}(x)\sigma_l(x) \quad (m=2, 3, \ldots). \quad (4.2.13)$$

Hence, from Eqns (4.2.6) and (4.2.11) to (4.2.13), the conserved quantities are obtained as

$$C_{2j+1} = \frac{1}{(2i)^{2j+1}} \int_{-\infty}^{\infty} \sigma_{2j+1} dx. \quad (4.2.14)$$

For example, for $j = 0, 1, 2$ the conserved quantities become, respectively,

$$\int_{-\infty}^{\infty} u \, dx = -\frac{1}{\pi} \int_{-\infty}^{\infty} \log(1-|r(k)|^2) dk - 4 \sum_{m}^{N} \kappa_m \quad (4.2.15a)$$

$$\int_{-\infty}^{\infty} u^2 \, dx = -\frac{4}{\pi} \int_{-\infty}^{\infty} k^2 \log(1-|r(k)|^2) dk + \frac{16}{3} \sum_{m}^{N} \kappa_m^3 \quad (4.2.15b)$$

$$\int_{-\infty}^{\infty} \left(u^3 + \frac{u_{,x}^2}{2}\right) dx = -\frac{8}{\pi} \int_{-\infty}^{\infty} k^4 \log(1-|r(k)|^2) dk - \frac{32}{5} \sum_{m}^{N} \kappa_m^5. \quad (4.2.15c)$$

Thus the integrals of σ_{2j+1}, which are polynomials of u and its derivatives, are conserved quantities and there exist an infinite number of them. Also, the above expressions demonstrate that the conserved quantities comprise the sum of the discrete spectrum (κ_m), corresponding to solitons, and an integral over the continuous spectrum, corresponding to the non-soliton part, that is, to the wave train. This implies that each conserved quantity is separable in terms of a spectral decomposition of the scattering data. It is also to be noticed that, as was shown at the end of the last section, the wave train decays asymptotically as $t \to \infty$. Nevertheless, $|r(k)|^2$ is time-independent; that is, the contribution from the wave-train to the integral of the conserved quantities is constant and does not decrease in time.

In order to show this result more explicitly, and in another way, we first write the KdV equation in canonical form with the Hamiltonian given by the left-hand side of Eq. (4.2.15c); that is the conserved quantity $\int_{-\infty}^{\infty} \sigma_5 \, dx$. Let the Lagrangian density be given by

$$\mathscr{L} \equiv \tfrac{1}{2}\phi_{,x}\phi_{,t} - \phi_{,x}^3 - \tfrac{1}{2}(\phi_{,xx})^2, \tag{4.2.16}$$

and ϕ be the potential for u, so that

$$u = \phi_{,x}. \tag{4.2.17}$$

Then the Lagrange equation

$$\frac{\partial}{\partial t}\left(\frac{\partial \mathscr{L}}{\partial \phi_{,t}}\right) + \frac{\partial}{\partial x}\left(\frac{\partial \mathscr{L}}{\partial \phi_{,x}}\right) - \frac{\partial^2}{\partial x^2}\left(\frac{\partial \mathscr{L}}{\partial \phi_{,xx}}\right) = 0 \tag{4.2.18}$$

becomes the KdV equation (4.1.3). Hence, introducing the momentum density by

$$\pi \equiv \frac{\partial \mathscr{L}}{\partial \phi_{,t}} = \frac{\phi_{,x}}{2} = \frac{u}{2}, \tag{4.2.19}$$

we have the Hamiltonian density

$$\mathscr{H} \equiv \pi \phi_{,t} - \mathscr{L} = \phi_{,x}^3 + \tfrac{1}{2}(\phi_{,xx})^2 = u^3 + \tfrac{1}{2}(u_{,x})^2, \tag{4.2.20}$$

which is given in Eq. (4.2.15c). The KdV equation can be written in the canonical form

$$u_{,t} = \frac{\partial}{\partial x}\left(\frac{\delta \bar{\mathscr{H}}}{\delta u(x)}\right), \tag{4.2.21}$$

where the Hamiltonian $\bar{\mathscr{H}}$ is the integral of the Hamiltonian density $\mathscr{H}$ over x from $-\infty$ to $+\infty$, and $\delta\bar{\mathscr{H}}/\delta u(x)$ is the functional derivative

$$\frac{\delta \bar{\mathscr{H}}}{\delta u(x)} \equiv \lim_{\varepsilon \to 0} \frac{1}{\varepsilon} \int_{-\infty}^{\infty} \left[\mathscr{H}\left(u(x') + \varepsilon\delta(x'-x), u_{,x'} + \varepsilon\frac{\partial}{\partial x'}\delta(x'-x)\right) - \mathscr{H}(u(x'), u_{,x'}) \right] dx'. \tag{4.2.22}$$

4.2 CONSERVATION LAWS AND CANONICAL FORM

The Hamiltonian $\bar{\mathcal{H}}$ is, of course, equal to the conserved quantity given by the left-hand side of Eq. (4.2.15c). However, as is obvious from Eq. (4.2.19), the canonical variable ϕ and its momentum density $\pi(\equiv \phi_{,x}/2)$ are not independent. This is simply because of the fact that the KdV equation is first order in the time-derivative, so that u and $u_{,t}$ cannot be given independently. Hence, a direct application of the ordinary canonical formalism, which assumes ϕ and π to be independent canonical variables, encounters various ambiguities. For similar systems there exist canonical formalisms with subsidiary conditions. This method was used by Watanabe (1974) with the KdV equation. Here we employ a different method given by Gardner (1971).

For periodic boundary conditions we assume that the solutions of the KdV equation are periodic functions with period 2π, then u may be expanded as the Fourier series,

$$u = \sum_{n=-\infty}^{\infty} u_n e^{inx} \quad (n \text{ an integer}). \tag{4.2.23}$$

In general, let $\bar{F} = \int_0^{2\pi} F(u(x, \alpha)) \, dx$ be a functional of a function $u(x, \alpha)$ with variable x and parameter α. Then its functional derivative $\delta \bar{F}/\delta u$ is defined by

$$\frac{d}{d\alpha} \bar{F} \equiv \int_0^{2\pi} \frac{\delta \bar{F}}{\delta u(x)} \frac{\partial u}{\partial \alpha} \, dx. \tag{4.2.24}$$

It is readily seen that this is, in fact, consistent with Eq. (4.2.22). If F is chosen as the Hamiltonian density $\mathcal{H}$ in Eq. (4.2.20), the functional $\bar{\mathcal{H}}$ becomes a function of the variables u_n. Hence partial differentiation of $\bar{\mathcal{H}}$ with respect to u_n is given via Eqns (4.2.23), (4.2.24) as

$$\frac{\partial \bar{\mathcal{H}}}{\partial u_n} = \int_0^{2\pi} \frac{\delta \bar{\mathcal{H}}}{\delta u} e^{inx} \, dx. \tag{4.2.25}$$

Consequently, the functional derivative $\delta \bar{\mathcal{H}}/\delta u$ may be obtained through the inverse Fourier expansion

$$\frac{\delta \bar{\mathcal{H}}}{\delta u} = \frac{1}{2\pi} \sum_{n=-\infty}^{\infty} \frac{\partial \bar{\mathcal{H}}}{\partial u_{-n}} e^{inx}. \tag{4.2.26}$$

Substituting this expression into Eq. (4.2.21) yields

$$\frac{du_n}{dt} = \frac{i}{2\pi} n \frac{\partial \bar{\mathcal{H}}}{\partial u_{-n}}. \tag{4.2.27}$$

Therefore, if for $n > 0$ the canonical variables and the Hamiltonian are introduced by

$$p_n \equiv u_{-n}, \quad q_n \equiv \frac{u_n}{n}, \quad H \equiv \frac{i}{2\pi} \bar{\mathcal{H}}, \tag{4.2.28}$$

then Eq. (4.2.27) becomes the canonical equations

$$\frac{dq_n}{dt} = \frac{\partial H}{\partial p_n}, \quad \frac{dp_n}{dt} = -\frac{\partial H}{\partial q_n}. \tag{4.2.29}$$

Also, the Poisson bracket of the functionals $\bar{F}, \bar{G}$ of the canonical variables p_n, q_n is defined by

$$(\bar{F}, \bar{G}) \equiv \frac{i}{2\pi} \sum_{n=1}^{\infty} \left(\frac{\partial \bar{F}}{\partial q_n} \frac{\partial \bar{G}}{\partial p_n} - \frac{\partial \bar{F}}{\partial p_n} \frac{\partial \bar{G}}{\partial q_n} \right) \tag{4.2.30}$$

which, in view of Eq. (4.2.28), takes the form

$$(\bar{F}, \bar{G}) = \frac{i}{2\pi} \sum_{n=1}^{\infty} \left(n \frac{\partial \bar{F}}{\partial u_n} \frac{\partial \bar{G}}{\partial u_{-n}} - n \frac{\partial \bar{F}}{\partial u_{-n}} \frac{\partial \bar{G}}{\partial u_n} \right)$$

$$= \frac{i}{2\pi} \sum_{n=-\infty}^{\infty} \left(n \frac{\partial \bar{F}}{\partial u_n} \frac{\partial \bar{G}}{\partial u_{-n}} \right).$$

Moreover, by means of Eq. (4.2.26), we finally obtain

$$(\bar{F}, \bar{G}) = \int_0^{2\pi} \frac{\delta \bar{F}}{\delta u} \frac{\partial}{\partial x} \left(\frac{\delta \bar{G}}{\delta u} \right) dx. \tag{4.2.31}$$

In particular, if $\bar{F}, \bar{G}$ are taken to be p_n, q_n, respectively, the well-known relations for the canonical conjugate variables

$$(p_n, q_{n'}) = -\frac{i}{2\pi} \delta_{nn'}, \quad (p_n, p_{n'}) = (q_n, q_{n'}) = 0 \tag{4.2.32}$$

are obtained.

So far it has been shown that for the periodic solutions the canonical equation (4.2.29) is valid for the canonical variables $p_n(=u_{-n})$, $q_n(=u_n/n)$. Hence, it will be possible to introduce similarly the canonical variables for the solutions which decrease rapidly at $|x|=\infty$. Since, in this case, we have seen that the solutions are given by the scattering data, it seems plausible to make use of the scattering data corresponding to u_n, u_{-n} which are the coefficients of the Fourier expansion. Then we can consider introducing the canonical variables given by the scattering data corresponding to p_n, q_n. For this approach the canonical variables, which are to be expressed in terms of $(k, r(k))$ (c_m, κ_m), are introduced by writing

$$\begin{aligned} P(k) &= P(k, r(k)) \quad (-\infty < k < \infty), \\ p_m &= p_m(\kappa_m, c_m) \quad (m = 1, 2, \ldots, N). \end{aligned} \tag{4.2.33}$$

Then, for the given Hamiltonian $\mathcal{H}$, we will find the canonical variables $Q(k)$, q_m conjugate to $P(k)$, p_m, respectively. Since, by Eqns (4.2.20),

4.2 CONSERVATION LAWS AND CANONICAL FORM

(4.2.15c), the Hamiltonian $\bar{\mathcal{H}}$ takes the form

$$\bar{\mathcal{H}} = \int_{-\infty}^{\infty} \left(u^3 + \frac{1}{2} u_{,x}^2 \right) dx$$

$$= -\left[\frac{8}{\pi} \int_{-\infty}^{\infty} k^4 \log(1-|r(k)|^2) \, dk + \frac{32}{5} \sum_{m=1}^{N} \kappa_m^5 \right], \quad (4.2.34a)$$

$P(k), p_m$ may be assumed to be

$$P(k) = \alpha_k \log(1-|r(k)|^2), \qquad p_m = \kappa_m^{\alpha_m}, \quad (4.2.35)$$

where α_k, α_m are constants to be determined in due course. Since $|r(k)|$, κ_m are constants of motion, $P(k), p_m$ are also constants of motion, and the Hamiltonian $\bar{\mathcal{H}}$ is given by a function of $P(k), p_m$ only. Hence, the $Q(k), q_m$ which are canonically conjugate to $P(k), p_m$ can be obtained easily by integrating the canonical equation

$$\frac{dQ(k)}{dt} = \frac{\delta \bar{\mathcal{H}}}{\delta P(k)}, \qquad \frac{dq_m}{dt} = \frac{\delta \bar{\mathcal{H}}}{\delta p_m}. \quad (4.2.36)$$

That is,

$$Q(k) = -\frac{8}{\pi} \frac{k^4}{\alpha_k} t + Q(k, 0),$$

$$q_m = -\frac{32 \kappa_m^{5-\alpha_m} t}{\alpha_m} + q_m(0). \quad (4.2.37)$$

On the other hand, from Eqns (4.1.16a), (4.1.14b) it follows that

$$8k^3 t = [\arg b(k,t) - \arg b(k,0)],$$
$$4\kappa_m^3 t = \log c_m(t) - \log c_m(0). \quad (4.2.38)$$

Therefore, putting

$$\alpha_k = -\frac{k}{\pi}, \qquad \alpha_m = 2, \quad (4.2.39)$$

we obtain the canonically conjugate variables given by the scattering data

$$P(k) = -\frac{k}{\pi} \log(1-|r(k)|^2), \quad (4.2.40a)$$

$$Q(k,t) = \arg b(k,t), \qquad Q(k,0) = \arg b(k,0), \quad (4.2.40b)$$

$$p_m = \kappa_m^2, \qquad q_m(t) = -4 \log c_m(t), \qquad q_m(0) = -4 \log c_m(0), \quad (4.2.40c)$$

and the Hamiltonian $\bar{\mathcal{H}}$ becomes

$$\bar{\mathcal{H}} = 8 \int_{-\infty}^{\infty} k^3 P(k) \, dk - \frac{32}{5} \sum_{m=1}^{N} p_m^{5/2}, \quad (4.2.34b)$$

which is given by the part consisting of the momentum p_m, corresponding to solitons, and the non-soliton part given by $P(k)$, corresponding to the wave-train. Thus the Hamiltonian is separable, and it demonstrates that the KdV equation is a completely integrable system for the canonical variables P, p_m and Q, q_m which are the action and angle variables, respectively. The canonical equations may be summarized as follows:

$$\frac{dP}{dt} = -\frac{\delta H}{\delta Q} = 0, \qquad \frac{dp_m}{dt} = -\frac{\delta H}{\delta q_m} = 0,$$
$$\frac{dQ}{dt} = \frac{\delta H}{\delta P} = 8k^3, \qquad \frac{dq_m}{dt} = \frac{\delta H}{\delta p_m} = -16p_m^{3/2}.$$
(4.2.41)

If Eq. (4.2.40) is used in Eq. (4.2.21), we arrive at the equations to determine the evolution of the scattering data which were obtained in the previous section. Here we derive in an heuristic way Eqns (4.2.34b), (4.2.40). For a systematic derivation we refer to the work of Zakharov and Faddeev (1971).

4.3 Periodic solutions and lattice dynamics

In Section 4.1 it has been shown that the initial value problem for the KdV equation can be solved by means of the inverse scattering method. There, by means of the assumption that the solutions to the KdV equation decay sufficiently rapidly as $|x| \to \infty$, the evolution of the scattering data, for example, Eqns (4.1.14) to (4.1.16) were obtained. Hence under periodic boundary conditions the method does not apply. However, as is obvious from Eq. (4.1.8), the eigenvalues λ of the Schrödinger equation (4.1.7) are time-independent constants for periodic boundary conditions. Since the eigenvalues correspond to the amplitudes and propagation velocities of solitons, even under periodic boundary condition the properties of solitons can be observed, provided the system length (the distance between the boundaries) is sufficiently large in comparison with the widths of the solitons. In this sense, a system of infinite length $|x| \to \infty$ can be well approximated by periodic boundary conditions. However, in such a system, after a sufficiently long time, characteristics of the periodicity can appear. In particular, the recurrence of the solution is one of the most interesting phenomena due to the periodicity of the boundary conditions.

As was already shown in Section 3.4, this was also discovered numerically by Zabusky and Kruskal (see Fig. 3.2). In order to explain this by means of a qualitative argument, we consider the initial condition $u_0 = \cos \pi x$ (the period $L = 2$).

4.3 PERIODIC SOLUTIONS AND LATTICE DYNAMICS

Then the eigenvalues of Eq. (4.1.7) can be estimated by expanding the potential u_0 about its lowest state by writing

$$-u_0 \simeq -1 + \frac{\pi^2}{2} x^2 + \cdots, \tag{4.3.1}$$

to give

$$\lambda_n \simeq -1 + \sqrt{2}\,(n + \tfrac{1}{2})\pi \qquad (n = 1, 2, 3, \ldots). \tag{4.3.2}$$

Consequently, it will be supposed that the solution breaks up into a number of solitons corresponding to these eigenvalues. In fact, the numerical results of Zabusky and Kruskal obtained for periodic boundary conditions show eight solitons corresponding to these eigenvalues. Those solitons repeat their mutual collisions and then, at the time $t_R/2$ all the solitons coalesce at two separate places. Finally, at the time t_R, the solution $u(x, t_R)$ returns to the initial condition $u_0(x, 0)$. In view of Eq. (4.3.2), the velocity difference betwen two neighbouring solitons is the same, and is equal to

$$\Delta v = 4(\lambda_{n+1} - \lambda_n) = 4\sqrt{2}\,\pi. \tag{4.3.3}$$

Since the velocity of a soliton does not change as the result of a collision, under periodic boundary conditions the relative distance between two solitons with velocity difference Δv becomes equal at every time T equal to $L/\Delta v$, provided the phase-shift given by Eq. (4.1.59c) due to the collision is neglected. Therefore the recurrence time t_R may be estimated roughly as $t_R \sim L/\Delta v = 1/(2\sqrt{2}\pi)$. As can be seen in Fig. 3.2, in the numerical computation of Zabusky and Kruskal, the collisions take place frequently, so that the phase-shifts due to the collisions are not negligible. Nevertheless, in order of magnitude, t_R is in agreement with the above estimate.

This recurrence phenomenon has been observed in experiments. Here we cite the experimental result obtained by Ikezi (1973) for an ion wave in a plasma. As was shown in the Example 3.4.3, the propagation of a weakly nonlinear ion wave of long wavelength can be described by the KdV equation. In the experiment a wave with frequency ω_0 is steadily excited at a point and the spatial variation of the wave is observed. The results are shown in Fig. 4.3(a), (b).

Figure 4.3(a) demonstrates that the wave steepens as it propagates, and breaks up into two solitons at a distance 5 cm from the excitor, and again becomes a sinusoidal wave at a distance 9 cm. Figure 4.3(b) shows the spatial variation of the amplitudes of the primary excited wave and the second and third harmonic waves. Near the excitor, the excited wave damps and the second harmonic wave grows, whilst from the 5 cm point

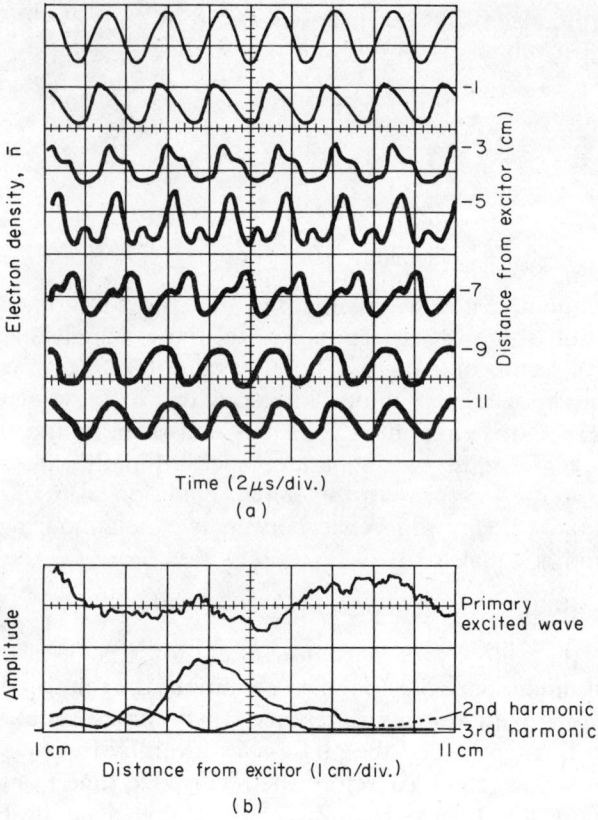

Fig. 4.3 (a) The time evolution of the electron density at distances 1 cm, 2 cm, ..., 11 cm from the excitation point; (b) The spatial variation of the amplitudes of the excited wave, the second and third harmonics (Ikezi (1973)).

the excited wave with frequency ω_0 grows and the second harmonic wave damps. At the 9 cm point the second harmonic wave entirely disappears and the recurrence is observed.

In this experiment the nonlinear effect is not strong in comparison with the dispersive effect, and consequently the wave with frequency ω_0 and the second harmonic wave are dominant. Hence the distance of the recurrence point may be estimated by considering those two waves only. Since the experiment shows the spatial development of waves, the KdV equation given in Section 3.4.3 governing the time evolution is transformed into one for the spatial development and, furthermore, to the laboratory system, which then takes the form

$$n_{,x} + n_{,t} - nn_{,t} - \tfrac{1}{2} n_{,ttt} = 0. \tag{4.3.4}$$

4.3 PERIODIC SOLUTIONS AND LATTICE DYNAMICS

Let the excited wave be given by

$$n_1(t, x) = a_1 e^{i(k_0 x - \omega_0 t)}, \tag{4.3.5}$$

where k_0, ω_0 are the wave number and the frequency which satisfy the linear dispersion relation of Eq. (4.3.4); that is, $k_0 = \omega_0 + (1/2)\omega_0^3$. Representing the second harmonic wave by

$$n_2(t, x) = a_2(x) e^{-i2\omega_0 t}, \tag{4.3.6}$$

and introducing these expressions into Eq. (4.3.4), we have the equation for $a_2(x)$:

$$a_{2,x} - ik_2 a_2 + i\omega_0 a_1^2 e^{i2k_0 x} = 0, \tag{4.3.7}$$

where k_2 is given by the linear dispersion relation for $2\omega_0$; namely, by $k_2 = 2\omega_0 + (1/2)(2\omega_0)^3$. By means of the boundary condition $a_2(0) = 0$, Eq. (4.3.7) may be integrated to give

$$a_2(x) = \frac{a_1^2}{3\omega_0^2} [e^{i2k_0 x} - e^{ik_2 x}], \tag{4.3.8}$$

which shows that the second harmonic wave consists of the forced oscillating term n_1 (the first term) and the second term satisfying the dispersion relation. Therefore, these two waves give rise to a spatial beat, so that at the point

$$x_R = l \frac{2\pi}{3\omega_0^3} \quad (l = 1, 2, 3, \ldots) \tag{4.3.9}$$

a_2 vanishes; that is $a_2(x_R) = 0$. The recurrence distance x_R thus obtained is approximately equal to the experimentally observed one.

We have mentioned two examples of the recurrence which can occur in a solution of the KdV equation. Historically, however, the recurrence was first discovered by Fermi *et al.* (1955) in a numerical computation for a model of a nonlinear string. This has been called the FPU paradox. As is well-known, a model for the small vibration of a string is provided by a one-dimensional system of N particles, linearly interacting with neighbouring ones, and executing small oscillations about the equilibrium points. In such a linear system the motion is given by the superpositions of eigen-oscillations, and there is no energy exchange amongst the eigen-oscillations; consequently the energy of each mode does not change with time.

However, as was shown in Section 3.8, in the presence of a nonlinear effect, there arise interactions amongst the eigen-modes and energy exchange then takes place. In view of this idea it has been thought that if nonlinear interaction exists, then starting from any initial condition the energy will eventually become equally distributed amongst all modes, so that the system will approach thermal equilibrium. In other words ergodicity will be achieved by means of the nonlinear effects. In order to examine the validity of this conjecture Fermi, Pasta and Ulam used a system with nonlinear forces proportional to the square or the cube of the displacement; that is, they integrated numerically the following coupled system of 64 equations:

$$\frac{d^2 y_n}{dt^2} = (y_{n+1} - 2y_n + y_{n-1}) + c[(y_{n+1} - y_n)^\alpha - (y_n - y_{n-1})^\alpha] \quad (\alpha = 2 \text{ or } 3), \tag{4.3.10}$$

in which y_n ($n = 1, 2, \ldots, 64$) is the displacement from the equilibrium point and y_1 and y_{64} are fixed. As a result they found a remarkable recurrence: if, initially, the energy is given to the eigen-mode with the lowest frequency (in the linear approximation), then several higher modes become excited by the nonlinear effects, but after a suitable lapse of time during which mode–mode interactions take place, almost all the energy returns to the mode comprising the initial state. That is, the system does not approach thermal equilibrium by means of the equi-partition of energy due to the nonlinear effects, but recurrence is observed and, consequently, the non-ergodicity of the system is demonstrated. The time variation of the energy in modes 1, 3, 5, and 7 is shown in Fig. 4.4.

In connection with the ergodicity of nonlinear systems, and to understand mathematically the FPU paradox, Toda (1967, 1980) proposed an exactly solvable model of a nonlinear lattice, which is given by the exponential potential

$$\phi(r) = \frac{a}{b} e^{-br} + ar \quad (ab > 0), \tag{4.3.11}$$

in which r is the displacement from the equilibrium point. If $a, b > 0$, for $r > 0$ the first term gives the repulsive force and the second the attractive force, whilst for $r < 0$ the first term gives the strong attractive force and the second term the repulsive force (Fig. 4.5). If $a, b < 0$, the same result holds for $r < 0$ and $r > 0$, respectively.

In the limit $b \to 0$, while ab is kept constant,

$$(a/b)e^{-br} + ar - (a/b) = (ab/2)\{r^2 - (b/3)r^3 + \cdots\},$$

and hence it becomes a harmonic lattice with the lattice constant $k = ab$. (Eq. (4.3.10) used by Fermi, Pasta, and Ulam corresponds to the case $c = -b/2$.) Also, in the limit $b \to \infty$, it reduces to the so-called potential with a hard shell. Under the potential (4.3.11), the equation of

4.3 PERIODIC SOLUTIONS AND LATTICE DYNAMICS

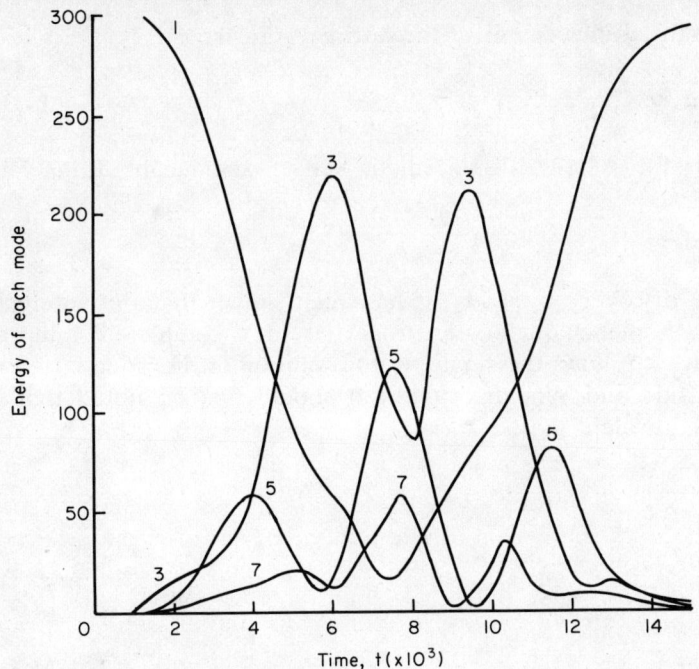

Fig. 4.4 The time variation of the energy of each mode (1, 3, 5, 7) (Fermi *et al.* (1955)).

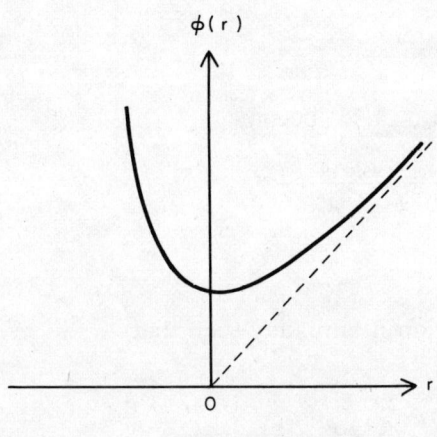

Fig. 4.5

motion for the displacement of the lattice y_n is given by

$$\frac{d^2 r_n}{dt^2} = a(2e^{-br_n} - e^{-br_{n-1}} - e^{-br_{n+1}}), \qquad (4.3.12)$$

where r_n is the relative displacement betwen the neighbouring lattice points, that is,

$$r_n = y_n - y_{n-1}.$$

Since the discovery by Toda, it has been shown that this interesting model has all the characteristic properties of a completely integrable system. In what follows the single soliton solution of the Toda lattice will be given (Toda and Wadati (1973)). It should first be noted that Eq. (4.3.12) is equivalent to

$$\log\left(1 + \frac{\ddot{s}_n}{a}\right) = b(s_{n-1} - 2s_n + s_{n+1}), \qquad (4.3.13)$$

where r_n and s_n are related by

$$r_n = -(s_{n-1} + s_{n+1} - 2s_n), \qquad (4.3.14)$$

$$\exp(-br_n) - 1 = \frac{\ddot{s}_n}{a}. \qquad (4.3.15)$$

The left-hand side of Eq. (4.3.15) is equivalent to the force f_n between the neighbouring lattice points, namely,

$$f_n = -\frac{1}{a}\frac{d\phi}{dr_n} = \exp(-br_n) - 1. \qquad (4.3.16)$$

Hence, in terms of the function φ_n introduced through the expression

$$s_n = \frac{1}{b}\log \varphi_n, \qquad (4.3.17)$$

Eqns. (4.3.13) and (4.3.15) become

$$1 + \frac{1}{ab}\frac{\ddot{\varphi}_n \varphi_n - \dot{\varphi}_n^2}{\varphi_n^2} = \frac{\varphi_{n+1}\varphi_{n-1}}{\varphi_n^2}, \qquad (4.3.18)$$

$$f_n = \frac{1}{ab}\frac{\ddot{\varphi}_n \varphi_n - \dot{\varphi}_n^2}{\varphi_n^2}. \qquad (4.3.19)$$

By direct substitution it is readily seen that

$$\varphi_n = 1 + Ae^{\kappa n - \beta t},$$

$$\beta^2 = 4ab \sinh^2\frac{\kappa}{2} \qquad (A, \kappa > 0), \qquad (4.3.20)$$

4.3 PERIODIC SOLUTIONS AND LATTICE DYNAMICS

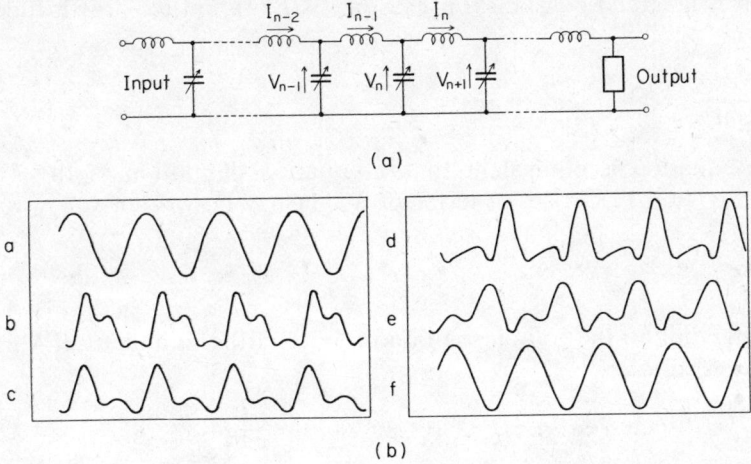

Fig. 4.6 (b) Voltage change at various points of the nonlinear circuit shown in (a) (Hirota and Suzuki (1970)).

satisfies Eq. (4.3.18). Substituting the above solution for φ_n in Eq. (4.3.19), we obtain the soliton solution of the Toda lattice

$$f_n = e^{-br_n} - 1 = \frac{\beta^2}{4ab} \operatorname{sech}^2 \left[\tfrac{1}{2}(\kappa n - \beta t + \delta)\right] \qquad (\delta \equiv \log a). \qquad (4.3.21)$$

The recurrence of the solution of the Toda lattice was confirmed by Hirota and Suzuki (1970) using a nonlinear circuit, which is shown in Fig. 4.6(a). The capacity of the condenser varies with the applied voltage, and is given by

$$C(V) = C_0 V_0 V^{-1} \log(1+V). \qquad (4.3.22)$$

Hereafter, for simplicity, we set $C_0 = V_0 = 1$. Consequently, when the voltage applied to the condenser C_n is denoted by V_n, the accumulated charge Q_n is

$$Q_n = \pm \log(1+V_n). \qquad (4.3.23)$$

The transmission equations for the circuit are

$$L \frac{\partial I_n(t)}{\partial t} = V_n(t) - V_{n+1}(t), \qquad (4.3.24a)$$

$$\frac{\partial Q_n(t)}{\partial t} = I_{n-1}(t) - I_n(t), \qquad (4.3.24b)$$

where I_n is the current flowing through the coil, and the inductance L of the coil is assumed constant. From Eqns (4.3.23) and (4.3.24) it follows that

$$L\frac{\partial^2 Q_n}{\partial t^2} = e^{Q_{n+1}} - 2e^{Q_n} + e^{Q_{n-1}}. \tag{4.3.25}$$

This equation is equivalent to the equation of motion of the Toda lattice Eq. (4.3.12). Also, in terms of V_n, it may be written

$$L\frac{\partial^2}{\partial t^2}\log(1+V_n) = V_{n+1} - 2V_n + V_{n-1}. \tag{4.3.26}$$

Corresponding to the soliton solution (4.3.21) of the Toda lattice, V_n has the soliton solution

$$V_n = \beta^2 \operatorname{sech}^2(\kappa n - \beta\tau), \qquad \beta = \sinh\kappa, \ \tau = \frac{t}{L^{1/2}}. \tag{4.3.27}$$

Figure 4.5(b) shows the waveform of the voltage observed at various points of the circuit and displays the recurrence phenomenon.

Finally, we show that in the continuous limit, Eq. (4.3.12) reduces to the Boussinesq equation, and for a wave propagating in one direction it further reduces to the KdV equation. Let $r(n, t) = r_n(t)$, then Eq. (4.3.12) becomes

$$\frac{\partial^2 r}{\partial t^2} = 2a\left[1 - \cosh\left(\frac{\partial}{\partial n}\right)\right]e^{-br}, \tag{4.3.28}$$

where

$$\exp\left(\pm\frac{\partial}{\partial n}\right)r(n) = r(n \pm 1). \tag{4.3.29}$$

Expanding the exponential part and retaining terms up to and including the order $\partial^4(br)/\partial n^4$, $\partial^2(br)^2/\partial n^2$, we see that Eq. (4.3.28) may be approximated by

$$\frac{\partial^2 r}{\partial t^2} = a\left[\frac{\partial^2(br)}{\partial n^2} + \frac{1}{12}\frac{\partial^4(br)}{\partial n^4} - \frac{1}{2}\frac{\partial^2(br)^2}{\partial n^2}\right]. \tag{4.3.30}$$

Then, introducing the transformation

$$t' = (12ab)^{1/2}t, \qquad x = 12^{1/2}n, \qquad u = -\frac{br}{12}, \tag{4.3.31}$$

and writing t' in place of t, we obtain the Boussinesq equation

$$u_{,tt} - u_{,xx} - 6(u^2)_{,xx} - u_{,xxxx} = 0. \tag{4.3.32}$$

As can be seen from the soliton solution (4.3.21) of Eq. (4.3.19), the Boussinesq equation admits the soliton solution

$$u(x, t) = \tfrac{1}{4}\kappa^2 \operatorname{sech}^2[\tfrac{1}{2}(\kappa x - \beta t + \delta)], \qquad \beta^2 = \kappa^2 + \kappa^4. \tag{4.3.33}$$

However, it should be noticed that for a plane wave of infinitesimal amplitude the Boussinesq equation yields the dispersion relation

$$\omega^2 = k^2 - k^4.$$

Consequently, the plane wave grows for large k, and therefore the above reduction to the Boussinesq equation is valid only for small k and not for large k.

For long wavelength waves, the Boussinesq equation can be reduced further to the KdV equation. That is, if we use the transformation

$$u \to \varepsilon u, \qquad \xi = \varepsilon^{1/2}(x \pm t), \qquad \tau = \varepsilon^{3/2} t, \tag{4.3.34}$$

for a wave propagating in one direction, it is approximated by the KdV equation

$$\pm u_{,\tau} - 6 u u_{,\xi} - \tfrac{1}{2} u_{,\xi\xi\xi} = 0. \tag{4.3.35}$$

As was already mentioned, the Toda lattice can approximate the FPU model, and therefore we can assert that the recurrence phenomenon presented as the FPU paradox is closely related to that of the KdV equation.

The mathematical theory of the KdV equation under periodic boundary condition has been studied extensively in recent years (Novikov (1974)) by means of an approach based on the inverse spectrum method. The same method of solution also applies to the Toda Lattice (Toda (1980)). However, these approaches require mathematical techniques which are beyond the level of the present book. We merely note here that Hirota's direct method can apply also to the periodic case.

4.4 The theory of the Lax pair

In Section 4.1, the equivalence of a nonlinear evolution equation and a set of linear equations was obtained on an *ad hoc* basis for the KdV equation. This relationship for the KdV equation was formulated by Lax in a very general theory. Before proceeding to the general theory, we first consider the special case of Section 4.1 in more detail.

As is well-known in quantum mechanics, the eigenvalues of the Schrödinger equation

$$L\psi \equiv \left(-\frac{\partial^2}{\partial x^2} + u\right)\psi = \lambda\psi \tag{4.4.1}$$

are invariant under a unitary transformation. Namely, let U be the unitary operator so that $UU^\dagger = 1$, where $U^\dagger$ is the adjoint of U, then multiplying Eq. (4.4.1) by U from the left gives

$$ULU^{-1}U\psi = \lambda U\psi. \tag{4.4.2}$$

Since $(ULU^{-1})^\dagger = UL^\dagger U^{-1}$, by means of the self-adjointness of the Schrödinger operator L $(L = L^\dagger)$, ULU^{-1} is also self-adjoint and the wave function $\psi' = U\psi$ becomes the eigenfunction of the transformed operator $L' \equiv ULU^\dagger$ for the same eigenvalue λ as for L. Namely, if the Schrödinger operator is given, the spectrum (that is, the set of eigenvalues) is determined uniquely; however, systems transformed by unitary transformations have the same eigenvalues, so these systems are said to be unitarily equivalent.

We now notice that in Eq. (4.4.1), when $u(x, 0)$ is given at $t = 0$, the eigenvalues λ do not change for $u(x, t)$, $t > 0$. Hence we see that for the parameter t, the systems given by $L(t) = -\partial^2/\partial x^2 + u(x, t)$ are unitarily equivalent for all t, that is

$$L(t) = U(t)L(0)U^\dagger(t), \tag{4.4.3a}$$

$$\psi(t) = U(t)\psi(0). \tag{4.4.3b}$$

The time-evolution of ψ is determined by that of u, and consequently by the KdV equation (4.1.3). This is, of course, given by Eq. (4.1.10). Since the transformation connecting $\psi(0)$ to $\psi(t)$ is Eq. (4.4.3b), which is unitary, the norm of ψ is conserved. Hence, if we write the equation for the time-evolution of ψ as

$$i\psi_{,t} = A\psi, \tag{4.4.4}$$

as is obvious from the analogy with quantum mechanics, A must be self-adjoint. On the other hand $\psi_{,t} = U_{,t}\psi(0)$ and, consequently, A and U are related by

$$iU_{,t} = AU, \tag{4.4.5}$$

which determines the time-development of U. Therefore the evolution of L in time is obtained as follows:

$$L_{,t} = U_{,t}L(0)U^\dagger + UL(0)U^\dagger_{,t} = -i[A, L], \tag{4.4.6}$$

which corresponds to a transformation to the Heisenberg representation from the Schrödinger representation in quantum mechanics.

These results were formulated by Lax in terms of the following general theorem.

Theorem Let a nonlinear evolution equation be given by

$$u_{,t} = K(u), \tag{4.4.7}$$

4.4 THE THEORY OF THE LAX PAIR

in which K is a nonlinear operator acting on a scalar function u. Suppose that a pair of self-adjoint linear operators L and A (the Lax pair) subject to solutions $u(x, t)$ of Eq. (4.4.7) exist and satisfy the operator equation

$$iL_{,t} = [A, L] = AL - LA. \tag{4.4.8}$$

Then, the eigenvalues λ of the operator L given by

$$L\psi = \lambda\psi \tag{4.4.9}$$

are time-independent. Also, the time-evolution of the eigenfunction ψ is determined by

$$i\psi_{,t} = A\psi. \tag{4.4.10}$$

Proof At $t = 0$, λ and $\psi(0)$ are determined by

$$L(0)\psi(0) = \lambda\psi(0), \tag{4.4.9'}$$

and the operator U introduced by

$$iU_{,t} = AU, \qquad U(0) = I, \tag{4.4.11}$$

is unitary. The operator $L(t)$ at $t > 0$ is then obtained by means of the unitary transformation

$$L(t) = U(t)L(0)U^{\dagger}(t). \tag{4.4.12}$$

This is readily seen from Eq. (4.4.11) which makes it possible to express $L(t)_{,t}$ in terms of A, so that Eq. (4.4.8) is valid. Since Eq. (4.4.12) yields $L(t)U(t) = U(t)L(0)$, multiplying Eq. (4.4.9') by $U(t)$ from the left, we have

$$L(t)U(t)\psi(0) = \lambda U(t)\psi(0). \tag{4.4.13}$$

Therefore, the eigenvalue of $L(t)$ is independent of t, and the eigenfunction of $L(t)$, namely $\psi(t)$, is given by

$$\psi(t) = U(t)\psi(0). \tag{4.4.14}$$

Differentiating Eq. (4.4.14) with respect to t and using Eq. (4.4.11), we find Eq. (4.4.10) which is the evolution equation for ψ. This completes our proof.

By way of example, if L is the Schrödinger operator we have

$$L_{,t} = u_{,t}. \tag{4.4.15}$$

Hence Eq. (4.4.8) requires that A be such that the commutator $[A, L]$ must be $iu_{,t}$, namely a factor which is to multiply an ordinary function without involving a differential operator. Such an operator is given by

$$A = iD, \qquad D \equiv \frac{\partial}{\partial x}. \tag{4.4.16}$$

Since D and D^2 commute,

$$[A, L] = [iD, u] = iu_{,x} \tag{4.4.17}$$

and Eq. (4.4.8) becomes

$$u_{,t} = u_{,x}. \tag{4.4.18}$$

That is, if u changes subject to Eq. (4.4.18), the Schrödinger operator $L(t) \equiv -\partial^2/\partial x^2 + u(x, t)$ is unitarily equivalent. This conclusion is obvious from the fact that the general solution to Eq. (4.4.18) is $u = f(x+t)$.

As an example of a non-trivial case we consider

$$A = -4i(D^3 + bD + Db), \tag{4.4.19}$$

where b will be determined later. The commutation relation with L is

$$[A, L] = -4i[3u_{,x}D^2 + 3u_{,xx}D + u_{,xxx} + 4b_{,x}D^2 \\ + 4b_{,xx}D + b_{,xxx} + 2bu_{,x}]. \tag{4.4.20}$$

Therefore, if

$$b = -\tfrac{3}{4}u, \tag{4.4.21}$$

the coefficients of D and D^2 vanish, so that $[A, L]$ becomes multiplication. Hence, if u evolves subject to

$$u_{,t} = -u_{,xxx} + 6uu_{,x}, \tag{4.4.22}$$

the operator $L(t)$ is unitarily equivalent and this equation is just the KdV equation.

The results obtained so far are valid even if U is not unitary provided U^{-1} exists. That is, Eq. (4.4.13) is valid and λ is time-independent provided $L(t)$ is transformed by the rule $L(t) = U(t)L(0)U^{-1}(t)$. In this case, L and A are, of course, not self-adjoint. (For the case of the nonlinear Schrödinger equation which will be discussed in the next section, L is not self-adjoint.)

4.5 Application of the inverse scattering method to the nonlinear Schrödinger equation

In Section 3.6 it was shown that the nonlinear Schrödinger equation is also an important equation with wide applications. We also know that solutions of this equation are remarkably different for the two cases that arise when the product of the group velocity dispersion ω_0'' and the coefficient of the nonlinear term Q is either positive or negative. Here, we consider the following nonlinear Schrödinger equation which corres-

4.5 THE NONLINEAR SCHRÖDINGER EQUATION

ponds to the first case when the product is positive; namely

$$iu_{,t} + u_{,xx} + Q|u|^2 u = 0 \quad (Q>2). \tag{4.5.1}$$

The envelope-soliton solution of this equation is given by Eq. (3.6.29), that is,

$$u(x,t) = 2\left(\frac{2}{Q}\right)^{1/2} \kappa \operatorname{sech}\left[2\kappa(x-x_0) + 8\iota\kappa t\right] \tag{4.5.2}$$
$$\times \exp\left[-4i(\iota^2 - \kappa^2)t - i\iota x + i\theta\right].$$

In contrast to the KdV soliton, this soliton involves two independent parameters, so that the amplitude (and the width) and the velocity can be given independently. It was shown by Zakharov and Shabat (1972) that Eq. (4.5.1) can be solved exactly by the inverse scattering method, if $|u|$ decays sufficiently rapidly as $|x| \to \infty$. They made the following choice for the Lax pair,

$$L = i\begin{bmatrix} 1+\beta & 0 \\ 0 & 1-\beta \end{bmatrix}\frac{\partial}{\partial x} + \begin{bmatrix} 0 & u^* \\ u & 0 \end{bmatrix}, \tag{4.5.3}$$

$$A = -\beta\begin{bmatrix} 1 & 0 \\ 0 & 1 \end{bmatrix}\frac{\partial^2}{\partial x^2} + \begin{bmatrix} |u|^2/(1+\beta) & iu_{,x}^* \\ -iu_{,x} & -|u|^2/(1-\beta) \end{bmatrix}, \quad \beta^2 = 1 - \frac{2}{Q}. \tag{4.5.4}$$

The operators L and A obviously satisfy the operator equation (4.4.8). Hence the two equations for the ψ with two components

$$L\psi = \lambda\psi, \qquad \psi = \begin{pmatrix} \psi_1 \\ \psi_2 \end{pmatrix}, \tag{4.5.5a}$$

$$i\psi_{,t} = A\psi, \tag{4.5.5b}$$

are required. (Equation (4.5.5a) is the eigenvalue equation for ψ with two components, and consequently is sometimes said to be of Dirac-type.) By virtue of this set of equations the eigenvalues λ of L are invariant, and the evolution of u is determined by the inverse scattering method in the same way as for the KdV equation. However, in order to follow the inverse scattering procedure in a way closely parallel to that in the case of the KdV equation, Zakharov and Shabat introduced a function $\tilde{\psi}$ by setting

$$\psi = S \exp(-i\tilde{\lambda}x)\tilde{\psi},$$
$$S = \begin{pmatrix} 0 & (1-\beta)^{1/2} \\ (1+\beta)^{1/2} & 0 \end{pmatrix}, \qquad \tilde{\lambda} = \frac{\lambda}{1-\beta^2} = \frac{Q}{2}\lambda, \tag{4.5.6}$$

and transformed the eigenvalue equation (4.5.5a) to

$$\tilde{\psi}_{1,x} - v\tilde{\psi}_2 = -ik\tilde{\psi}_1, \tag{4.5.7a}$$

$$\tilde{\psi}_{2,x} + v^*\tilde{\psi}_1 = ik\tilde{\psi}_2, \tag{4.5.7b}$$

$$v = \frac{iu}{(1-\beta^2)^{1/2}} = i\left(\frac{Q}{2}\right)^{1/2} u, \quad k = \tilde{\lambda}\beta. \tag{4.5.7c}$$

Here the matrix S is not unitary, and hence the set of equations (4.5.7) does not give a self-adjoint operator. For example, if the equations are written as

$$i\sigma_3\tilde{\psi}_{,x} + P\tilde{\psi} = k\tilde{\psi},$$

$$\sigma_3 = \begin{pmatrix} 1 & 0 \\ 0 & -1 \end{pmatrix}, \quad P = \begin{pmatrix} 0 & -iv \\ -iv^* & 0 \end{pmatrix}, \tag{4.5.8}$$

P is not Hermitian, and consequently the eigenvalues k are complex. Nevertheless the inverse scattering method of Eqns (4.5.7) closely parallels that of the Schrödinger equation. In what follows, ψ will be written in place of $\tilde{\psi}$, unless otherwise stated. Let the eigenfunctions for the eigenvalues k_a and k_b be denoted by $\psi^{(a)}$ and $\psi^{(b)}$, respectively. Then from Eqns (4.5.7) it follows that

$$\frac{d}{dx}(\psi_1^{(a)}\psi_2^{(b)} - \psi_1^{(b)}\psi_2^{(a)}) + i(k_a - k_b)(\psi_1^{(a)}\psi_2^{(b)} + \psi_2^{(a)}\psi_1^{(b)}) = 0. \tag{4.5.9}$$

If ψ is the eigenfunction for the complex eigenvalue $k = \iota + i\kappa$ (ι, κ are real), the wave function $\bar{\psi}$ defined by

$$\bar{\psi} \equiv \begin{pmatrix} \psi_2^* \\ -\psi_1^* \end{pmatrix} = i\sigma_2\psi^*, \quad i\sigma_2 = \begin{pmatrix} 0 & 1 \\ -1 & 0 \end{pmatrix}, \tag{4.5.10}$$

becomes the eigenfunction for the eigenvalue $k^* = \iota - i\kappa$. That is, it satisfies the equation

$$i\sigma_3\bar{\psi}_{,x} + P\bar{\psi} = k^*\bar{\psi}, \tag{4.5.11}$$

in which we have used the relationships $\sigma_2\sigma_3 = -\sigma_3\sigma_2$, $\sigma_2 P^* = P\sigma_2$.

The Jost functions are introduced in correspondence to Eq. (4.1.26) as follows. φ and χ are solutions to Eq. (4.5.8) for Im $k \equiv \kappa = 0$ and satisfy asymptotically the conditions

$$\varphi \to \begin{pmatrix} 1 \\ 0 \end{pmatrix} e^{-ikx} \quad (x \to -\infty, \kappa = 0), \tag{4.5.12a}$$

$$\chi \to \begin{pmatrix} 0 \\ 1 \end{pmatrix} e^{ikx} \quad (x \to \infty, \kappa = 0). \tag{4.5.12b}$$

Since the solutions $\chi, \bar{\chi}$ constitute a complete set of solutions, φ is

4.5 THE NONLINEAR SCHRODINGER EQUATION

expressible as a linear combination of $\bar{\chi}, \chi$, so that

$$\varphi = a(k)\bar{\chi} + b(k)\chi. \tag{4.5.13}$$

Applying to the solutions $\varphi, \bar{\varphi}$ of Eq. (4.5.8) the relation (4.5.9) yields

$$|a(k)|^2 + |b(k)|^2 = 1. \tag{4.5.14}$$

It can be shown in a similar way to that used for the Schrödinger equation that analytic continuation into the upper half of the complex k-plane is possible for the functions φ and χ given by Eq. (4.5.12). On the other hand, from Eq. (4.5.9), for Im $k = 0$ it follows that

$$a(k) = (\varphi_1\chi_2 - \varphi_2\chi_1). \tag{4.5.15}$$

Hence $a(k)$ is also analytic for Im $k > 0$. In the limit as $|k| \to \infty$, in Eqns (4.5.7) $v \approx 0$, so that as

$$|k| \to \infty \ (\text{Im } k \geq 0), \qquad a(k) \to 1. \tag{4.5.16}$$

For Im $k > 0$, the eigenvalues k_m of Eq. (4.5.8) are given by $a(k_m) = 0$ ($m = 1, 2, \ldots, N$), corresponding to which the eigenfunctions are given by

$$\varphi(x, k_m) = \bar{c}_m \chi(x, k_m) \qquad (m = 1, 2, \ldots, N). \tag{4.5.17}$$

In contrast to the KdV equation, the eigenvalues k_m are not necessarily purely imaginary. However, if $v = v^*$, then for Im $k = 0$

$$\varphi(x, -k) = \varphi^*(x, k), \qquad \chi(x, -k) = \chi^*(x, k). \tag{4.5.18}$$

Consequently,

$$\begin{aligned} a(k) &= a^*(-k) & (\text{Im } k = 0), \\ a(k) &= a^*(-k^*) & (\text{Im } k > 0), \end{aligned} \tag{4.5.19}$$

and therefore k_m is purely imaginary.

Substituting for ψ in Eq. (4.5.5b) the $\tilde{\psi}$ given by Eq. (4.5.6), we find, as $|x| \to \infty$,

$$i\tilde{\psi}_{,t} = \frac{k^2}{\beta}\tilde{\psi} + 2ik\tilde{\psi}_{,x} - \beta\tilde{\psi}_{,xx}. \tag{4.5.20}$$

By making use of a similar procedure to that used in Section 4.1, the time evolution of the scattering data can be determined by this equation. However, the (left) Jost function φ given by Eq. (4.5.13) does not satisfy Eq. (4.5.20), because as is expressed by Eq. (4.5.12a) the asymptotic form of φ as $x \to -\infty$ does not have any time dependence. Therefore, we put

$$\tilde{\psi} \equiv \varphi e^{i\alpha t} \tag{4.5.21a}$$

to determine a, b, and α.

Using the asymptotic form Eq. (4.5.12a) of φ at $x = -\infty$, we have

$$\alpha = -\left\{\frac{1}{\beta} + 2 + \beta\right\}k^2. \tag{4.5.21b}$$

Also, as $x \to \infty$, introducing Eq. (4.5.13) for φ into Eq. (4.5.20) and using Eq. (4.5.21b), we get

$$a_{,t} = 0, \qquad b_{,t} = i4k^2 b. \tag{4.5.22}$$

Therefore, the scattering data $a(k, t)$, $b(k, t)$ for the continuous spectrum are found to be

$$a(k, t) = a(k, 0), \qquad b(k, t) = b(k, 0)e^{i4k^2 t}. \tag{4.5.23}$$

The same procedure applies to the discrete spectrum k_m for Im $k > 0$. Namely, from Eqns (4.5.12a) and (4.5.17), we have as $x \to \infty$

$$\psi = \tilde{c}_m(t)\begin{pmatrix}0\\1\end{pmatrix}e^{ik_m x + i\alpha_m t} \qquad (\text{Im } k_m > 0), \tag{4.5.24a}$$

and as $x \to -\infty$

$$\psi = \begin{pmatrix}1\\0\end{pmatrix}e^{-ik_m x + i\alpha_m t} \qquad (\text{Im } k_m > 0). \tag{4.5.24b}$$

Consequently, substituting these expressions for $\hat{\psi}$ in place of $\tilde{\psi}$ in Eq. (4.5.20), we obtain as $x \to -\infty$

$$\alpha_m = -\left[\frac{1}{\beta} + 2 + \beta\right]k_m^2, \tag{4.5.25}$$

and as $x \to \infty$

$$\tilde{c}_{m,t} = i4k_m^2 \tilde{c}_m, \tag{4.5.26}$$

and hence

$$\tilde{c}_m(t) = \tilde{c}_m(0)e^{i4k_m^2 t}. \tag{4.5.27}$$

In order to obtain $u(x, t)$ from the scattering data $a(k, 0)$, $b(k, t)$ and $\tilde{c}_m(t)$ by means of the inverse scattering method, corresponding to the function given by Eq. (4.1.36) in Section 4.1, we introduce the function $\Psi(k, x)$:

$$\Psi(k, x) = \begin{cases} a^{-1}(k)\varphi(k, x)e^{ikx} & (\text{Im } k > 0), \\ \bar{\chi}(k^*, x)e^{ikx} & (\text{Im } k < 0). \end{cases} \tag{4.5.28}$$

(Here $\bar{\chi}$ is defined by Eq. (4.5.10).) From Eq. (4.5.13) it is obvious that $\Psi(k, x)$ also has a jump $\phi(\iota)$ across the real axis.

$$\phi(\iota) \equiv \Psi(\iota + i0) - \Psi(\iota - i0) = r(\iota)\chi(\iota, x)\exp(i\iota x) \tag{4.5.29}$$

4.5 THE NONLINEAR SCHRÖDINGER EQUATION

where

$$r(\iota) = \frac{b(\iota)}{a(\iota)}. \tag{4.5.30}$$

$\Psi(k, x)$ has poles at $a(k_m) = 0$ in the upper half-plane and the jump across the real axis $r(\iota)$ corresponds to the reflection coefficient. Hence, in the same way as was done in Section 4.1, considering the integral (4.1.38), for Im $k < 0$ we have

$$i\pi \begin{pmatrix} 1 \\ 0 \end{pmatrix} + \int_{-\infty}^{\infty} \frac{\Psi(k' + i\delta)}{k' - k} \, dk' = 2\pi i \sum_n \frac{A_n}{k_n - k},$$

$$A_n \equiv C_n \chi_n(x) e^{ik_n x}, \qquad C_n \equiv \frac{\bar{c}_n}{a'(k_n)}, \tag{4.5.31}$$

where

$$a'(k_n) = \frac{da(k)}{dk}\bigg|_{k=k_n}.$$

Also, corresponding to Eq. (4.1.41b), the integral over the lower half-plane gives

$$i\pi \begin{pmatrix} 1 \\ 0 \end{pmatrix} - \int_{-\infty}^{\infty} \frac{\Psi(k' - i\delta)}{k' - k} \, dk' = 2\pi i \overline{\chi(k^*, x)} e^{ikx}. \tag{4.5.32}$$

Equations (4.5.29), (4.5.31), and (4.5.32) may be combined to give, for Im $k < 0$,

$$\overline{\chi(k^*, x)} e^{ikx} = \begin{pmatrix} 1 \\ 0 \end{pmatrix} + \frac{1}{2\pi i} \int_{-\infty}^{\infty} \frac{r(k')\chi(k', x)}{k' - k} e^{ik'x} \, dk' - \sum_n^N \frac{C_n \chi_n(x)}{k_n - k} e^{ik_n x}. \tag{4.5.33}$$

Taking the complex conjugate of Eq. (4.5.33) and rewriting k^* as k yields the equation for Im $k > 0$. Then, using the transformation Eq. (4.5.10), and noting that $\overline{(\bar{\chi})} = -\chi$, we obtain, for Im $k > 0$,

$$\chi_1(k, x) e^{-ikx} = \frac{1}{2\pi i} \int_{-\infty}^{\infty} \frac{r^*(k')\chi_2^*(k', x)}{k' - k} e^{-ik'x} \, dk' + \sum_{n=1}^{N} \frac{C_n^* \chi_{n2}^*(x)}{k_n^* - k} e^{-ik_n^* x}, \tag{4.5.34a}$$

$$\chi_2(k, x) e^{-ikx} = 1 - \frac{1}{2\pi i} \int_{-\infty}^{\infty} \frac{r^*(k')\chi_1^*(k', x)}{k' - k} e^{-ik'x} \, dk'$$
$$- \sum_{n=1}^{N} \frac{C_n^* \chi_{n1}^*(x)}{k_n^* - k} e^{-ik_n^* x}. \tag{4.5.34b}$$

Substituting for k in this equation first $k = \iota + i\delta$ and then $k = k_m$ enables us to obtain $2(N+1)$ equations connecting the scattering data $(a(\iota, 0),$

$b(\iota, t)$, $\tilde{c}_m(t)$) and the solutions to the Dirac type Eq. (4.5.8), $\chi_1(\iota, x)$, $\chi_2(\iota, x)$, $\chi_{m1}(x)$ and $\chi_{m2}(x)$.

The potential $v(x, t)$ of Eq. (4.5.7) can be obtained from the asymptotic behaviour in the limit $|k| \to \infty$. Let χ_1, χ_2 be given by

$$\chi_1 \equiv f_1 e^{ikx}, \qquad \chi_2 \equiv f_2 e^{ikx}, \qquad (4.5.35)$$

then Eq. (4.5.17) gives

$$f_{1,x} + 2ikf_1 = vf_2, \qquad f_{2,x} = -v^* f_1. \qquad (4.5.36)$$

In view of Eq. (4.5.12b), $\chi_1 \to 0$, $\chi_2 \to e^{ikx}$ as $x \to \infty$, and hence $f_1 \to 0$, $f_2 \to 1$ as $x \to \infty$. Consequently for $|k| \gg 1$, f_1, f_2 may be expanded in powers of k^{-1} as

$$\begin{aligned} f_1 &= k^{-1} f_1^{(1)} + k^{-2} f_1^{(2)} + \cdots, \\ f_2 &= 1 + k^{-1} f_2^{(1)} + k^{-2} f_2^{(2)} + \cdots, \end{aligned} \qquad (4.5.37)$$

which may be substituted for f_1, f_2 in Eqns (4.5.36) to determine successively the expansion coefficients

$$f_1^{(1)} = \frac{1}{2i} v, \qquad f_2^{(1)} = \frac{1}{2i} \int_x^\infty |v|^2 \, dx'. \qquad (4.5.38)$$

Hence Eq. (4.5.35) becomes

$$\chi e^{-ikx} = \begin{pmatrix} 0 \\ 1 \end{pmatrix} + \frac{1}{2ik} \begin{pmatrix} v \\ \int_x^\infty |v|^2 \, dx' \end{pmatrix} + \cdots. \qquad (4.5.39)$$

Expanding the right-hand side of Eq. (4.5.34) in powers of k^{-1} in similar fashion, and comparing with Eq. (4.5.39), we have

$$v = -\frac{1}{\pi} \int_{-\infty}^\infty r^*(k') \chi_2^*(k', x) e^{-ik'x} \, dk' - 2i \sum_n^N C_n^* \chi_{n2}^* e^{-ik_n^* x}, \qquad (4.5.40)$$

$$\int_x^\infty |v|^2 \, dx' = \frac{1}{\pi} \int_{-\infty}^\infty r(k') \chi_1(k', x) e^{ik'x} \, dk' - 2i \sum_n^N C_n \chi_{n1} e^{ik_n x}. \qquad (4.5.41)$$

Therefore the initial value problem of the nonlinear Schrödinger equation (4.5.1) can be solved as follows.

Obtain the scattering data for the initial condition $u(x, 0)$, $(a, (\iota, 0)$, $b(\iota, 0)$, $a(k_m) = 0$, $\tilde{c}_m(0)$ $(m = 1, 2, \ldots, N)$) from the eigenvalue equation (4.5.7). The scattering data at the arbitrary time $t > 0$ is given by Eqns (4.5.23) and (4.5.27), from which we may obtain the wave functions at time t from the $2(N+1)$ equations (4.5.34). Next, we introduce these wave functions into Eq. (4.5.40) to get $v(x, t)$; that is, $u(x, t)$.

The inverse scattering problem for the nonlinear Schrödinger equation

4.5 THE NONLINEAR SCHRÖDINGER EQUATION

can also be solved by the G–L–M equation in the same way as for the KdV equation in Section 4.1. That is, corresponding to Eq. (4.1.49), introduce functions $K_1(x, y)$ and $K_2(x, y)$ by

$$K_1(x, y) = i \sum_{n=1}^{N} C_n^* \chi_{n2}^* e^{-ik_n^* y} + \frac{1}{2\pi} \int_{-\infty}^{\infty} r^*(k) \chi_2^*(k, x) e^{-iky} \, dk \quad (4.5.42a)$$

$$K_2(x, y) = -i \sum_{n=1}^{N} C_n^* \chi_{n1}^* e^{-ik_n^* y} - \frac{1}{2\pi} \int_{-\infty}^{\infty} r^*(k) \chi_1^*(k, x) e^{-iky} \, dk \quad (4.5.42b)$$

respectively. Then, from Eq. (4.5.34), it follows that

$$\int_x^{\infty} K_1(x, y) e^{iky} \, dy = \chi_1(k, x), \quad (4.5.43a)$$

$$\int_x^{\infty} K_2(x, y) e^{iky} \, dy = \chi_2(k, x) - e^{ikx}. \quad (4.5.43b)$$

Now we introduce the function $B(x + y, t)$ given in terms of the scattering data by

$$B(x + y; t) = \frac{1}{2\pi} \int_{-\infty}^{\infty} r(k, t) e^{ik(x+y)} \, dk - i \sum_{n=1}^{N} C_n e^{ik_n(x+y)}. \quad (4.5.44)$$

Then it is readily seen, by substitution of Eqns (4.5.43) for χ_1 and χ_2 in Eqns (4.5.34), that the kernels $K_1(x, y)$ and $K_2(x, y)$ satisfy the G–L–M equation

$$K_1(x, y; t) = B^*(x + y; t) + \int_x^{\infty} B^*(y + z; t) K_2^*(x, z; t) \, dz \quad (4.5.45a)$$

$$K_2^*(x, y; t) = -\int_x^{\infty} B(y + z; t) K_1(x, z; t) \, dz \quad (4.5.45b)$$

Also, comparing Eq. (4.5.42) with Eqns (4.5.40) and (4.5.41), we can express $v(x, t)$ and $\int_x^{\infty} |v|^2 \, dx'$ by means of the solutions $K_1(x, x)$, $K_2(x, x)$ of the G–L–M equation (4.5.45) as

$$v(x, t) = -2K_1(x, x), \quad (4.5.46a)$$

$$\int_x^{\infty} |v|^2 \, dx' = -2K_2^*(x, x). \quad (4.5.46b)$$

Let us obtain the single soliton solution. When $r(k) = 0$, $N = 1$, Eq. (4.5.34) reduces to

$$\chi_1 e^{-ik_1 x} = \frac{C_1^* \chi_2^* e^{-ik_1^* x}}{k_1^* - k_1},$$

$$\chi_2 e^{-ik_1 x} = 1 - \frac{C_1^* \chi_1^* e^{-ik_1^* x}}{k_1^* - k_1}.$$

$(4.5.47)$

As can be seen from Eq. (4.5.40), $v(x, t)$ depends only on χ_2, and hence eliminating χ_1 from Eq. (4.5.47), and using Eqns (4.5.27) and (4.5.31), we have

$$C_1^* \chi_2^* e^{-ik_1^* x} = \frac{\gamma^* e^{-4i(\iota^2-\kappa^2)t - 2i\iota x}}{|\gamma|(2\kappa)^{-2} e^{-8\iota\kappa t - 2\kappa x} + e^{8\iota\kappa t + 2\kappa x}}, \qquad (4.5.48)$$

where $k_1 = \iota + i\kappa$, $\gamma = \tilde{c}_1(0)/a'(k_1)$. Introducing this equation into Eq. (4.5.40), and using the relation between u and v given by Eq. (4.5.7c), we obtain the single soliton solution Eq. (4.5.2), with

$$\theta = \arg \gamma \quad \text{and} \quad x_0 = (2\kappa)^{-1} \log(|\gamma|/2\kappa).$$

The N-soliton solution can be obtained in almost the same way as for the KdV equation. In Eq. (4.5.34), putting $k = k_m$, $\chi_1 = \chi_{m1}$, $\chi_2 = \chi_{m2}$, and multiplying Eq. (4.5.34a) and the complex conjugate of Eq. (4.5.34b) by $(C_m)^{1/2} e^{ik_m x}$ and $(C_m^*)^{1/2} e^{-ik_m^* x}$, respectively, we obtain an algebraic equation corresponding to Eq. (4.1.55b)

$$(I+D)\Psi = E, \qquad (4.5.49)$$

in which

$$\Psi = \begin{pmatrix} \varphi_{11} \\ \vdots \\ \varphi_{N1} \\ \varphi_{12}^* \\ \vdots \\ \varphi_{N2}^* \end{pmatrix}, \quad \varphi_{m1} = (C_m)^{1/2} \chi_{m,1}, \quad \varphi_{m2}^* = (C_m^*)^{1/2} \chi_{m2}^*$$

$$D \equiv \begin{pmatrix} 0 & D^{(12)} \\ D^{(21)} & 0 \end{pmatrix}, \quad D_{mn}^{(12)} = -D_{mn}^{(21)*} = (C_m C_n^*)^{1/2} \frac{e^{i(k_m - k_n^*)x}}{k_m - k_n^*}$$

$$(4.5.50)$$

$$E = \begin{pmatrix} 0 \\ \vdots \\ 0 \\ (C_1^*)^{1/2} e^{-ik_1^* x} \\ \vdots \\ (C_N^*)^{1/2} e^{-ik_N^* x} \end{pmatrix}.$$

Consequently,

$$\Psi_n = \Delta^{-1} \sum_{m}^{2N} E_m Q_{mn} = \Delta^{-1} \sum_{m=N+1}^{2N} E_m Q_{mn}, \qquad (4.5.51)$$

$$\Delta \equiv \det(I+D) = \det(I + D^{(12)*} D^{(12)}),$$

where Q_{mn} is the cofactor of the element (m, n) of the matrix $(I+D)$.

4.5 THE NONLINEAR SCHRÖDINGER EQUATION

Also, multiplying the complex conjugate of Eq. (4.5.34a) by $(C_m^*)^{1/2} e^{-2ik_m^* x}$ and Eq. (4.5.34b) by $(C_m)^{1/2} e^{ik_m^* x}$, in the same way we obtain an algebraic equation

$$(I+D)\bar{\Psi} = \bar{E}. \tag{4.5.52}$$

Here

$$\bar{\Psi} = \begin{pmatrix} \varphi_{12} \\ \vdots \\ \varphi_{N2} \\ -\varphi_{11}^* \\ \vdots \\ -\varphi_{N1}^* \end{pmatrix}, \quad \bar{E} = \begin{pmatrix} (C_1)^{1/2} e^{ik_1 x} \\ \vdots \\ (C_N)^{1/2} e^{ik_N x} \\ 0 \\ \vdots \\ 0 \end{pmatrix}, \tag{4.5.53}$$

and hence

$$\bar{\Psi}_n = \frac{1}{\Delta} \sum_m^{2N} \bar{E}_m Q_{mn} = \frac{1}{\Delta} \sum_{m=1}^{N} \bar{E}_m Q_{mn}. \tag{4.5.54}$$

On the other hand, Eq. (4.5.41) becomes

$$\int_x^\infty |v|^2 \, dx' = -2i \sum_{n=1}^N C_n \chi_{n1} e^{ik_n x}$$
$$= -i \left\{ \sum_{n=1}^N C_n \chi_{n1} e^{ik_n x} - \sum_{n=1}^N C_n^* \chi_{n1}^* e^{-ik_n^* x} \right\}, \tag{4.5.55}$$

where in deriving this last relation we have used the fact that the left-hand side is real. Since

$$\varphi_{n1} = (C_n)^{1/2} \chi_{n1}, \quad \sum_{n=1}^N C_n \chi_{n1} e^{ik_n x} = \bar{E}_n \Psi_n$$

and

$$\sum_{n=1}^N C_n^* \chi_{n1}^* e^{-ik_n^* x} = -\sum_{m=N+1}^{2N} \bar{E}_m \bar{\Psi}_m,$$

substituting for Ψ and $\bar{\Psi}$ in Eqns (4.5.55), (4.5.51) and (4.5.54), we obtain

$$\int_x^\infty |v|^2 \, dx = -\frac{1}{\Delta} \left\{ \sum_{n=1}^N \sum_{m=N+1}^{2N} \left(\frac{d}{dx} D_{mn}^{(21)} \right) Q_{mn} \right.$$
$$\left. + \sum_{n=N+1}^{2N} \sum_{m=1}^N \left(\frac{d}{dx} D_{mn}^{(12)} \right) Q_{mn} \right\}$$
$$= -\frac{1}{\Delta} \sum_{n=1}^{2N} \sum_{m=1}^{2N} \left(\frac{d}{dx} D_{mn} \right) Q_{mn} = -\frac{1}{\Delta} \frac{d}{dx} \Delta = -\frac{d}{dx} (\log \Delta), \tag{4.5.56}$$

which, by means of Eq. (4.5.7c), gives the expression for the N-soliton solution corresponding to Eq. (4.1.57):

$$|u(x,t)|^2 = \frac{2}{Q}\frac{d^2}{dx^2}(\log \Delta). \tag{4.5.57}$$

If the real parts ι_n of all the eigenvalues k_n are different, then as $|t| \to \infty$, the solution $v(x,t)$ yields N solitons which all are sufficiently separated from each other to be identifiable and which proceed with velocities $-4\iota_n$ in precisely the same way as for the KdV equation. However, their amplitudes and widths are determined by the imaginary parts κ_n of the eigenvalues k_n. Also, because of collisions, each soliton changes its phase θ and the position of its centre x_0 (see Eq. (4.5.2)) (Zakharov and Shabat (1972)). Since the velocity of a soliton is given by the real part of the corresponding eigenvalue, two solitons with the same ι do not separate, but form a bound state of two-solitons. Moreover, in contrast to the eigenvalues of the Schrödinger equation, the eigenvalues of the Dirac-type equation (4.5.17) may be degenerate, and there then exists a bound state corresponding to the degenerate eigenstate (Zakharov and Shabat (1972)). Also, if $\iota = 0$, as is obvious from Eq. (4.5.2), the soliton oscillates with the angular frequency $\omega = 4\kappa^2$.

The conservation laws can also be derived in the same way as for the KdV equation. The asymptotic forms of $\varphi(k,x)$ as $x \to \pm \infty$ may be derived from Eqns (4.2.12) and (4.2.13), respectively, as

$$\begin{aligned}\varphi_1(k,x) &\sim a(k)e^{-ikx} \quad (x \to \infty), \\ \varphi_1(k,x) &\sim e^{-ikx} \quad (x \to -\infty).\end{aligned} \tag{4.5.58}$$

Hence, similar to Eqns (4.2.3) and (4.2.5), introducing θ and δ by

$$\varphi_1 = e^{\theta}, \quad \sigma \equiv \theta_{,x} + ik, \tag{4.5.59}$$

respectively, we have

$$\int_{-\infty}^{\infty} \sigma \, dx = \log a(k). \tag{4.5.60}$$

Substituting φ in Eqns (4.5.7) in place of $\tilde{\psi}$, where φ is given by Eqns (4.5.58) and (4.5.59), we find that σ satisfies the Riccati equation

$$\sigma_{,x} + \sigma^2 - \sigma\frac{v_{,x}}{v} + |v|^2 - 2ik\sigma = 0. \tag{4.5.61}$$

Therefore, as in Section 4.2, the asymptotic expansion of σ for large $|k|$ is given by

$$\sigma(k,x) = \sum_{m=1}^{\infty} \frac{\sigma_m}{(2ik)^m},$$

$$\sigma_m = v\left(\frac{\sigma_{m-1}}{v}\right)_{,x} + \sum_{l=1}^{m-2} \sigma_{m-l-1}\sigma_l, \quad \sigma_1 = |v|^2, \tag{4.5.62}$$

while for Im $k \geq 0$, $|k| \to \infty$, $\log a(k) \to 0$. Consequently $\log a(k)$ can also be expanded in powers of $1/k$, that is,

$$\log a(k) = \sum_{m=1}^{\infty} \frac{C_m}{k^m}. \tag{4.5.63}$$

Then, from Eqns (4.5.62), (4.5.63), it follows that

$$(2i)^m C_m = \int_{-\infty}^{\infty} \sigma_m(x)\, dx \quad (m = 1, 2, \ldots). \tag{4.5.64}$$

As is readily seen from Eq. (4.5.62), the conservation laws (4.5.64) involve v and $v_{,x}$. For example, for $m = 1, 2, 3$,

$$2iC_1 = \frac{Q}{2} \int_{-\infty}^{\infty} |u(x,t)|^2\, dx, \tag{4.5.65a}$$

$$(2i)^2 C_2 = -\frac{Q}{4} \int_{-\infty}^{\infty} (u^* u_{,x} - u u^*_{,x})\, dx, \tag{4.5.65b}$$

$$(2i)^3 C_3 = -\frac{Q}{2} \int_{-\infty}^{\infty} \left(|u_{,x}|^2 - \frac{Q}{2} |u|^4 \right) dx. \tag{4.5.65c}$$

In the same way as in Section 4.2, the nonlinear Schrödinger equation is also given by a canonical form with a completely integrable Hamiltonian (Zakharov and Manakov (1974)).

4.6 The Sine–Gordon equation and the Bäcklund transformation

As was shown in Section 2.1, the reason why the method of characteristics is useful with the linear wave equation (2.1.1), namely $\phi_{tt} - \phi_{xx} = 0$, is because there exist the Riemann invariants $\phi_{,\xi} = r(\xi)$, $\phi_{,\eta} = s(\eta)$ along the characteristics $\xi = x - t$, $\eta = x + t$, respectively. If the Klein–Gordon equation (2.3.13)

$$\phi_{,tt} - \phi_{,xx} + m^2 \phi = 0$$

is written in terms of the characteristic coordinates $\xi = (x-t)/2$, $\eta = (x+t)/2$, we arrive at the result

$$\phi_{,\xi\eta} = m^2 \phi, \tag{4.6.1}$$

when the integrals along the characteristics

$$\phi_{,\eta} = m^2 \int \phi\, d\xi + s(\eta), \tag{4.6.2a}$$

$$\phi_{,\xi} = m^2 \int \phi\, d\eta + r(\xi), \tag{4.6.2b}$$

involve integrals either with respect to ξ or η. Hence, in general, no local relationship exists along a characteristic amongst the derivatives of ϕ, such as occurred in Eq. (2.1.5). However, there will be particular cases in which such relationships hold.

We shall first examine under what conditions such relationships can exist. Let us assume that $\phi_{,\xi}$, $\phi_{,\eta}$ are capable of expression by means of the local relations

$$\phi_{,\xi} = P(\phi), \qquad (4.6.3a)$$

$$\phi_{,\eta} = Q(\phi), \qquad (4.6.3b)$$

and then look for the functions $P(\phi)$ and $Q(\phi)$. Differentiate Eq. (4.6.3a) with respect to η and use Eqns (4.6.1) (4.6.3b) to obtain

$$QP_{,\phi} = m^2\phi. \qquad (4.6.4a)$$

Similarly, from Eq. (4.6.3b), we obtain

$$PQ_{,\phi} = m^2\phi. \qquad (4.6.4b)$$

From Eq. (4.6.4) we have

$$Q = a^2 P, \qquad (4.6.5)$$

where a is an arbitrary constant. Introducing Eq. (4.6.5) into Eq. (4.6.4), and integrating with respect to ϕ, we obtain

$$P^2 = \frac{m^2\phi^2 + 2C}{a^2} = \left(\frac{Q}{a^2}\right)^2. \qquad (4.6.6)$$

For simplicity, let us set the integration constant C equal to zero, then Eq. (4.6.3) becomes

$$\phi_{,\xi} = \pm\left(\frac{m}{a}\right)\phi, \qquad (4.6.7a)$$

$$\phi_{,\eta} = \pm(ma)\phi, \qquad (4.6.7b)$$

which can be integrated to give

$$\phi = e^{\pm m\zeta + b} = A e^{\pm m(Kx - \Omega t)}, \qquad (4.6.8)$$

where

$$\zeta \equiv \frac{\xi}{a} + a\eta, \qquad (4.6.9)$$

$$K \equiv \frac{1+a^2}{2a}, \quad \Omega \equiv \frac{1-a^2}{2a}, \qquad (4.6.10)$$

and b (and $A = e^b$) is an integration constant. Hence, if $a^2 < 0$, by writing

4.6 THE SINE–GORDON EQUATION

$ik \equiv mK$ and $i\omega \equiv m\Omega$, Eq. (4.6.8) then represents a plane wave with the dispersion relation $\omega^2 - k^2 = m^2$. If $a^2 > 0$, Eq. (4.6.8) represents a solution which grows or decreases exponentially at $|x| = \infty$. If the integration constant C is finite, and non-zero, the exponential function in Eq. (4.6.8) becomes either an hyperbolic function or a trigonometric function, corresponding to $a^2 > 0$ or $a^2 < 0$, respectively. The above-mentioned method of solution can readily be extended to a nonlinear equation of the form

$$\phi_{,tt} - \phi_{,xx} = -F(\phi) = \frac{\partial V}{\partial \phi}. \tag{4.6.11}$$

Functions P, Q which satisfy Eq. (4.6.3) are given by, for example (for $a^2 > 0$),

$$P = \pm \frac{[2(C-V)]^{1/2}}{a} = \frac{Q}{a^2}, \tag{4.6.12}$$

and when introduced into Eq. (4.6.3) they lead to the formal solution

$$\phi = K^{-1}(\zeta + b). \tag{4.6.13a}$$

Here K^{-1} is the function inverse to

$$K(\phi) \equiv \pm \int [2(C-V)]^{-1/2} \, d\phi, \tag{4.6.13b}$$

b is an integration constant and ζ is given by Eq. (4.6.9).

If ϕ is a function of ζ only, Eq. (4.6.11) reduces to $\phi_{,\zeta\zeta} = -\partial V/\partial \phi$; and consequently, by regarding ϕ, ζ, and V as the coordinate, the time and the potential of a particle, respectively, the ζ dependence of ϕ can be deduced by analogy with the motion of a particle in a potential V.

For example, in order that an oscillating solution (a wave-train) exists, $V(\phi)$ is required to have a minimum. In general, the solution involves the two parameters C, a and the integration constant b. Also, if V has a maximum as well as a minimum, such as would be given by the function $V = -\phi^2(1-\phi)^n$ (n a positive integer), solitary-wave like solutions and kink-like solutions can exist which approach a uniform state as $|x| \to \infty$. Since in this case C is determined by the potential, apart from the integration constant b, the solution depends on the single parameter a only.

We now show that amongst these equations, the Sine–Gordon equation introduced in Section 3.9 has a particular property like that of a linear equation, in so far as the superposition of solutions is possible. Let two solutions of Eq. (4.6.11) be $\phi^{(0)}$, $\phi^{(1)}$, then for each of these we have

$$\phi^{(i)}_{,\xi} = \int F(\phi^{(i)}) \, d\eta, \qquad \phi^{(i)}_{,\eta} = \int F(\phi^{(i)}) \, d\xi \qquad (i = 0, 1). \tag{4.6.14}$$

Now we assume that the local relation Eq. (4.3.6) is valid for the superposition of the solutions, so that

$$\phi^{(1)}_{,\xi} + \alpha\phi^{(0)}_{,\xi} = \int \{F(\phi^{(1)}) + \alpha F(\phi^{(0)})\}\, d\eta = P(\phi^{(1)}, \phi^{(0)}), \qquad (4.6.15a)$$

$$\phi^{(1)}_{,\eta} + \beta\phi^{(0)}_{,\eta} = \int \{F(\phi^{(1)}) + \beta F(\phi^{(0)})\}\, d\xi = Q(\phi^{(1)}, \phi^{(0)}), \qquad (4.6.15b)$$

in which α, β are constants. Differentiating Eqns (4.6.15a, b) with respect to η, ξ, respectively, yields

$$\phi^{(1)}_{,\eta} P_{,\phi^{(1)}} + \phi^{(0)}_{,\eta} P_{,\phi^{(0)}} = F(\phi^{(1)}) + \alpha F(\phi^{(0)}), \qquad (4.6.16a)$$

$$\phi^{(1)}_{,\xi} Q_{,\phi^{(1)}} + \phi^{(0)}_{,\xi} Q_{,\phi^{(0)}} = F(\phi^{(1)}) + \beta F(\phi^{(0)}). \qquad (4.6.16b)$$

Equations (4.6.15) and (4.6.16) can be considered as the algebraic equations for $\phi^{(i)}_{,\xi}$ and $\phi^{(i)}_{,\eta}$. Hence, requiring that the set of equations (4.6.15a) and (4.6.16b) and the set of equations (4.6.15b) and (4.6.16a) are, respectively, equivalent, we have proportionality amongst their coefficients, which can be summarized in the form of the following equations:

$$PQ_{,\phi^{(1)}} - QP_{,\phi^{(1)}} = (\beta - \alpha)F^{(0)}, \qquad (4.6.17a)$$

$$PQ_{,\phi^{(0)}} - QP_{,\phi^{(0)}} = (\alpha - \beta)F^{(1)}, \qquad (4.6.17b)$$

$$\alpha Q_{,\phi^{(1)}} - Q_{,\phi^{(0)}} = 0,$$
$$\beta P_{,\phi^{(1)}} - P_{,\phi^{(0)}} = 0, \qquad (4.6.17c)$$

$$F^{(i)} \equiv F(\phi^{(i)}) \qquad (i = 0, 1). \qquad (4.6.17d)$$

From Eqns (4.6.17c, d) it follows that P and Q are functions of $v \equiv \phi^{(1)} + \beta\phi^{(0)}$ and $u \equiv \phi^{(1)} + \alpha\phi^{(0)}$, respectively. Assuming $\alpha \neq \beta$, differentiating Eq. (4.6.17a) with respect to $\phi^{(1)}$ and noticing that $F^{(0)}$ is a function of $\phi^{(0)}$, we obtain

$$PQ_{,uu} - QP_{,vv} = 0. \qquad (4.6.18)$$

Consequently,

$$P_{,vv} + \kappa^2 P = 0, \qquad Q_{,uu} + \kappa^2 Q = 0, \qquad (4.6.19)$$

where κ is a constant.

If $\kappa = 0$, it is readily seen that P, Q are linear functions of u and v, and that F must also be a linear function of ϕ. For example, in the case of the Klein–Gordon equation (4.6.1),

$$P = \pm\frac{m}{a} v, \qquad Q = \pm mau, \qquad (4.6.20)$$

4.6 THE SINE–GORDON EQUATION

which when substituted for P, Q in Eqns (4.6.15) give

$$\phi_{,\xi}^{(1)} + \alpha\phi_{,\xi}^{(0)} = \pm\frac{m}{a}(\phi^{(1)} + \beta\phi^{(0)}), \tag{4.6.21a}$$

$$\phi_{,\eta}^{(1)} + \beta\phi_{,\eta}^{(0)} = \pm ma(\phi^{(1)} + \alpha\phi^{(0)}). \tag{4.6.21b}$$

Differentiating Eq. (4.6.21a) with respect to η, and making use of Eq. (4.6.21b), it is easily seen that when $\phi^{(0)}$ is a solution of Eq. (4.6.1), then $\phi^{(1)}$ is also a solution. (Equation (4.6.21) may be regarded as an equation for $\phi^{(1)}$ for a given $\phi^{(0)}$; then similar arguments show that the integrability condition for this equation is satisfied when $\phi^{(0)}$ satisfies Eq. (4.6.1).) In particular, if $\alpha = \beta$, Eq. (4.6.21) is merely the principle of superposition. In this sense, Eq. (4.6.21) can be considered as an extension of the principle of superposition. On the other hand, Eq. (4.6.21) is a relation connecting the two integral surfaces $\phi^{(0)}$, $\phi^{(1)}$, and consequently it can be regarded as a transformation from $\phi^{(0)}$ to $\phi^{(1)}$ (or vice versa). However, the transformation involves not only ϕ, but also its derivatives (the gradients of the integral surface), so that it is not a point transformation but a contact transformation, and it belongs to a special class of Bäcklund transformations which are to be discussed later. These are called the restricted Bäcklund transformations (McLaughlin, Scott (1973)). In what follows, such transformations will be called simply Bäcklund transformations. We remark that, for a general Bäcklund transformation, $\phi^{(0)}$ need not be a solution of Eq. (4.6.1), and it is necessary only that $\phi^{(0)}$ is a sufficiently smooth function of ξ, η, in which case the integrability condition is $\alpha = \beta$.

From Eqns (4.6.21), when $\alpha, \beta, \phi^{(0)}$ are given, $\phi^{(1)}$ is obtained as a solution involving the single parameter a and one integration constant. For example, for the trivial solution $\phi^{(0)} = 0$ (the vacuum solution), $\phi^{(1)}$ becomes Eq. (4.6.8). Since Eqns (4.6.21) hold for a set of two arbitrary solutions, further solutions can be obtained successively by means of this process. In general, replacing $\phi^{(0)}$ and $\phi^{(1)}$ by $\phi^{(n-1)}$ and $\phi^{(n)}$, respectively, and the parameter a by a_n, from Eqns (4.6.21) we have

$$\phi^{(n-1)} = \sum_{j=1}^{n-1} A_j^{(n-1)} e^{\pm m\zeta_j}, \tag{4.6.22a}$$

$$\phi^{(n)} = \sum_{j=1}^{n} A_j^{(n)} e^{\pm m\zeta_j}, \tag{4.6.22b}$$

$$\zeta_j \equiv \frac{\xi}{a_j} + a_j \eta,$$

where

$$A_j^{(n)} = \gamma_j^{(n)} A_j^{(n-1)}, \tag{4.6.23a}$$

and

$$\gamma_j^{(n)} \equiv \frac{a_j \beta - a_n \alpha}{a_n - a_j} \qquad (a_n \neq a_j). \tag{4.6.23b}$$

Hence, for $\alpha = \beta$, the argument proceeds parallel to that for a Fourier expansion. The explicit forms of the first few terms constructed from the vacuum solution are as follows:

$$\begin{aligned}
\phi^{(1)} &= A^{(1)} e^{\pm m\zeta_1}, \\
\phi^{(2)} &= A^{(2)} e^{\pm m\zeta_2} + \gamma_1^{(2)} A^{(1)} e^{\pm m\zeta_1}, \\
\phi^{(3)} &= A^{(3)} e^{\pm m\zeta_3} + \gamma_2^{(3)} A^{(2)} e^{\pm m\zeta_2} + \gamma_1^{(3)} \gamma_1^{(2)} A^{(1)} e^{\pm m\zeta_1}.
\end{aligned} \tag{4.6.24}$$

These can be written formally by means of the increment operator B_j as

$$\phi^{(n)} = B_n \phi^{(n-1)} = B_n B_{n-1} \cdots B_1 \phi^{(0)}. \tag{4.6.25}$$

As is clearly seen from the above explicit forms Eq. (4.6.24), for α, β given, the B_j depends on one parameter a_j and one integration constant $A^{(j)}$. The increment operators are not necessarily commutative. For example, $B_3 B_2 \phi^{(1)} \neq B_2 B_3 \phi^{(1)}$. However, the integration constant $A^{(j)}$ of the operator B_j may be chosen arbitrarily, and hence it is always possible to choose the $A^{(2)'}$, $A^{(3)'}$, appropriately, so that

$$B_2(a_2, A^{(2)'}) B_3(a_3, A^{(3)'}) = B_3(a_3, A^{(3)}) B_2(a_2, A^{(2)})$$

holds ($A^{(2)'} = \gamma_2^{(3)} A^{(2)}$, $A^{(3)'} = A^{(3)}/\gamma_3^{(2)}$). That is to say, by means of an appropriate choice of $A^{(j)}$, the B_j can be made mututally commutative. This is often represented in the form of the so-called Lamb diagram (Fig. 4.7).

For the Klein–Gordon equation, α, β are arbitrary constants and are not uniquely determined. However, as will be shown in the following discussion, when $\kappa \neq 0$, α, β are not arbitrary but are in fact determined uniquely, corresponding to which F becomes nonlinear and its functional form is also determined.

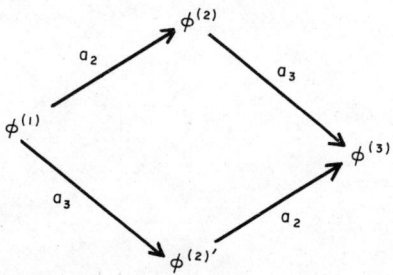

Fig. 4.7

4.6 THE SINE–GORDON EQUATION

First let us assume that $\kappa^2 > 0$, so that κ is real. Then, differentiating Eq. (4.6.17b) with respect to $\phi^{(0)}$, in the same way as was used to get Eq. (4.6.19), we obtain

$$P_{,vv} + \iota^2 \alpha^2 P = 0, \qquad Q_{,uu} + \iota^2 \beta^2 Q = 0. \tag{4.6.26}$$

Here ι is constant, and from Eqns (4.6.19) and (4.6.26) it follows that $\alpha^2 = \beta^2$, whilst $\alpha \neq \beta$ has been assumed, so that we are led to the result

$$\alpha = -\beta. \tag{4.6.27}$$

By means of Eqns (4.6.19) and (4.6.26), P and Q then take the forms

$$Q = q \sin(\kappa u + \theta), \tag{4.6.28a}$$

$$P = p \sin(\kappa v + \chi), \tag{4.6.28b}$$

where p, q, θ, χ are constants.

Substituting, for ϕ, P, Q in Eqns (4.6.17a, b), equations (4.6.28) then make it possible to determine $F^{(0)}, F^{(1)}$ as

$$F^{(0)} = \frac{\kappa pq}{2\alpha} \sin(2\kappa\alpha\phi^{(0)} + \theta - \chi), \tag{4.6.29a}$$

$$F^{(1)} = \frac{\kappa pq}{2} \sin(2\kappa\phi^{(1)} + \theta + \chi), \tag{4.6.29b}$$

which require the further conditions $\alpha = 1$, $\chi = 0$. We thus find that

$$F = \frac{\kappa pq}{2} \sin(2\kappa\phi + \theta), \tag{4.6.30}$$

$$Q = q \sin(\kappa u + \theta), \qquad P = p \sin \kappa v, \tag{4.6.31}$$

By virtue of Eq. (4.6.31), Eqns (4.6.15) become

$$\phi^{(1)}_{,\xi} + \phi^{(0)}_{,\xi} = p \sin \kappa(\phi^{(1)} - \phi^{(0)}), \tag{4.6.32a}$$

$$\phi^{(1)}_{,\eta} - \phi^{(0)}_{,\eta} = q \sin \kappa(\phi^{(1)} + \phi^{(0)} + \theta). \tag{4.6.32b}$$

These results show that in order that Eqns (4.6.15) hold, F cannot be an arbitrary function of ϕ, but (for real κ) it must agree with the right-hand side of the Sine–Gordon equation (3.9.13). For this equation, in the same way as for the linear equation (4.6.21), the Bäcklund transformation (4.6.32) which links two solutions $\phi^{(1)}$, $\phi^{(0)}$ is valid. It should, however, be noticed that in this case α, β are no longer arbitrary, but are fixed with $\alpha = -\beta = 1$.

In particular, putting $\theta = 0$, $\kappa = \frac{1}{2}$, $pq = \pm 4$, we find

$$F = \pm \sin \phi \tag{4.6.33}$$

and Eq. (4.6.11) becomes identical with Eq. (3.9.13); in which case, for

$q = \pm 2a$, $p = 2/a$, the Bäcklund transformation for the Sine–Gordon equation is given by

$$\phi^{(1)}_{,\xi} + \phi^{(0)}_{,\xi} = \frac{2}{a} \sin \frac{\phi^{(1)} - \phi^{(0)}}{2}, \tag{4.6.34a}$$

$$\phi^{(1)}_{,\eta} - \phi^{(0)}_{,\eta} = \pm 2a \sin \frac{\phi^{(1)} + \phi^{(0)}}{2}, \tag{4.6.34b}$$

where the $\pm$ signs on the right-hand sides correspond to the $\pm$ signs of F, respectively.

In Eq. (4.6.34), if $\phi^{(0)}$ is given as a known solution of the Sine–Gordon equation, another soluton $\phi^{(1)}$ is obtained. It can be proved that the solution $\phi^{(1)}$ thus obtained depends on only the single parameter a and one integration constant. This is due to the fact that $d\phi^{(1)}$ is a perfect differential of $d\xi$ and $d\eta$. For more details refer, for example, to Seeger et al. (1953). Consequently, the wave-train solution for Eq. (4.6.13) which contains three constants cannot be obtained and the solution is restricted like a soliton.

For example, for the trivial solution of the S–G equation, $\phi^{(0)} = 0$, Eq. (4.6.34) becomes

$$\frac{d(\phi^{(1)}/2)}{\sin(\phi^{(1)}/2)} = \frac{d\xi}{a} = \pm a \, d\eta, \tag{4.6.35}$$

so that

$$\phi^{(1)} = 4 \tan^{-1} e^{\zeta + b},$$

$$\zeta \equiv \pm a\eta + a^{-1}\xi = \frac{\pm a + a^{-1}}{2}\left[x - \frac{1 \mp a^2}{1 \pm a^2} t\right], \tag{4.6.36}$$

where b is an integration constant and, for simplicity, will hereafter be set equal to zero. If instead of a, the parameter u is introduced by $u = (1 \mp a^2)/(1 \pm a^2)$, then for the $+$ sign of F, e.g. in the case of the positive type, $a^2 = (1-u)/(1+u)$ giving $|u| < 1$, and corresponding to $a = \pm[(1-u)/(1+u)]^{1/2}$, ϕ takes the form

$$\phi = 4 \tan^{-1}[\exp\{\pm(1-u^2)^{-1/2}(x - ut)\}], \tag{4.6.37}$$

which is merely the solution to Eq. (3.9.18a).

$\phi^{(0)} = \pi$ is also a trivial solution of the S–G equation (3.9.15), in which case the Bäcklund transformation (4.6.34) becomes

$$\phi^{(1)}_{,\xi} = \frac{2}{a} \sin \frac{\phi^{(1)} - \pi}{2}, \tag{4.6.38a}$$

$$\phi^{(1)}_{,\eta} = \pm 2a \sin \frac{\phi^{(1)} + \pi}{2}. \tag{4.6.38b}$$

4.6 THE SINE–GORDON EQUATION

In the above equations, writing $\phi^{(2)}$ in place of $\phi^{(1)} - \pi$ causes a reversal of the sign of the right-hand side of Eq. (4.6.38b), to give $\mp 2a \sin \phi^{(2)}$. Hence the solution $\phi_+^{(1)}$ of the equation of the positive type creates from $\phi^{(2)} = \phi_+^{(1)} - \pi$ the solution $\phi^{(2)}$ of the equation of the negative type, which implies that Eq. (4.6.38) is a transformation connecting solutions of the equations of the positive and negative types.

In the same way as was used for the linear Klein–Gordon equation (cf. Fig. 4.7), starting from a known solution, we construct the Bäcklund transformation Eq. (4.6.34) to give $\phi^{(1)}$ and $\phi^{(2)}$ corresponding to arbitrary constants $a = a_1$ and $a = a_2$ ($a_1 \neq a_2$). From $\phi^{(1)}$, $\phi^{(2)}$ we then obtain $\phi^{(3)}$, $\phi^{(3)'}$, corresponding to $a = a_2$, and $a = a_1$. Then, by means of an appropriate choice of the integration constants, we can choose $\phi^{(3)}$ such that $\phi^{(3)} = \phi^{(3)'}$. (As was already explained for the Klein–Gordon equation, the Bäcklund transformation can be made commutative by adjusting the integration constants. For the case of nonlinear equations we refer the reader to McLaughlin and Scott (1973) and Seeger et al. (1953).) We thus arrive at the results

$$\phi^{(1)} = B_1 \phi^{(0)}, \qquad \phi^{(2)} = B_2 \phi^{(0)}, \qquad \phi^{(3)} = B_2 B_1 \phi^{(0)} = B_1 B_2 \phi^{(0)},$$

(4.6.39)

where B_j denotes the increment operator corresponding to the constant a_i.

For example, for each transformation Eq. (4.6.34a) takes the form (in what follows only the positive type will be considered):

$$\frac{(\phi^{(1)} + \phi^{(0)})_{,\xi}}{2} = a_1^{-1} \sin \frac{\phi^{(1)} - \phi^{(0)}}{2}, \tag{4.6.40a}$$

$$\frac{(\phi^{(2)} + \phi^{(0)})_{,\xi}}{2} = a_2^{-1} \sin \frac{\phi^{(2)} - \phi^{(0)}}{2}, \tag{4.6.40b}$$

$$\frac{(\phi^{(3)} + \phi^{(1)})_{,\xi}}{2} = a_2^{-1} \sin \frac{\phi^{(3)} - \phi^{(1)}}{2}, \tag{4.6.40c}$$

$$\frac{(\phi^{(3)} + \phi^{(2)})_{,\xi}}{2} = a_1^{-1} \sin \frac{\phi^{(3)} - \phi^{(2)}}{2}. \tag{4.6.40d}$$

Subtracting the sum of Eqns (4.6.40b) and (4.6.40c) from the sum of Eqns (4.6.40a) and (4.6.40d) yields

$$a_1 \sin \left\{ \frac{(\phi^{(3)} - \phi^{(0)}) - (\phi^{(1)} - \phi^{(2)})}{4} \right\} = a_2 \sin \left\{ \frac{(\phi^{(3)} - \phi^{(0)}) + (\phi^{(1)} - \phi^{(2)})}{4} \right\}.$$

(4.6.41)

Using Eq. (4.6.34b) leads to the same result. Therefore,

$$\tan\frac{\phi^{(3)}-\phi^{(0)}}{4} = \frac{a_1+a_2}{a_1-a_2}\tan\frac{\phi^{(1)}-\phi^{(2)}}{4}. \tag{4.6.42}$$

The above formula is called the addition theorem for the Bäcklund transformation. This relation makes it possible to obtain algebraically the new solution $\phi^{(3)}$ from the known solutions $\phi^{(0)}$, $\phi^{(1)}$, $\phi^{(2)}$, without involving any integration. Repeating this process we can obtain successively new solutions by means of purely algebraic manipulations.

By way of example, we now show how the solution which decouples asymptotically to a kink and an anti-kink (the possible two-kink solution) can be obtained. Let $\phi^{(0)}$ be given by $\phi^{(0)} = 0$. Then $\phi^{(1)}$, $\phi^{(2)}$ becomes the kink solutions Eq. (4.6.36) corresponding to

$$\zeta_i = \pm\gamma_i(x-u_it), \qquad \gamma_i = (1-u_i^2)^{-1/2} \qquad (i=1,2), \tag{4.6.43}$$

respectively. Here the $\pm$ signs correspond to those of $a_i = \pm\{(1-u_i)/(1+u_i)\}^{1/2}$. Consequently, by means of the addition theorem Eq. (4.6.42), we find

$$\phi^{(3)} = 4\tan^{-1}\left[\operatorname{sgn}(\Gamma)\exp(\log|\Gamma|)\frac{\sinh\{(\zeta_1-\zeta_2)/2\}}{\cosh\{(\zeta_1+\zeta_2)/2\}}\right],$$
$$\Gamma = \frac{a_1+a_2}{a_1-a_2}. \tag{4.6.44}$$

In the case $a_1 > 0$, $a_2 > 0$, for simplicity we shall assume $u_1 > u_2 > 0$, so that then $a_1 < a_2$ and $\Gamma < -1$, and also

$$\zeta_1 \pm \zeta_2 = \gamma^{(\pm)}(x - v^{(\pm)}t), \tag{4.6.45}$$

where

$$\gamma^{(\pm)} \equiv \gamma_1 \pm \gamma_2, \tag{4.6.46a}$$

$$v^{(\pm)} \equiv \frac{1+u_1u_2 \mp \{(1-u_1^2)(1-u_2^2)\}^{1/2}}{u_1+u_2}. \tag{4.6.46b}$$

By assumption, the following inequalities for $\gamma^{(\pm)}$, $v^{(\pm)}$ hold,

$$\gamma^{(\pm)} > 0, \qquad u_2 < v^{(+)} < u_1 < v^{(-)}. \tag{4.6.47}$$

If $t < 0$ and $|t| \gg 1$, by means of the inequalities (4.6.47), in the region $x > v^{(+)}t$, the approximations $\sinh\{(\zeta_1-\zeta_2)/2\} \sim (1/2)e^{(\zeta_1-\zeta_2)/2}$ $\cosh\{(\zeta_1+\zeta_2)/2\} \sim (1/2)e^{(\zeta_1+\zeta_2)/2}$ are valid. Therefore, Eq. (4.6.44) becomes, approximately,

$$\phi^{(3)} \sim 4\tan^{-1}(-e^{-\zeta_2+\log|\Gamma|})$$
$$= 4\tan^{-1}(-e^{-\gamma_2(x-u_2t-\Delta_2)}) \qquad (x > v^{(+)}t) \tag{4.6.48a}$$

where $\Delta_i \equiv \gamma_i^{-1}\log|\Gamma| > 0$.

4.6 THE SINE–GORDON EQUATION

In the same way, in the region $v^{(+)}t > x > v^{(-)}t$

$$\phi^{(3)} \sim 4\tan^{-1}(-e^{\gamma_1(x-u_1t+\Delta_1)}) \qquad (v^{(+)}t > x > v^{(-)}t). \qquad (4.6.48b)$$

Also, in the region $v^{(-)}t > x$,

$$\phi^{(3)} \sim 4\tan^{-1}(e^{\gamma_2(x-u_2t+\Delta_2)}) \qquad (v^{(-)}t > x). \qquad (4.6.48c)$$

Consequently, by virtue of the inequality (4.6.47), in the limit as $t \to -\infty$, for $x > u_2t + \Delta_2$, Eq. (4.6.48a) yields

$$\phi^{(3)} \to 0. \qquad (4.6.49a)$$

Also, in the same way, for $u_2t + \Delta_2 > x > u_1t - \Delta_1$, from Eqns (4.6.48a, b) it follows that

$$\phi^{(3)} \to -2\pi, \qquad (4.6.49b)$$

while for $x < u_1t - \Delta_1$, Eqns (4.6.48b, c) give

$$\phi^{(3)} \to 0. \qquad (4.6.49c)$$

We thus find that the two-kink solution $\phi^{(3)}$ comprises asymptotically the anti-kink solution with width γ_1 and velocity u_1, and the kink solution with width γ_2 and velocity u_2 (Cf. Fig. 4.8).

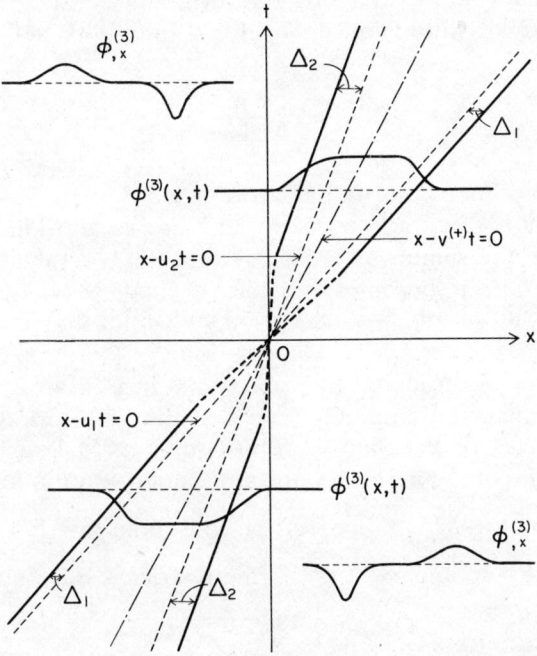

Fig. 4.8

Similarly, for $t \gg 0$, in the regions $x > v^{(+)}t_x$ and $x < v^{(+)}t$, we have

$$\phi^{(3)} \sim 4\tan^{-1}(e^{-\gamma_1(x-u_1t-\Delta_1)}) \qquad (x > v^{(+)}t), \tag{4.6.50a}$$

$$\phi^{(3)} \sim 4\tan^{-1}(e^{\gamma_2(x-u_2t+\Delta_2)}) \qquad (x < v^{(+)}t), \tag{4.6.50b}$$

respectively, and hence in the limit as $t \to \infty$, the solution reduces to

$$\phi^{(3)} \to 0 \qquad (x > u_1t + \Delta_1), \tag{4.6.51a}$$

$$\to 2\pi \qquad (u_1t + \Delta_1 > x > u_2t - \Delta_2), \tag{4.6.51b}$$

$$\to 0 \qquad (u_2t - \Delta_2 > x). \tag{4.6.51c}$$

Therefore, for $t \to \infty$, the two-kink solution again comprises the anti-kink and the kink solutions, the widths and velocities of which are given at $t = -\infty$ (cf. Fig. 4.8). As is also shown in Fig. 4.8, we notice that the phases of these kinks shift by $2\Delta_1$ and $2\Delta_2$ as a result of the overtaking process. The behaviour of the kink and the anti-kink solutions can be more easily understood in relation to solitons by considering $\phi_{,x}$. For example, both of the two-kink solutions given by Eqns (4.6.48a) and (4.6.50b) can be expressed in terms of $\phi^{(3)}_{,x}$ as

$$\frac{\partial \phi^{(3)}}{\partial x} \sim 2\gamma_2 \operatorname{sech}[\gamma_2(x - u_2t \mp \Delta_2)], \tag{4.6.52a}$$

which exhibits a compressive soliton (a bright pulse). On the other hand, for the anti-kink solutions given by Eqns (4.6.48b) and (4.6.50a), $\phi^{(3)}_{,x}$ becomes

$$\frac{\partial \phi^{(3)}}{\partial x} \sim -2\gamma_1 \operatorname{sech}[\gamma_1(x - u_1t \pm \Delta_1)]. \tag{4.6.52b}$$

giving an expansion soliton (a dark soliton).

The result of the interaction of two kink or two anti-kink solutions can be obtained in the same way. Also, by repeatedly applying the addition theorem (4.6.42), it becomes possible to find N-kink solutions. For various solutions of the S-G equation, we refer the reader to Lamb (1971).

So far, we have considered the case $\kappa^2 > 0$ in Eq. (4.6.19). If $\kappa^2 < 0$ (κ is purely imaginary), it is readily seen that the Sinh–Gordon equation is obtained in place of the Sine–Gordon equation. McLaughlin and Scott (1973) showed that the necessary and sufficient condition for the equation

$$\phi_{,\xi\eta} + c_1\phi_{,\xi} + c_2\phi_{,\eta} = F(\phi) \qquad (c_1, c_2, \text{constants}) \tag{4.6.53}$$

to admit for its solutions $\phi^{(n)}$, $\phi^{(n-1)}$ the restricted Bäcklund transformation

$$\begin{aligned}\phi^{(n)}_{,\xi} &= f(\phi^{(n)}, \phi^{(n-1)}, \phi^{(n-1)}_{,\xi}),\\ \phi^{(n)}_{,\eta} &= g(\phi^{(n)}, \phi^{(n-1)}, \phi^{(n-1)}_{,\eta}),\end{aligned} \tag{4.6.54}$$

is as follows: (a) F is linear, or (b) $c_1 = c_2 = 0$ and $d^2 F/d\phi^2 = \kappa F$ (κ is an arbitrary constant).

4.7 Bäcklund transformation and the inverse scattering method

We may suppose that in general, when the second-order equation

$$K(\phi_{xx}, \phi_{xt}, \phi_{tt}, \phi_x, \phi_t, \phi; x, t) = 0 \tag{4.7.1}$$

is given, the Bäcklund transformation for its two solutions $\phi^{(n)}$ and $\phi^{(n-1)}$ will take the form

$$\phi_x^{(n)} = P(\phi^{(n)}, \phi^{(n-1)}, \phi_x^{(n-1)}, \phi_t^{(n-1)}; x, t), \tag{4.7.2a}$$

$$\phi_t^{(n)} = Q(\phi^{(n)}, \phi^{(n-1)}, \phi_x^{(n-1)}, \phi_t^{(n-1)}; x, t), \tag{4.7.2b}$$

where the $\phi^{(n)}, \phi^{(n-1)}$ related by the Bäcklund transformation need not satisfy the same equation. That is, the Bäcklund transformation may be a transformation from a solution of one equation to that of another. As an example, suppose Eq. (4.6.11) takes the form

$$\phi_{,xt} = F(\phi) \equiv e^{\phi}, \tag{4.7.3}$$

then the Bäcklund transformation Eq. (4.7.2) becomes

$$\phi_{,x} = \phi'_{,x} - ae^{(\phi+\phi')/2}, \tag{4.7.4a}$$

$$\phi_{,t} = -\phi'_{,t} - \frac{2}{a} e^{(\phi-\phi')/2}, \tag{4.7.4b}$$

where a is an arbitrary constant. The integrability condition for Eqns (4.7.4) requires

$$\phi'_{,xt} = 0. \tag{4.7.5}$$

Hence, the Bäcklund transformation Eq. (4.7.4) is a transformation connecting solutions ϕ of Eq. (4.7.3) with the solutions ϕ' of Eq. (4.7.5). This shows that the solutions ϕ of the nonlinear equation (4.7.3) can be obtained from the solutions ϕ' of the linear equation (4.7.5) by means of a Bäcklund transformation. The Cole–Hopf transformation Eq. (1.2.18) can be regarded as a Bäcklund transformation which links the solutions u of the Burgers equation (1.2.17) and the solutions ψ of the diffusion equation (1.2.15), provided it is given in the form

$$\psi_{,x} = -\frac{\alpha}{2\mu} u\psi, \tag{4.7.6a}$$

$$\psi_{,t} = -\frac{\alpha}{2} u_{,x} + \frac{\alpha^2}{4\mu} u. \tag{4.7.6b}$$

Moreover, the Miura transformation Eq. (4.1.2) can also be regarded as a Bäcklund transformation connecting the solutions of the KdV equation and those of the modified KdV equation. It has already been stated there that if u is given, the Miura transformation Eq. (4.1.2) becomes the Riccati equation for v, and by means of the transformation (4.1.5) the Schrödinger equation (4.1.7) was derived. This suggests that in general the Riccati equation can be derived by the Bäcklund transformation, and that the transformation of the Riccati equation to a linear equation leads to the eigenvalue equation for the inverse scattering method. In view of this scheme, we shall attempt to derive the Riccati equation and the linear equation from the Bäcklund transformation of the S–G equation.

For the S–G equation of the positive type Eq. (3.9.15), the Bäcklund transformation Eq. (4.6.34) takes the form

$$\frac{\phi_{,t} + \phi'_{,t}}{2} = a^{-1} \sin \frac{\phi - \phi'}{2}, \tag{4.7.7a}$$

$$\frac{\phi_{,x} - \phi'_{,x}}{2} = a \sin \frac{\phi + \phi'}{2}. \tag{4.7.7b}$$

Then, introducing Γ by writing

$$\Gamma = \tan \frac{\phi + \phi'}{4}, \tag{4.7.8}$$

these equations become the respective Riccati equations,

$$\Gamma_{,t} - \frac{\sin \phi}{2a}(1 - \Gamma^2) + \frac{\cos \phi}{a} \Gamma = 0, \tag{4.7.9a}$$

$$\Gamma_{,x} - \frac{\phi_{,x}}{2}(1 + \Gamma^2) + a\Gamma = 0. \tag{4.7.9b}$$

As is well-known, the Riccati equations may be transformed to linear equations for ψ_1 and ψ_2 by the change of variable

$$\Gamma = \frac{\psi_2}{\psi_1} \tag{4.7.10}$$

That is, Eqns (4.7.9) are transformed, respectively, to

$$\psi_{1,t} = \frac{1}{2a}(\psi_1 \cos \phi + \psi_2 \sin \phi), \tag{4.7.11a}$$

$$\psi_{2,t} = \frac{1}{2a}(\psi_1 \sin \phi - \psi_2 \cos \phi), \tag{4.7.11b}$$

and

$$\psi_{1,x} + \frac{\phi_{,x}}{2}\psi_2 = \frac{a}{2}\psi_1, \tag{4.7.12a}$$

$$\psi_{2,x} - \frac{\phi_{,x}}{2}\psi_1 = -\frac{a}{2}\psi_2. \tag{4.7.12b}$$

It can be shown that the set of equations (4.7.12) are the eigenvalue equations for the inverse scattering method for the S–G equation (3.9.15), and that the parameter a of the Bäcklund transformation Eq. (4.7.7) corresponds to the eigenvalue. Moreover, the set of equations (4.7.11) are the equations for the time-evolution of ψ_1 and ψ_2 (Ablowitz et al. (1973a)). Conversely, starting from the equations for the inverse scattering method (the eigenvalue equation and the evolution equation for the eigenfunction), we can obtain the Riccati equation and, further, the Bäcklund transformation. In order to show those results in a general form, we use the Ablowitz, Kaup, Newell and Segur (AKNS) formalism (Ablowitz et al. (1973b), (1974)), which is an extension of the theory by Zakharov and Shabat for the nonlinear Schrödinger equation given in Section 4.5.

Consider the eigenvalue equation for the linear operator L

$$L\psi = \zeta\psi \tag{4.7.13}$$

and the evolution equation for ψ,

$$i\psi_{,t} = A\psi. \tag{4.7.14}$$

Here the linear operator L is the 2×2 matrix,

$$L \equiv \begin{pmatrix} i\,\partial/\partial x & -iq(x,t) \\ ir(x,t) & -i\,\partial/\partial x \end{pmatrix}, \tag{4.7.15}$$

ψ is the vector

$$\psi \equiv \begin{pmatrix} \psi_1 \\ \psi_2 \end{pmatrix} \tag{4.7.16}$$

and A is the matrix given by

$$A \equiv \begin{pmatrix} \mathcal{A}(x,t;\zeta) & \mathcal{B}(x,t;\zeta) \\ \mathcal{C}(x,t;\zeta) & -\mathcal{A}(x,t;\zeta) \end{pmatrix}, \tag{4.7.17}$$

in which $\mathcal{A}, \mathcal{B}, \mathcal{C}$ are functions of $x, t,$ and ζ. Since L and A are not self-adjoint, the eigenvalue ζ is complex, but it will be assumed to be time-independent. Also, $q(x,t)$, $r(x,t)$, the elements of L, are solutions to a nonlinear evolution equation which will be specified later, so that these do not explicitly depend on ζ. The linear operators L and A may be regarded as a generalization of the operators L and A (Eqns (4.5.3) and (4.5.4)) for the nonlinear Schrödinger equation (4.5.1). (In what follows, the system of equations (4.7.13) to (4.7.17) will be called the AKNS equations.)

The integrability condition for Eqns (4.7.13) and (4.7.14) is obtained by cross-differentiation of these equations. In order that the eigenvalue ζ be time-independent, the following relationships between $q(x, t)$, $r(x, t)$ and the elements $\mathcal{A}(x, t; \zeta)$, $\mathcal{B}(x, t; \zeta)$, $\mathcal{C}(x, t; \zeta)$ must hold:

$$\mathcal{A}_{,x} = q\mathcal{C} - r\mathcal{B}, \tag{4.7.18a}$$

$$\mathcal{B}_{,x} + 2i\zeta\mathcal{B} = iq_{,t} - 2\mathcal{A}q, \tag{4.7.18b}$$

$$\mathcal{C}_{,x} - 2i\zeta\mathcal{C} = ir_{,t} + 2\mathcal{A}r. \tag{4.7.18c}$$

By way of illustration, let $\mathcal{A}$ be given by a cubic polynomial of ζ,

$$\mathcal{A}(x, t; \zeta) = 4\zeta^3 + 2qr\zeta + irq_{,x} - iqr_{,x}, \tag{4.7.19}$$

then by Eqns (4.7.18), $\mathcal{B}$ and $\mathcal{C}$ are determined successively, in descending powers of ζ, by

$$\mathcal{B}(x, t; \zeta) = 4iq\zeta^2 - 2q_{,x}\zeta + i(2q^2r - q_{,xx}), \tag{4.7.20a}$$

$$\mathcal{C}(x, t; \zeta) = 4ir\zeta^2 + 2r_{,x}\zeta + i(2qr^2 - r_{,xx}). \tag{4.7.20b}$$

In particular, the terms independent of ζ (i.e. involving ζ^0) yield the evolution equations

$$q_{,t} - 6rqq_{,x} + q_{,xxx} = 0, \tag{4.7.21a}$$

$$r_{,t} - 6rqr_{,x} + r_{,xxx} = 0. \tag{4.7.21b}$$

If we choose

$$r = 1, \qquad q(x, t) = u(x, t), \tag{4.7.22}$$

Eq. (4.7.21) reduces to the KdV equation (4.1.3), while the eigenvalue equation (4.7.13) becomes the Schrödinger equation when we set $\psi_2 = \psi$, $\zeta^2 = \lambda$. Also, for the choice

$$r(x, t) = q(x, t) = v(x, t), \tag{4.7.23}$$

Eq. (4.7.21) becomes the modified KdV equation (4.1.1) (Wadati, 1972)).

If $\mathcal{A}$ is quadratic in ζ, so that

$$\mathcal{A}(x, t; \zeta) = 2\zeta^2 + qr, \tag{4.7.24}$$

($\mathcal{B}$ and $\mathcal{C}$ are linear in ζ) then by Eq. (4.7.18) we have

$$\mathcal{B}(x, t; \zeta) = 2iq\zeta - q_{,x}, \tag{4.7.25a}$$

$$\mathcal{C}(x, t; \zeta) = 2ir\zeta + r_{,x}. \tag{4.7.25b}$$

4.7 BÄCKLUND TRANSFORMATION

and

$$iq_{,t} + q_{,xx} - 2q^2 r = 0, \quad (4.7.26a)$$
$$ir_{,t} - r_{,xx} + 2qr^2 = 0. \quad (4.7.26b)$$

Therefore, for

$$r(x, t) = -\frac{Q}{2} q^*(x, t), \qquad q(x, t) = u(x, t), \quad (4.7.27)$$

Eqns (4.7.26) yield the nonlinear Schrödinger equation (4.5.1). On the other hand, if $\mathcal{A}$ is given as an inverse power of ζ, say by

$$\mathcal{A}(x, t; \zeta) = -\frac{\cos \phi}{4\zeta}, \quad (4.7.28)$$

Eqns (4.7.18) give

$$\mathcal{B}(x, t; \zeta) = \frac{q_{,t}}{2\zeta}, \quad (4.7.29a)$$

$$\mathcal{C}(x, t; \zeta) = -\frac{r_{,t}}{2\zeta}, \quad (4.7.29b)$$

and

$$(\cos \phi)_{,x} = 2(qr)_{,t}, \quad (4.7.30a)$$
$$q_{,xt} = q \cos \phi, \quad (4.7.30b)$$
$$r_{,xt} = r \cos \phi. \quad (4.7.30c)$$

Consequently, putting

$$r(x, t) = -q(x, t) = \frac{\phi_{,x}}{2}, \quad (4.7.31)$$

Eq. (4.7.30) becomes the Sine–Gordon equation (3.9.15), while $\mathcal{B}$ and $\mathcal{C}$ are given by

$$\mathcal{B}(x, t; \zeta) = \mathcal{C}(x, t; \zeta) = -\frac{\sin \phi}{4\zeta}. \quad (4.7.32)$$

Here it should be remarked that, as is seen from these examples, $\mathcal{A}$, $\mathcal{B}$, $\mathcal{C}$ obey the boundary conditions

$$\mathcal{A}(x, t, \zeta) \to \mathcal{A}_0(\zeta),$$
$$\mathcal{B}(x, t, \zeta) \to 0,$$
$$\mathcal{C}(x, t, \zeta) \to 0,$$

as $|x| \to \infty$, and $\mathcal{A}_0$ is determined by the linear dispersion relation of the evolution equation. Namely, noting that q, r decay (exponentially) rapidly as $|x| \to \infty$, we find from Eq. (4.7.19) that for the KdV equation and the modified KdV equation $\mathcal{A}_0 \propto \zeta^3$ and for the nonlinear Schrödinger equation and the S–G equation, $\mathcal{A}_0 \propto \zeta^2$ from Eq. (4.7.24) and $\mathcal{A}_0 \propto \zeta^{-1}$ from Eq. (4.7.28), respectively. It was shown by Ablowitz et al. (1974) that the following coupled nonlinear evolution equation can be solved by the inverse scattering method of the AKNS equations (4.7.13) and (4.7.14):

$$\begin{pmatrix} r_{,t} \\ -q_{,t} \end{pmatrix} + 2\mathcal{A}_0(L^+)\begin{pmatrix} r \\ q \end{pmatrix} = 0, \qquad (4.7.33)$$

where L^+ is the integro-differential operator

$$L^+ \equiv \frac{1}{2i}\begin{pmatrix} \dfrac{\partial}{\partial x} - 2r\int_{-\infty}^{x} dy\, q & 2r\int_{-\infty}^{r} dy\, r \\ -2q\int_{-\infty}^{x} dy\, q & -\dfrac{\partial}{\partial x} + 2q\int_{-\infty}^{x} dy\, r \end{pmatrix}.$$

It is clear that $A_0(\zeta)$ is directly related to the dispersion relation of the linearized version of equation (4.7.33). In this regard, the method of solution of the nonlinear evolution equation by means of the inverse scattering method is called by those authors (Ablowitz et al.) the nonlinear Fourier transformation. For a linear equation the process illustrated in Fig. 4.1 is replaced by the usual procedure based on the Fourier and the inverse Fourier transformations. For a nonlinear equation, a spectrum of a certain linear operator must be used in place of the plane wave so that the time-evolution equation for each spectrum component becomes separable. The crucial point exhibited by Eq. (4.7.33) is that the nonlinear equation which can be solved in this way has a correspondence to the linearized equation.

As a by-product of Eq. (4.7.33), Ablowitz et al. also showed that the class of nonlinear equations for which the Schrödinger equation is the appropriate scattering problem is given by

$$q_{,t} + C_0(L_s^+)q_{,x} = 0,$$

with

$$L_s^+ \equiv -\frac{1}{4}\frac{\partial^2}{\partial x^2} - q + \tfrac{1}{2}q_{,x}\int_{x}^{\infty} y,$$

and $C_0(k^2)$ is related to the phase speed in the linearized problem.

We now return to our attempt to obtain the Bäcklund transformation from the inverse scattering method. Introducing Eqns (4.7.28), (4.7.31), and (4.7.32) into the AKNS equations we obtain the eigenvalue equation (and the evolution equation for the scattering data) for the S–G equation (3.9.15). These are found to be in agreement with the linear equations

(4.7.11) and (4.7.12) derived from the Bäcklund transformation, in which case, the parameter a of the Bäcklund transformation is related to the eigenvalue by $a = -2i\zeta$.

That is to say, for the S–G equation, we have proved that:

Bäcklund transformation → Riccati equations → AKNS equations.

It is possible to reverse the process of proof. By the transformations

$$\Gamma_1 = \psi_2/\psi_1 \quad \text{and} \quad \Gamma_2 = \psi_1/\psi_2, \tag{4.7.34}$$

the AKNS equations (4.7.13), (4.7.14) are transformed into the Riccati equations

$$\Gamma_{1,x} = 2i\zeta\Gamma_1 + r - q\Gamma_1^2, \tag{4.7.35a}$$

$$\Gamma_{2,x} = -2i\zeta\Gamma_2 + q - r\Gamma_2^2, \tag{4.7.35b}$$

and

$$\Gamma_{1,t} = i(2\mathscr{A}\Gamma_1 - \mathscr{C} + \mathscr{B}\Gamma_1^2), \tag{4.7.36a}$$

$$\Gamma_{2,t} = -i(2\mathscr{A}\Gamma_2 + \mathscr{B} - \mathscr{C}\Gamma_2^2), \tag{4.7.36b}$$

respectively. Hence, for the S–G equation, by Eq. (4.7.35) and the transformation Eq. (4.7.8) we obtain the Bäcklund transformation. That is, we have proved the inverse of the above result, namely:

AKNS equations → Riccati equations
 → the Bäcklund transformation.

In general, for the evolution equation (4.7.33) which is solvable by the inverse scattering method and the AKNS equations, the Bäcklund transform can be derived by a suitable transformation, such as Eq. (4.7.8), from the Riccati equations (4.7.35) and (4.7.36) (Chen (1974), Ablowitz et al. (1974)). Thus, we find that through the Riccati equation, the inverse scattering method and the Bäcklund transformation are closely related to each other. On the other hand, as was remarked upon in Section 4.2, the infinite number of conservation laws of the KdV equation can be derived from the Miura transformation. Hence, if the connection of the Miura transformation with the Riccati equation is taken into account, it suggests that the Riccati equations (4.7.35a, b) derived from the AKNS equations will make it possible to derive an infinite number of conservation laws.

In fact, from Eqns (4.7.18a, b), the conservation law

$$(q\Gamma_1)_{,t} + i(\mathscr{A} + \mathscr{B}\Gamma_1)_{,x} = 0 \tag{4.7.37a}$$

follows at once. Similarly, we have

$$(r\Gamma_2)_{,t} + i(-\mathscr{A} + \mathscr{B}\Gamma_2)_{,x} = 0. \tag{4.7.37b}$$

Hence, $q\Gamma_1$, $r\Gamma_2$ are conserved densities. Now Eq. (4.7.35a) can be written as

$$q\Gamma_1 = \frac{1}{2i\zeta}[(q\Gamma_1)^2 - qr + q\Gamma_{1,x}], \qquad (4.7.38)$$

and therefore, expanding $q\Gamma_1$ in powers of ζ^{-1}, we find

$$q\Gamma_1 = \sum_{n=1}^{\infty} f_n \zeta^{-n}, \qquad (4.7.39)$$

which yields the recurrence formula for f_n,

$$f_{n+1} = \frac{1}{2i}\left[\sum_{k=1}^{n-1} f_k f_{n-k} - (rq)\,\delta_{n,0} + q\left(\frac{f_n}{q}\right)_{,x}\right]. \qquad (4.7.40)$$

Substituting for $q\Gamma_1$ in Eq. (4.7.37a) from the expression (4.7.39), for each power of ζ^{-1} we obtain an infinite number of conserved quantities. In this way, the inverse scattering method, the Bäcklund transformation and the conservation laws are mutually related through the Riccati equation. The relation given in Fig. 4.9 is thus proved, on the basis of the AKNS equations, for the KdV equation, the S–G equation and the modified KdV equation (Wadati et al. (1975)). As can be seen from the discussion of the KdV equation in Section 4.2, the existence of an infinite number of conservation laws implies that such a system is completely integrable; that is, that the Hamiltonian is separable, and this has been shown, for example, by McLaughlin (1975a, b), Kodama (1975), and others. We finally note that by extending the dimension of the matrices L and A in the AKNS equations, the equation for three-wave interactions (3.8.54) (Zakharov and Manakov (1976)) and the equation for self-induced transparency (the Bloch equations (3.9.8) and (3.9.9)) (Lamb (1973)) can be solved similarly by means of the inverse scattering method.

However, to get only the N-soliton solutions, the inverse scattering method need not be used, and the Bäcklund transformation may be used instead. However, we mention that the method due to Hirota is particularly useful (Hirota (1976); Hirota and Satsuma (1976)). As will be shown in the next section, this method makes it possible to get multiple-soliton

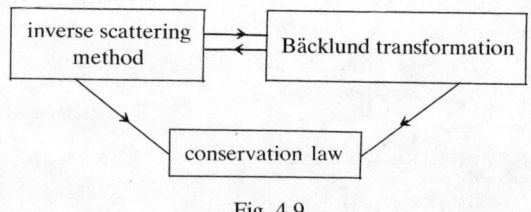

Fig. 4.9

solutions for a system which has not been solved by the inverse scattering method such as the Boussinesq equation (Jeffrey and Kawahara (1982)).

4.8 Hirota's method

A method of solution which yields multi-soliton solutions by directly solving nonlinear evolution equations has been established by Hirota and is called Hirota's method. The method can be applied not only to equations solvable by the inverse scattering method, but also to some equations which are not completely integrable. The method is also being extended so that it is applicable to initial value problems, periodic solutions and multi-dimensional problems.

In order to show the basic idea of this unique method, following Hirota we first consider the one-soliton solution of the modified KdV equation

$$u_{,t} + \alpha(u^3)_{,x} + u_{,xxx} = 0 \qquad (\alpha > 0). \tag{4.8.1}$$

By means of the standard method the following soliton solution is readily obtained:

$$u = (2/\alpha)^{1/2} p \,\text{sech}\, [p(x - \lambda t) + \delta], \tag{4.8.2}$$

where p and δ are constant and $\lambda = p^2$. As was already noted in Section 4.1 for the KdV equation, as $|x| \to \infty$ the solution damps exponentially so that the linear approximation to Eq. (4.8.1) is valid, for example, at $x = -\infty$. We thus have $u \propto e^\theta$ with $\theta = px - \Omega t + \delta$, where p is positive and $\Omega = p^3$. We remark that the above asymptotic form is obtained if k is replaced by $-ip$ in the plane wave solution $\exp(ikx - iwt)$.

Hirota noticed that when a solution decays exponentially as $|x| \to \infty$, then at $x = -\infty$, for example, the exact solution may be obtained by successive approximation, starting from the linear solution, in the form

$$u = \varepsilon u_1 + \varepsilon^2 u_2 + \cdots, \tag{4.8.3}$$

where ε is a parameter specifying the smallness of u (given by Eq. (4.8.2)) due to the exponential decay of u as $|x| \to \infty$. Then, substituting for u in Eq. (4.8.1), and setting

$$\mathscr{L} \equiv \frac{\partial}{\partial t} + \frac{\partial^3}{\partial x^3}$$

we have as a result of collecting terms involving the same power of ε,

$$\mathscr{L} u_1 = 0,$$
$$\mathscr{L} u_3 = -\alpha (u_1^3)_{,x}$$
$$\mathscr{L} u_5 = -3\alpha (u_1^2 u_3)_{,x}.$$

This set of equations can be integrated successively to give

$$u_1 = e^\theta,$$
$$u_3 = -(\alpha/8p^2)e^{3\theta},$$
$$u_5 = (\alpha/8p^2)^2 e^{5\theta},$$
$$\cdots$$
$$u_{2n+1} = (-\alpha/8p^2)^n e^{(2n+1)\theta}.$$
$$\cdots$$

These can be summed to give

$$u = \frac{\varepsilon e^\theta}{1 + \varepsilon^2 (\alpha/8p^2) e^{2\theta}}, \qquad (4.8.4)$$

which for $\varepsilon = 1$ is merely the one-soliton solution Eq. (4.8.2). Namely, for $p > 0$, although each term u_{2n+1} diverges as $x \to +\infty$, summing all the terms yields the correct result for all x; that is, analytic continuation from $x \to -\infty$ to $x \to +\infty$ is achieved in this way for any time. (For example, $f(x) = 1 + x + x^2 + x^3 + \cdots$ diverges for $|x| > 1$, but if it is summed then $f(x) = 1/(1-x)$, which is well defined for $|x| > 1$.)

This procedure applies, in principle, also to the KdV equation (4.1.3). Expanding u in the same manner as in Eq. (4.8.3), we have the system

$$\left. \begin{array}{l} \mathscr{L} u_1 = 0 \\ \mathscr{L} u_2 = 6 u_1 u_{1,x} \\ \mathscr{L} u_3 = 6(u_2 u_{1,x} + u_1 u_{2,x}) \\ \cdots, \end{array} \right\} \qquad (4.8.5)$$

which can be integrated successively to give the infinite series

$$u = \sum_{n=1}^{\infty} a_n (\varepsilon e^\theta)^n.$$

In this case the coefficients $\{a_n \mid n = 1, 2, \ldots\}$ do not take on simple forms as in Eq. (4.8.3), so that it is not easy to sum the infinite series to obtain the one-soliton solution Eq. (4.1.11). However, noting that Eq. (4.1.11) can also be written as an entire function of e^θ similar to Eq. (4.8.4), that is, as

$$u = -\frac{2p^2 e^\theta}{(1+e^\theta)^2} \qquad (p = 2\kappa), \qquad (4.8.6)$$

Hirota assumed the following form for u:

$$u = G/F, \qquad (4.8.7)$$

where G and F are polynomials of e^θ, to be determined such that u

4.8 HIROTA'S METHOD

satisfies the KdV equation. (In this sense, Hirota's method may be regarded as a Padé approximation.) By using such an approach he succeeded in obtaining not only the one-soliton solution but also the N-soliton solution.

In what follows, for the sake of better understanding, we first use the potential ϕ introduced in Eq. (4.2.17), the equation for which becomes

$$\phi_{,t} - 3(\phi_{,x})^2 + \phi_{,xxx} = 0, \tag{4.8.8}$$

where we have assumed that $\phi \to 0$ as $x \to -\infty$. (We note that ϕ corresponds to $K(x, x)$ and Eq. (4.1.57) shows that $\phi = -2\Delta_{,x}/\Delta$, so that ϕ is related to Δ by the Cole–Hopf transformation Eq. (1.2.18).) Corresponding to Eq. (4.8.7) we put

$$\phi = g/f, \tag{4.8.9}$$

then Eq. (4.8.8) becomes

$$\begin{aligned}(g_{,t}f - gf_{,t})/f^2 - 3(g_{,x}f - gf_{,x})^2/f^4 \\ + (g_{,xxx}f - 3g_{,xx}f_{,x} - 3g_{,x}f_{,xx} - gf_{,xxx})/f^2 \\ + 6(fg_{,x}f_{,x}^2 + fgf_{,xx}f_{,x} - gf_{,x}^3)/f^4 = 0,\end{aligned} \tag{4.8.10}$$

where the last two terms give the dispersion $\phi_{,xxx}$ and the second term is the nonlinear term.

The first procedure in Hirota's method involves a decoupling, that is, the developing of Eq. (4.8.10) into two equations which are given by the bilinear form which will be explained later. Finding the decoupling is, in general, heuristic, but in this case from Eq. (4.1.57) it is readily deduced that setting $g = -2f_{,x}$ leads to decoupling. The nonlinear term then becomes

$$-12(f_{,xx}f - f_{,x}^2)^2/f^4.$$

Another term containing the factor f^{-4} is the last term which reduces to

$$12(-2ff_{,xx}f_{,x}^2 + f_{,x}^4)/f^4.$$

Hence, by adding to this the term $12f^2f_{,xx}^2/f^4$, we cancel the nonlinear term.

Now we must subtract from the remaining part of Eq. (4.8.10) this same term which is equivalent to changing the sign of $-3g_{,x}f_{,xx}/f^2$ in the third term of Eq. (4.8.10). We thus obtain the result

$$f(f_{,t} + f_{,xxx})_{,x} - f_{,x}(f_{,t} + f_{,xxx}) - 3(f_{,xxx}f_{,x} - f_{,xx}^2) = 0. \tag{4.8.11}$$

This equation can be solved for f by means of the expansion at $x = -\infty$.

$$f = f_0 + \varepsilon f_1 + \varepsilon^2 f_2 + \cdots. \tag{4.8.12}$$

Namely, we have the successive results

$$f_0 = \text{constant}$$
$$\mathcal{L}f_{1,x} = 0$$
$$\mathcal{L}f_{2,x} = 3(f_{1,xxx}f_{1,x} - f_{1,xx}^2) \qquad (4.8.13)$$
$$\mathcal{L}f_{3,x} = -(f_1 \mathcal{L}f_{2,x} - f_{1,x}\mathcal{L}f_2)$$
$$\qquad + 3(f_{2,xxx}f_{1,x} + f_{1,xxx}f_{2,x} - 2f_{1,xx}f_{2,xx}),$$

and so on.

Since Eq. (4.8.11) is invariant with respect to multiplication of f by a constant, without loss of generality f_0 may be equated to unity. To obtain the one-soliton solution we take the special solution for f_1, $f_1 = e^\theta$. Then it is readily seen that f_2 can be equated to zero and all the higher order terms f_n ($n \geq 2$) can also be made equal to zero, so that $f_1 = 1 + \varepsilon e^\theta$. Consequently, putting $\varepsilon = 1$, we get

$$\phi = -\frac{2pe^\theta}{1+e^\theta},$$

and differentiation of this immediately yields Eq. (4.8.6). The two-soliton solution can be obtained in similar fashion. Let f_1 be given as $f_1 = e^{\theta_1} + e^{\theta_2}$, where $\theta_i = p_i t - \Omega_i x + \delta_i$, $\Omega_i = p_i^3$, so that $\mathcal{L}e^{\theta_i} = 0$. Then we have an equation for f_2 with the forcing term $3p_1 p_2 (p_1 - p_2)^2 e^{\theta_1 + \theta_2}$, which may be integrated to give

$$f_2 = [(p_1 - p_2)^2 / (p_1 + p_2)^2] e^{\theta_1 + \theta_2}. \qquad (4.8.14)$$

Introducing f_1 and f_2 thus obtained into the third order equation for f_3, after some tedious but straightforward computations we find that f_3 can be equated to zero, as can all the higher order terms.

However, for the three-soliton solution the computations become laborious. In order to overcome this difficulty, following Hirota we rewrite Eq. (4.8.11) in the following bilinear form:

$$D_x(D_t + D_x^3)f \cdot f = 0. \qquad (4.8.15)$$

Here the operator D_x is defined by

$$D_x a \cdot b = \left(\frac{\partial}{\partial x} - \frac{\partial}{\partial x'}\right) a(x) b(x')\big|_{x=x'} \qquad (4.8.16)$$
$$= a_{,x} b - a b_{,x}.$$

Consequently,

$$D_x^n a \cdot b = \left(\frac{\partial}{\partial x} - \frac{\partial}{\partial x'}\right)^n a(x) b(x')\big|_{x=x'}, \qquad (4.8.16)'$$

4.8 HIROTA'S METHOD

from which it follows that $D_x^n a \cdot a = 0$ if n is odd. D_t is defined similarly, and hence

$$D_x D_t f \cdot f = (\partial_x - \partial_{x'})(\partial_t - \partial_{t'}) f(x, t) f(x', t')\big|_{\substack{x=x' \\ t=t'}}$$
$$= 2(f f_{,xt} - f_{,x} f_{,t})$$

and Eq. (4.8.16)' yields (for $n = 4$)

$$D_x \cdot D_x^3 f \cdot f = 2(f_{,xxxx} f - 4 f_{,xxx} f_{,x} + 3 f_{,xx}^2).$$

From these formulae we find the equivalence of Eqns (4.8.11) and (4.8.15). It is worth noticing that the bilinear form Eq. (4.8.15) is expressible as

$$D_x \mathscr{L}(D) f \cdot f = 0 \qquad (4.8.15)'$$

where $\mathscr{L}(D)$ is now defined by replacing $\partial/\partial x$ and $\partial/\partial t$ in $\mathscr{L}(\partial)$ by D_x and D_t, respectively. Hereafter $D_x \mathscr{L}(D)$ will be written as $\bar{\mathscr{L}}(D)$. By definition, $D_x f \cdot 1 = \partial_x f$, and hence

$$\bar{\mathscr{L}}(D) f \cdot 1 = \bar{\mathscr{L}}(\partial) f. \qquad (4.8.17)$$

We now solve Eq. (4.8.15)' by means of the expansion Eq. (4.8.12). The equations corresponding to Eq. (4.8.13) become

$$\bar{\mathscr{L}}(D)(f_1 \cdot 1 + 1 \cdot f_1) = 0 \qquad (4.8.18a)$$
$$\bar{\mathscr{L}}(D)(f_2 \cdot 1 + f_1 \cdot f_1 + 1 \cdot f_2) = 0 \qquad (4.8.18b)$$
$$\bar{\mathscr{L}}(D)(f_3 \cdot 1 + f_2 \cdot f_1 + f_1 \cdot f_2 + 1 \cdot f_3) = 0, \qquad (4.8.18c)$$

and so on. By means of the formulae (4.8.17), we see immediately that Eq. (4.8.18a) is satisfied by $f_1 = e^\theta$. Then Eq. (4.8.18b) reduces to

$$\bar{\mathscr{L}}(\partial) f_2 + \bar{\mathscr{L}}(D) e^\theta \cdot e^\theta = 0.$$

It is readily seen that f_2 can be equated to zero by virtue of the property of the D operator

$$D_x^m e^{p_1 x} \cdot e^{p_2 x} = (p_1 - p_2)^m e^{(p_1 + p_2) x}. \qquad (4.8.19)$$

so that $\bar{\mathscr{L}}(D) e^\theta \cdot e^\theta = 0$. Thus we obtain the one-soliton solution $f_1 = e^\theta$. For the two-soliton solution, $f_1 = e^{\theta_1} + e^{\theta_2}$ similarly satisfies Eq. (4.8.18a) because of (4.8.17). Then Eq. (4.8.18b) becomes

$$2 \mathscr{L}(\partial) f_2 + \bar{\mathscr{L}}(D) [(e^{\theta_1} + e^{\theta_2}) \cdot (e^{\theta_1} + e^{\theta_2})] = 0,$$

which may be integrated to give $f_2 = e^{a_{12} + \theta_1 + \theta_2}$, where

$$e^{a_{12}} = e^{a_{21}} = -\frac{\bar{\mathscr{L}}(\Omega_1 - \Omega_2, p_1 - p_2)}{\bar{\mathscr{L}}(\Omega_1 + \Omega_2, p_1 + p_2)} = \frac{(p_1 - p_2)^2}{(p_1 + p_2)^2}. \qquad (4.8.20)$$

Introducing f_1 and f_2 into Eq. (4.8.18c), we obtain terms like

$$\bar{\mathcal{L}}(D)(e^{\theta_1+\theta_2} \cdot e^{\theta_1}),$$

which, by use of Eq. (4.8.19), are computed as $D_x \mathcal{L}(\Omega_2, p_2) e^{2\theta_1+\theta_2}$, and consequently vanish because of the dispersion relation $\mathcal{L}(\Omega_i, p_i) = 0$. Therefore, we see immediately that f_3 can be equated to zero, as can all the higher order f_n.

The above property of the bilinear form also enables us to obtain three-soliton solutions quite easily. That is, assuming

$$f_1 = e^{\theta_1} + e^{\theta_2} + e^{\theta_3}$$

gives

$$f_2 = a^{a_{12}+\theta_1+\theta_2} + e^{a_{23}+\theta_2+\theta_3} + e^{a_{31}+\theta_3+\theta_1},$$

since in Eq. (4.8.18c) all terms like $\bar{\mathcal{L}}(D) e^{\theta_1} \cdot e^{(\theta_1+\theta_2)}$ vanish because of (4.8.19).

In terms of the notation $\bar{\mathcal{L}}(p_1+p_2-p_3) \equiv \bar{\mathcal{L}}(\Omega_1+\Omega_2-\Omega_3, p_1+p_2-p_3)$, f_3 is given as

$$f_3 = [e^{a_{12}} \bar{\mathcal{L}}(p_1+p_2-p_3) \\
+ e^{a_{23}} \bar{\mathcal{L}}(p_2+p_3-p_1) \\
+ e^{a_{31}} \bar{\mathcal{L}}(p_3+p_1-p_2)]/\bar{\mathcal{L}}(p_1+p_2+p_3) \\
\times e^{\theta_1+\theta_2+\theta_3} = e^{a_{12}+a_{23}+a_{31}+\theta_1+\theta_2+\theta_3},$$

in which we have used the identity

$$(p_1-p_2)^2 (p_2+p_3)^2 (p_3+p_1)^2 \bar{\mathcal{L}}(p_1+p_2-p_3) \\
+ (p_2-p_3)^2 (p_3+p_1)^2 (p_1+p_2)^2 \bar{\mathcal{L}}(p_2+p_3-p_1) \\
+ (p_3-p_1)^2 (p_1+p_2)^2 (p_2+p_3)^2 \bar{\mathcal{L}}(p_3+p_1-p_2) \\
+ (p_1-p_2)^2 (p_2-p_3)^2 (p_3-p_1)^2 \bar{\mathcal{L}}(p_1+p_2+p_3) = 0. \quad (4.8.21)$$

This is proved as follows: let the left-hand side of Eq. (4.8.21) be denoted by Q, then Q is a symmetric homogeneous polynomial in p_1, p_2 and p_3 of degree 8, and vanishes if $p_1^2 = p_2^2$, $p_2^2 = p_3^2$ or $p_3^2 = p_1^2$, so that $Q \propto (p_1^2-p_2^2)^2 (p_2^2-p_3^2)^2 (p_3^2-p_1^2)^2$, which is a polynomial of degree 12, which contradicts the fact that it is of degree 8. Hence Q must vanish identically.

The identity can be generalized to the case of n variables $p_1, p_2, \ldots, p_n$ (Hirota, 1971):

$$\sum_{\sigma_1,\sigma_2,\ldots,\sigma_n} b(\sigma_1 p_1, \sigma_2 p_2, \ldots, \sigma_n p_n) \bar{\mathcal{L}}(\sigma_1 p_1, \sigma_2 p_2, \ldots, \sigma_n p_n) = 0.$$

$$(4.8.22)$$

Here, b is defined by

$$b(\sigma_1 p_1, \sigma_2 p_2, \ldots, \sigma_n p_n) = \prod_{k<l}^{(n)} (\sigma_k p_k - \sigma_l p_l)^2,$$

where (n) indicates the product of all possible combinations of the n elements (with the specified condition $k<l$, as indicated), the summation being over all possible combinations of $\sigma_1 = \pm 1, \sigma_2 = \pm 1, \ldots, \sigma_n = \pm 1$. By virtue of the identity, for the N-soliton solution f can be expressed as

$$f = \sum_{\substack{\mu_i = 0,1 \\ \text{for all } i}} \exp\left(\sum_{i<j}^{(N)} \mu_i \mu_j a_{ij} + \sum_{i=1}^{N} \mu_i \theta_i\right), \tag{4.8.23}$$

where the summation over $\mu_i = 0, 1$ for all i means summation for all possible combinations of $\mu_i = 1, 0$ (for example, for $N = 3$, all of μ_i $(i = 1, 2, 3) = 0$ gives 1, and the combination $\mu_1 = 1$, $\mu_2 = \mu_3 = 0$ gives e^{θ_1} etc.), and $\sum_{i>j}^{(N)}$ indicates summation over all pairs chosen from N elements.

In the derivation of N-soliton solution by means of Hirota's method, the explicit form of $\mathscr{L}$ is not used, but the identity Eq. (4.8.22) is crucial. In fact, it was proved by Hirota that in general a differential equation in the bilinear form

$$F(D_t, D_x, D_y, \ldots) f \cdot f = 0 \tag{4.8.24}$$

where F satisfies the conditions

$$F(0, 0, 0, \ldots) = 0, \qquad F(D_t, D_x, D_y, \ldots) = F(-D_t, -D_x, -D_y, \ldots),$$
$$F(\Omega_i, p_i, q_i, \ldots) = 0 \quad \text{for real sets of } \Omega_i, p_i, q_i, \ldots (i = 1, 2, \ldots, N)$$

admits a solution of the form Eq. (4.8.23), where $\theta_i = \Omega_i t + p_i x + q_i y + \cdots$, provided that the identity

$$\sum_{\sigma = \pm 1} F\left(\sum_{i=1}^{n} \sigma_i \Omega_i, \sum_{i=1}^{n} \sigma_i p_i, \ldots\right)$$
$$\times \prod_{i<j}^{(n)} F(\sigma_i \Omega_i - \sigma_j \Omega_j, \sigma_i p_i - \sigma_j p_j, \ldots) \sigma_i \sigma_j = 0 \tag{4.8.25}$$

is valid for $n = 1, 2, \ldots, N$.

If F is connected with solutions to a nonlinear evolution equation through F_x/F, the above solution gives the N-soliton solutions. Since Eq. (4.8.25) is trivial for $n = 1$ and 2, it follows that the two-soliton solution exists whenever Eq. (4.8.24) is valid.

We thus find that Hirota's method involves transforming a nonlinear evolution equation into the bilinear differential equation (4.8.24).

If this is possible, the two-soliton solution is obtained by setting
$$F = 1 + e^{\theta_1} + e^{\theta_2} + e^{a_{12}+\theta_1+\theta_2},$$
where
$$e^{a_{12}} = \frac{F(\Omega_1-\Omega_2, p_1-p_2, q_1-q_2, \ldots)}{F(\Omega_1+\Omega_2, p_1+p_2, q_1+q_2, \ldots)}.$$

Furthermore, if the identity Eq. (4.8.25) is valid, the N-soliton solution can be obtained by means of Eq. (4.8.23). Conversely, transforming the bilinear differential equation (4.8.23) with the identity Eq. (4.8.25) back to the usual differential equation, Hirota obtained a number of nonlinear evolution equations which admit N-soliton solutions. Here we mention the following three equations:

(i) The Boussinesq equation
$$u_{,tt} - u_{,xx} = (3u^2)_{,xx} + u_{,xxxx}$$
is transformed into
$$(D_t^2 - D_x^2 - D_x^4)f \cdot f = 0;$$

(ii) The Kadomtsev–Petviashvili equation (Kadomtsev, Petviashvili (1970)) (the weakly two-dimensional KdV equation)
$$u_{,tx} + 6(uu_{,x})_{,x} + u_{,xxxx} \pm u_{,yy} = 0$$
is transformed into
$$(D_t D_x + D_x^4 \pm D_y^2)f \cdot f = 0;$$

(iii) A model equation for shallow water
$$u_{,t} - u_{,xxt} - 3uu_{,t} + 3u_{,x} \int_x^\infty u_{,t}\, dx' + u_{,x} = 0$$
is transformed into
$$D_x(D_t - D_t D_x^2 + D_x)f \cdot f = 0.$$

In all of these three equations u and f are related by
$$u = 2(\log f)_{,xx}.$$

However physically important nonlinear equations, such as the modified KdV equation, the nonlinear Schrödinger equation and the Sine–Gordon equation, etc., cannot be transformed into a single bilinear equation but into a set of bilinear equations (cf. Problems 4.8.2, 3, 4). Even in those cases though, N-soliton solutions are obtained when the expansion terminates at a finite order. Therefore, in essence, the procedure is the same in all of those cases.

4.8 HIROTA'S METHOD

Finally, we briefly introduce the Bäcklund transformation in the context of Hirota's method. Let f and f' be two different solutions of a bilinear equation $F(D)f \cdot f = 0$. Then the bilinear relations relating the two solutions $F_1(D)f \cdot f' = 0$ and $F_2(D)f \cdot f' = 0$ are called by Hirota the Bäcklund transformation in bilinear form. These may be derived from the identity

$$[F(D)f \cdot f]f'f' - ff[F(D)f' \cdot f'] = 0. \tag{4.8.26}$$

For example, for the KdV equation $F(D) = \mathcal{L}(D)$, Eq. (4.8.26) can be transformed into

$$2D_x[\{D_t + 3\lambda D_x + D_x^3\}f' \cdot f] \cdot ff' + 6D_x[(D_x^2 - \lambda)f' \cdot f](D_x f \cdot f') = 0, \tag{4.8.26}'$$

where λ is an arbitrary constant. This may be done straightforwardly, but it is accomplished more easily by means of the exchange formulae given by Hirota (1974) (cf. Problem 4.8.7). Hence, we have the Bäcklund transformation

$$[D_t + 3\lambda D_x + D_x^3]f' \cdot f = 0, \tag{4.8.27a}$$

$$D_x^2 f' \cdot f = \lambda f' f. \tag{4.8.27b}$$

Transforming from f to ϕ by $\phi = -2(\log f)_{,x}$ then leads to the Bäcklund transformation for the KdV equation given first by Wahlquist and Estabrook (1973). On the other hand, putting $\psi = f'/f$, $u = -2(\log f)_{,xx}$ into Eqns (4.8.27) yields the set of equations in the inverse method.

Problem 4.8.1

Show that f given by Eq. (4.8.14) is equal to D of Eq. (4.1.23a).

Problem 4.8.2

Show that the modified KdV equation (4.8.1) is decoupled by Eq. (4.8.7) into

$$\mathcal{L}(D)G \cdot F = 0,$$
$$D_x^2 F \cdot F = \alpha G^2.$$

Obtain the two-soliton solution by means of expansions

$$F = 1 + \varepsilon^2 F_2 + \varepsilon^4 F_4,$$
$$G = \varepsilon G_1 + \varepsilon^3 G_3.$$

Show also that Eq. (4.8.1) is transformed by $u = (8/\alpha)^{1/2} \operatorname{Im} \partial/\partial x$

$\log (f + ig)$ into

$$(D_t + D_x^3) g \cdot f = 0,$$
$$D_x^2 (f \cdot f + g \cdot g) = 0,$$

and obtain the N-soliton solution (Hirota, 1976).

Problem 4.8.3

Decouple the nonlinear Schrödinger equation by setting $u = G/F$ where F is real, and obtain the N-soliton solution (Hirota, 1976).

Problem 4.8.4

Decouple the Sine–Gordon equation $\phi_{,xx} + \phi_{,yy} - \phi_{,tt} = \sin \phi$ by setting $u = 4 \tan^{-1} (g/f) = \text{Im } 4 \log (f + ig)$, and obtain the three-soliton solution (Hirota, 1976).

Problem 4.8.5

Solve the modified Volterra equation (Hirota (1976), Yoshikawa and Yamaguti (1974))

$$\left(\frac{\partial}{\partial t} + v_1 \frac{\partial}{\partial x} \right) N_1 = N_1 (a_1 - N_2),$$

$$\left(\frac{\partial}{\partial t} + v_2 \frac{\partial}{\partial x} \right) N_2 = -N_2 (a_2 - N_1),$$

(a_1, a_2, v_1, v_2 are constant). Show that if $a_1 = a_2 = 0$ the system becomes the two-wave interaction (Hashimoto (1974)).
Hint Put $N_1 = \phi_1 e^{a_1 t}$, $N_2 = \phi_2 e^{-a_2 t}$, $\phi_1 = G_1/F$ and $\phi_2 = G_2 F$.

Problem 4.8.6

Solve the three-wave interaction equation (3.8.54) (Hirota, 1976).

Problem 4.8.7

Prove the exchange formulae

$$e^{D_1}(e^{D_2} a \cdot b) \cdot (e^{D_3} c \cdot d) = e^{1/2(D_2 - D_3)} [e^{1/2(D_2 + D_3) + D_1} a \cdot d][e^{1/2(D_2 + D_3) - D_1} c \cdot b],$$

where $D_i = \varepsilon_i D_x + \delta_i D_t$, and then derive Eq. (4.8.26)'.
Hint $e^{\varepsilon_i D_x} a \cdot b = a(x + \varepsilon_i) b(x - \varepsilon_i)$; equate the same powers of $\varepsilon_i \delta_i$ to derive the various relations.

4.8 HIROTA'S METHOD

Problem 4.8.8

Derive the Bäcklund transformation of the KdV equation (4.8.8) from Eqns (4.8.27).

Hint Transform Eq. (4.8.27) into the equation for $\phi' - \phi = -2(\log f'/f)_{,x}$ and $\phi' + \phi = -2(\log f'f)_{,x}$.

Appendix Two-fluid model of a plasma (electron fluid and ion fluid)

The high-temperature plasma is a fully ionized gas, comprising electrons and ions, which are governed by the two hydrodynamic equations (cf. Section 2.2) for electrons and ions coupled with the Maxwell equations:
The continuity equations

$$\frac{\partial n_j}{\partial t} + \nabla \cdot (n_j \boldsymbol{v}_j) = 0 \qquad (j=\mathrm{i, e}); \tag{A.1a}$$

The equations of motion

$$m_j n_j \left[\frac{\partial \boldsymbol{v}_j}{\partial t} + (\boldsymbol{v}_j \cdot \nabla)\boldsymbol{v}_j\right] = q_j n_j \left[\boldsymbol{E} + \frac{\boldsymbol{v}_j}{c} \times \boldsymbol{B}\right] - \nabla p_j \qquad (j=\mathrm{i, e}); \tag{A.1b}$$

The Maxwell equations

$$\nabla \cdot \boldsymbol{E} = 4\pi(q_i n_i + q_e n_e), \tag{A.1c}$$

$$\nabla \cdot \boldsymbol{B} = 0, \tag{A.1d}$$

$$\frac{\partial \boldsymbol{B}}{\partial t} + c\nabla \times \boldsymbol{E} = 0, \tag{A.1e}$$

$$-\frac{\partial \boldsymbol{E}}{\partial t} + c\nabla \times \boldsymbol{B} = 4\pi(q_i n_i \boldsymbol{v}_i + q_e n_e \boldsymbol{v}_e); \tag{A.1f}$$

The equations of state

$$p_j = n_j T_j \qquad (j=\mathrm{i, e}). \tag{A.1g}$$

Here, $\boldsymbol{n}$, $\boldsymbol{v}$, p, $\boldsymbol{E}$, $\boldsymbol{B}$ are the density, the flow velocity, the pressure, the electric field vector, and the magnetic field vector, respectively. T is the product of the Boltzmann constant and the temperature; q, m are the charge and the mass, respectively; the subscripts i and e denote quantities related to the ions and the electrons, respectively. (The dissipation due to collisions is neglected.)

For an electrostatic wave, that is, a longitudinal wave ($\nabla \times \boldsymbol{E} = \boldsymbol{0}$, $\boldsymbol{B} = \boldsymbol{0}$), propagating in one dimension, setting $v_x \equiv u$, $q_i = -q_e \equiv e$, we have

$$\frac{\partial n_j}{\partial t} + \frac{\partial (n_j u_j)}{\partial x} = 0 \quad (j = i, e), \tag{A.2a}$$

$$\frac{\partial u_j}{\partial t} + u_j \frac{\partial u_j}{\partial x} = \pm \frac{e}{m_j} E - \frac{1}{m_j n_j} \frac{\partial (T_j n_j)}{\partial x} \quad (j = i, e), \tag{A.2b}$$

$$\frac{\partial E}{\partial x} = 4\pi e (n_i - n_e), \tag{A.2c}$$

$$\frac{\partial E}{\partial t} = 4\pi e (n_e u_e - n_i u_i). \tag{A.2d}$$

(Eqns (A.2a, d) and Eqns (A.2a, c) are equivalent.)

(1) *The electron plasma*

In this we let $m_i \to \infty$, $n_i = n_0$ (= constant) and $u_i = 0$, so that the ion motion is neglected. Hence, omitting the subscript e, we get

$$\frac{\partial n}{\partial t} + \frac{\partial (nu)}{\partial x} = 0, \tag{A.3a}$$

$$\frac{\partial u}{\partial t} + u \frac{\partial u}{\partial x} = -\frac{e}{m} E - \frac{1}{mn} \frac{\partial (Tn)}{\partial x}, \tag{A.3b}$$

$$\frac{\partial E}{\partial x} = 4\pi e (n_0 - n), \tag{A.3c}$$

$$\frac{\partial E}{\partial t} = 4\pi e n u, \tag{A.3d}$$

which is usually supplemented by the adiabatic law $p \propto n^\gamma$ ($\gamma = 2/m + 1$, where m is the degrees of freedom of the motion) or by the isothermal law $T = $ constant.

(2) *High-temperature electrons and low-temperature ions*

Here we let $m_e \to 0$ and $T_e = $ constant with $T_e \gg T_i$. Then we find that:

$$(A.2b) \to -eE - \frac{T_e}{n_e} \frac{\partial n_e}{\partial x} = 0 \tag{A.4}$$

$$(A.2a \sim c) \to \frac{\partial n_i}{\partial t} + \frac{\partial (n_i u_i)}{\partial x} = 0 \tag{A.5a}$$

$$\frac{\partial u_i}{\partial t} + u_i \frac{\partial u_i}{\partial x} = \frac{e}{m_i} E - \frac{1}{m_i n_i} \frac{\partial (n_i T_i)}{\partial x} \tag{A.5b}$$

$$\frac{\partial E}{\partial x} = 4\pi e (n_i - n_e). \tag{A.5c}$$

These are the equations for n_i, u_i, E, and n_e if T_i is a given constant, or by the adiabatic law so that $n_i T_i \propto n_i^\gamma$. Setting $E = -\partial\phi/\partial x$, with ϕ the electrostatic potential, Eq. (A.4) yields the Boltzmann distribution for electrons $n_e = n_0 \exp[e\phi/T_e]$.

Bibliography

Ablowitz, M. J., Kaup, D. J., Newell, A. C., and Segur, H., *Phys. Rev. Letters*, **30** (1973a), 1262; **31** (1973b), 125; *Studies Appl. Math.*, **53** (1974), 249.

Ablowitz, M. J., and Newell, A. C., *J. Math. Phys.*, **14** (1973), 1277.

Akhiezer, A. I., Lubarski, G. J., and Polovin, R. V., *Soviet Phys. JETP*, **8** (1959), 507.

Akhmanov, S. A., Sukhorukov, A. P., and Khokhlov, R. V., *Soviet Phys. JETP*, **23** (1966), 1025.

Asano, N., *J. Phys. Soc. Japan.*, **36** (1974a), 861.

Asano, N., *Prog. Theor. Phys. Suppl.*, No. 55 (1974b), 52.

Asano, N., Taniuti, T., and Yajima, N., *J. Math. Phys.*, **14** (1973), 1389.

Balanis, G. N., *J. Math. Phys.*, **13** (1972), 1001.

Bloch, B., *Phys. Rev.*, **70** (1946), 460.

Boillat, G., *La Propagation des Ondes*, Gauthier-Villars, Paris (1965).

Born, M., and Infeld, L., *Proc. Roy. Soc. (London)*, Ser. **A144** (1934), 425; **147** (1934), 522; **150** (1935), 141.

Boussinesq, J., *J. Math. Pures Appl.*, **17** (1872), 55.

Chen, H. H., *Phys. Rev. Letters*, **33** (1974), 925.

Coulson, C. A., and Jeffrey, A., *Waves* (2nd Edition), Longman, London & New York (1977).

Courant, R., and Friedrichs, K. O., *Supersonic Flow and Shock Waves*, Interscience, New York (1948).

Dafermos, C. M., *Nonlinear Waves*, Cornell Univ. Press, Ithaca (1974), p. 82.

Fermi, E., Pasta, J., and Ulam, S., *Los Alamos Report*, LA1940 (1955).

Flaschka, H., *Prog. Theor. Phys. (Kyoto)*, **51** (1974), 703.

Friedrichs, K. O., *Nichtlineare Differenzialgleichungen* (1955).

Friedrichs, K. O., and Kranzer, H., *New York University Rept.*, MH-8 (1958).

Gardner, C. S., *J. Math. Phys.*, **12** (1971), 1548.

Gardner, C. S., Greene, J. M., Kruskal, M. D., and Miura, R. M., *Phys. Rev. Letters*, **19** (1967), 1095.

Gardner, C. S., Greene, J. M., Kruskal, M. D., and Miura, R. M., *Commns. Pure and Appl. Math.*, **27** (1974), 97.

Gardner, C. S., and Morikawa, G. K., *Courant Inst. Math. Sci. Rept.*, NYO-9082 (1960), 1.

Gel'fand, I. M., and Levitan, B. M., *Amer. Math. Soc. Transl.*, **1** (1955), 253.

Grad, H., *Commns Pure and Appl. Math.*, **2** (1949), 331; **5** (1952), 257.

Hasegawa, A., *Phys. Rev.*, **A1** (1970), 1746.

Hasegawa, A., *Plasma Instabilities and Nonlinear Effects*, Springer-Verlag, Berlin, Heidelberg and New York (1975).

Hashimoto, H., *Proc. Japan Acad.*, **50** (1974), 623.

Heisenberg, W., and Euler, H., *Z. Physik*, **98** (1936), 714.

Hirota, R., *Phys. Rev. Letters*, **27** (1971), 1192.

Hirota, R., *Prog. Theor. Phys.*, **52** (1974), 1498.

Hirota, R., in *Bäcklund Transformations, The Inverse Scattering Method, Solitons, and Their Applications* (R. M. Miura (ed.), *Lecture Notes in Mathematics*, No. 515), Springer-Verlag, Berlin, Heidelberg and New York (1976).

Hirota, R., and Satsuma, J., *Prog. Theor. Phys. Suppl.*, No. 59 (1976), 64.

Hirota, R., and Suzuki, K., *J. Phys. Soc. Japan*, **28** (1970), 1366.

Ikezi, H., *Phys. Fluids*, **16** (1973), 1668.

Ikezi, H., Nishikawa, K., and Mima, K., *J. Phys. Soc. Japan*, **37** (1974), 766.

Inoue, Y., and Matsumoto, Y., *J. Phys. Soc. Japan*, **36** (1974), 1446.

Jeffrey, A., *Quasilinear Hyperbolic Systems and Waves, Research Note in Mathematics 5*, Pitman Publishing, London (1976).

Jeffrey, A., and Taniuti, T., *Nonlinear Wave Propagation*, Academic Press, New York and London (1964).

Jeffrey, A., and Kakutani, T., *SIAM Rev.*, **14** (1972), 582.

Jeffrey, A., and Kawahara, T., *Asymptotic Methods in Nonlinear Wave Theory*, Pitman Publishing, London, (1982).

Kadomtsev, B. B., and Petviashvili, V. I., *Sov. Phys. Dokl.*, **15** (1970), 539.

Kakutani, T., Ono, H., Taniuti, T., and Wei, C. C., *J. Phys. Soc. Japan*, **24** (1968), 1159.

Kakutani, T., and Sugimoto, N., *Phys. Fluids*, **17** (1974), 1617.

Karpman, V. I., *Phys. Letters*, **25A** (1967), 708.

Karpman, V. I., *Nonlinear Waves in Dispersive Media* (transl. by F. F. Cap), Pergamon Press, Oxford and New York (1975).

Karpman, V. I., and Kruskal, E. M., *Soviet Phys. JETP*, **28** (1969), 277.

Kato, Y., *Prog. Theor. Phys. Suppl.*, No. 55 (1974), 247.

Kaup, D. J., *Stud. Appl. Math.*, **54** (1975), 165.

Kawahara, T., *J. Phys. Soc. Japan*, **35** (1973), 805.

Kelley, P. L., *Phys. Rev. Letters*, **15** (1965), 1005.

Kodama, Y., *Prog. Theor. Phys. (Kyoto)*, **54** (1975), 669.

Kodama, Y., *J. Phys. Soc. Japan*, **45** (1978), 311.

Kodama, Y., and Taniuti, T., *J. Phys. Soc. Japan*, **47** (1979a), 1706; *J. Phys. Soc. Japan*, **45** (1978a), 298; *J. Phys. Soc. Japan*, **45** (1978b), 1465; *Physica Scripta*, **19** (1979b), 486.

Kruskal, M. D., Miura, R. M., and Gardner, C. S., *J. Math. Phys.*, **11** (1970), 952.

Ladyzhenskaya, A., *Doklady Akad. Nauk S.S.S.R.*, **111** (1956), 291.

Lamb, G. L., *Revs. Modern Phys.*, **43** (1971), 99.

Lamb, G. L., *Physica*, **66** (1973), 298.

Landau, L. D., and Lifshitz, E. M., *Quantum Mechanics, Nonrelativistic Theory*, Addison-Wesley, Reading, Mass. (1958).

Landau, L. D., and Lifshitz, E. M., *Electrodynamics of Continuous Media*, Pergamon Press, Oxford and New York (1960).

Lax, P. D., *Annales Math. Studies (Princeton)*, **33** (1954), 211.

Lax, P. D., *Commns. Pure and Appl. Math.*, **10** (1957), 537; **21** (1968), 467.

Lax, P. D., *Contributions to Nonlinear Functional Analysis*, Academic Press, New York (1971), p. 603.

Lax, P. D., *Commns. Pure and Appl. Math.*, **5** (1971), 280.

Lighthill, M. J., *Surveys in Mechanics*, Cambridge Univ. Press, London (1956).

Lutzky, M., and Toll, J. S., *Phys. Rev.*, **113** (1959), 1649.

Marchenko, V. A., *Doklady Akad. Nauk S.S.S.R.*, **104** (1955), 695.

McCall, S. L., and Hahn, E. L., *Phys. Rev.*, **183** (1969), 457.

McLaughlin, D. W., *J. Math. Phys.*, **16** (1975a), 96; **16** (1975b), 1704.

McLaughlin, D. W., and Scott, A. C., *J. Math. Phys.*, **14** (1973), 1817.

Miura, R. M., *J. Math. Phys.*, **9** (1968), 1202.

Miura, R. M., Gardner, C. S., and Kruskal, M. D., *J. Math. Phys.*, **9** (1968), 1204.

Mizohata, S., *The Theory of Differential Equations*, Cambridge Univ. Press (1973).

Møller, C., *The Theory of Relativity*, Oxford Univ. Press, Clarendon (1972).

Mott-Smith, H. M., *Phys. Rev.*, **82** (1951), 885.

Newton, R. G., *Scattering Theory of Waves and Particles*, McGraw-Hill, New York (1966).

Nishihara, K., *J. Phys. Soc. Japan*, **39** (1975), 803.

Nishikawa, K., Hojo, H., Mima, K., and Ikezi, H., *Phys. Rev. Letters*, **33** (1974), 148.

Novikov, S. P., *Funct. Annal. Appls.*, **8** (1974), 54.

Nozaki, K., *J. Phys. Soc. Japan*, **37** (1974a), 206; **37** (1974b), 1124.

Nozaki, K., and Taniuti, T., *J. Phys. Soc. Japan*, **34** (1973), 796.

Ohsawa, Y., and Nozaki, K., *J. Phys. Soc. Japan*, **36** (1974), 591.

Oikawa, M., and Yajima, N., *J. Phys. Soc. Japan*, **34** (1973), 1093.

Oikawa, M., and Yajima, N., *Prog. Theor. Phys. Suppl.*, No. 55 (1974), 36.

Oleinik, O. A., *Uspekhi Matem. Nauk*, **9** (1954), 231; **10** (1955), 229.

Oleinik, O. A., *Doklady Akad. Nauk S.S.S.R.*, **109** (1956), 1098.

Ostrovskii, L. A., *Soviet Phys. Tech. Phys.*, **8** (1964), 679.

Ostrovskii, L. A., *Soviet Phys. JETP*, **24** (1967), 797.

Polovin, R. V., *Soviet Phys. Uspekhi*, **3** (1961), 677.

Rudenko, O. V., and Soluyan, G. I., *Theoretical Foundations of Nonlinear Acoustics* (translated by R. T. Berger), Consultants Bureau, New York and London (1977).

Sagdeev, R. Z., *Review of Plasma Physics* (ed. by M. A. Leontovich), Vol. IV, Consultants Bureau, New York (1966).

Sagdeev, R. A., and Galeev, A. A., *Nonlinear Plasma Theory* (revised and ed. by T. M. O'Neil and D. L. Book), Benjamin, New York (1969).

Scott, A. C., Chu, F. Y. F., and McLaughlin, D. W., *Proc. IEEE*, **61** (1973), 1443.

Seeger, A., Donth, H., and Kochendorfer, A. Z., *Physik*, **134** (1953), 173.

Shvartsburg, A. B., *Nonlinear Electromagnetics*, Academic Press, New York and London (1980), 133–187.

Shvartsburg, A. B., *Fortschritte der Physik*, **28** (1980), 1–33.

Stewartson, K., and Stuart, J. I., *J. Fluid Mech.*, **48** (1971), 529.

Taniuti, T., *Prog. Theor. Phys.* (*Kyoto*), **10** (1953), 525; **17** (1957), 461; **20** (1958), 529; **28** (1962), 756.

Taniuti, T., *Prog. Theor. Phys. Suppl.*, No. 9 (1959), 69; No. 55 (1974), 1.

Taniuti, T., *J. Phys. Soc. Japan*, **18** (1963), 408.

Taniuti, T., and Wei, C. C., *J. Phys. Soc. Japan*, **24** (1968), 941.

Taniuti, T., and Yajima, N., *J. Math. Phys.*, **10** (1969), 1369; **14** (1973), 1389.

Tatsumi, T., and Tokunaga, H., *J. Fluid Mech.*, **65** (1974), 581.

Toda, M., *J. Phys. Soc. Japan*, **22** (1967), 431; **23** (1967), 501.

Toda, M., *Prog. Theor. Phys. Suppl.*, No. 45 (1970), 174.

Toda, M., and Wadati, M., *J. Phys. Soc. Japan*, **34** (1973), 18.

Toda, M., *Nonlinear Lattice Dynamics*, Springer, Berlin and New York (1980).

VanDam, J. W., *J. Phys. Soc. Japan*, **34** (1973), 1633.

Wadati, M., *J. Phys. Soc. Japan*, **32** (1972), 1681.

Wadati, M., and Toda, M., *J. Phys. Soc. Japan*, **32** (1972), 1403.

Wadati, M., Sanuki, H., and Konno, K., *Prog. Theor. Phys. (Kyoto)*, **53** (1975), 419.

Wahlquist, H. D., and Estabrook, F. B., *Phys. Rev. Letters*, **31** (1973), 1386.

Washimi, H., *J. Phys. Soc. Japan*, **34** (1973), 1373.

Washimi, H., and Taniuti, T., *Phys. Rev. Letters*, **17** (1966), 996.

Watanabe, Y., *J. Math. Phys.*, **15** (1974), 453.

Whitham, G. B., *Proc. Roy. Soc. (London)*, **A283** (1965a), 238.

Whitham, G. B., *J. Fluid Mech.*, **22** (1965b), 273.

Whitham, G. B., *Linear and Nonlinear Waves*, John Wiley & Sons, New York (1974).

Yajima, N., *Prog. Theor. Phys. (Kyoto)*, **52** (1974), 1066.

Yajima, N., Outi, A., and Taniuti, T., *Prog. Theor. Phys. (Kyoto)*, **40** (1968), 243.

Yajima, N., and Outi, A., *Prog. Theor. Phys. (Kyoto)*, **45** (1971), 1997.

Yoshikawa, M., and Yamaguti, M., *Publ. Res. Inst. Math. Sci.*, Kyoto Univ., **9** (1974), 577.

Zabusky, N. J., and Kruskal, M. D., *Phys. Rev. Letters*, **15** (1965), 240.

Zakharov, V. E., *Soviet Phys. JETP*, **35** (1972), 908.

Zakharov, V. E., and Faddeev, L. D., *Funct. Anal. Appls.*, **5** (1971), 280.

Zakharov, V. E., and Manakov, S. V., *Theor. Math. Phys.*, **19** (1974), 551.

Zakharov, V. E., and Manakov, S. V., *Soviet Phys. JETP*, **42** (1976), 842.

Zakharov, V. E., and Shabat, A. B., *Soviet Phys. JETP*, **34** (1972), 62.

Zakharov, V. E., and Shabat, A. B., *Func. Anal. Appl.*, **8** (1974), 43.

Zel'dovich, Ya. B., and Raizer, Yu. P., *Physics of Shock Waves and High-Temperature Hydrodynamic Phenomena* (ed. by W. D. Hayes and R. F. Probstein), Academic Press, New York (1966).

Index

Ablowitz, Kaup, Newell and Segur formalism, 225–230
   AKNS equation, 226, 228–230
   and inverse scattering method, 225–230
   relation with Riccati equation and Bäcklund transformation, 224–230
addition theorem for Bäcklund transformation, 220
adiabatic index, 33
adiabatic law, 56
Alfvén wave, 53, 55
anti-kink solution of Sine–Gordon equation, 157, 220–222

Bäcklund transformation, 211–231
   addition theorem for, 220
   Cole–Hopf transformation as, 223
   general, 215
   and Hirota's method, 239
   increment operator for, 219
   and inverse scattering method, 223–231
   Miura transformation as, 223
   N-soliton solution by, 220
   restricted, 215
   Ricatti equation and, 224–230
   Sine–Gordon equation and, 211–223
Bloch equation, 154, 230

blow-up type of instability, 52
Born–Infeld field, 40, 85
Boussinesq equation, 196, 197, 238
   Hirota's method for, 238
   soliton solution of, 197
bright pulse (compressive soliton), 124, 222
Burgers equation, 12, 14–20, 51, 95–99
   Cole–Hopf transformation of, 12
   shock solution of, 14, 68, 96
   similarity law of, 96

canonical equation, 186, 188
canonical form
   conservation law and, 180–188
   of KdV equation, 180–188
   of nonlinear Schrödinger equation, 211
canonical variable for KdV equation, 186–188
carrier wave, 119, 152
   see also plane wave
Cauchy problem, 74
centred expansion wave, 6, 20
Čerenkov radiation, 83
characteristic (curve), 2, 7*ff*, 23–32, 41–55, 75
   method of, 23–32
characteristic base curve, 8
characteristic equation, 7*ff*, 74

characteristic initial value problem, 51
characteristic ray, 75
characteristic speed, 16, 26, 47
characteristic surface, 74, 75
Cole–Hopf transformation, 12, 99, 160, 223
   as Bäcklund transformation, 223
compatibility condition, 120
completely exceptional system, 53
completely integrable Hamiltonian, 188, 211, 230
completely integrable system, 188, 230
   and infinite number of conservation laws, 180–188
compression (converging) wave, 4
compressive soliton (bright pulse), 102, 222
conservation of energy, 32, 109, 142
conservation of entropy, 33
conservation form, 19, 56, 72
conservation law, 19, 180
   and canonical form, 180–188
   infinite number of, 180, 229–230
   and inverse scattering method, 180–188
   inverse scattering method, Bäcklund transformation and, 230
   of KdV equation, 180–188
   of mass, 32
   of momentum, 32, 142
   of nonlinear Schrödinger equation, 210–211
conservation of mass, 32, 109
conservation of momentum, 9, 32, 109, 142
conserved density, 180, 230
conserved quantity, 180
   infinite number of, 180–184, 229–230
   inverse scattering method and infinite number of, 180–188
   see also conservation law
contact surface, 58, 86
contact transformation, 215

continuous spectrum as scattering data in inverse scattering method, 162*ff*, 204
coupling equation, 141
   mode, 141*ff*

dark pulse, 124
dark soliton, 222
decay-type instability, 144
   see also parametric excitation
diffusion equation, 12, 13, 51
Dirac's equation, 86
Dirac-type eigenvalue equation, 201
direct problem in inverse scattering method, 162*ff*
   see also inverse scattering method
dispersion, 13, 100
   reductive perturbation method for weakly dispersive system, 99–116
   reductive perturbation method for strongly dispersive system, 116–130
   strong, 100, 117
   weak, 100
dispersion relation, 13, 100, 117, 125, 139
   for electron plasma wave, 127
   for ion-sound wave, 111
   for KdV equation, 100
   for Klein–Gordon equation, 128
   linear, 117, 125, 139
   nonlinear, 125
dispersive medium, 137
dispersive shallow water wave, 113
dispersive string, 110
disruption of wave, 132
dissipation, 13, 95*ff*
   weak dissipation and shock wave, 14–22
dissipative system, 95–99, 107–114
   reductive perturbation method for, 107–114

divergent (expansion) wave, 4
domain of dependence, 49

eigenstate, as scattering data in inverse scattering method, 162*ff*
eigenvalue equation
  for inverse scattering method, 160–164, 201, 225–230
  for KdV equation, 160*ff*
  and Lax pair, 197–200
  for nonlinear Schrödinger equation, 201*ff*
  for Sine–Gordon equation, 225
electron plasma wave, 127–128
electrostatic wave, 243
  see also ion-sound wave; electron plasma wave
elliptic equation, 39, 51
energy density of wave, 142
enthalpy, 67
entropy, 16, 33
  law of increase of, 18, 60, 66, 69
entropy wave, 44, 53, 55
envelope cavity, 124
envelope soliton, 122, 123, 201
  of nonlinear Schrödinger equation, 122, 201
equation of hydrodynamics, 32
equation of magnetohydrodynamics, 54
equations for mode coupling, 141*ff*
evolutionary condition, 19, 63, 65
exceptional case
  linear, 98
  nonlinear, 53, 98, 110
expansion shock wave, 18
expansion soliton, 102, 123, 222
expansion wave (divergent wave), 4
  centred, 6, 20
exterior derivative, 45

far field, 89*ff*, 103, 116
  for hyperbolic equation, 89–95
  for linear wave modulation, 114–116

shock as invariant far field, 96
soliton as invariant far field, 103
fast magnetoacoustic wave, 55
Fermi–Pasta–Ulam paradox, 191, 192, 197
functional, 180, 186
functional derivative, 184, 185

Galilean invariance of KdV equation, 160
Gardner–Morikawa transformation, 94*ff*, 105, 113
gas dynamics, shock wave in, 55–72
Gauss' theorem, 19
Gelfand–Levitan–Marchenko (G–L–M) equation, 164–177, 207
general Bäcklund transformation, 215
generalized nonlinear Schrödinger equation, 121
generalized Rankine–Hugoniot relation, 63
genuinely nonlinear system, 53
group velocity, 52
  see also modulation of plane wave

Hamilton–Jacobi equation, 10, 76
Hamiltonian, 10, 76, 133, 184–188
  completely integrable, 188, 211, 230
  for KdV equation, 184–188
  for nonlinear Schrödinger equation, 211
Hamiltonian density, 184
Heisenberg–Euler equation, 85
Hirota's method, 230–241
  and Bäcklund transformation, 230
  for Boussinesq equation, 238
  for Kadomtsev–Petviashvili equation, 238
  for KdV equation, 232–237
  for modified KdV equation, 239–240
  for modified Volterra equation, 240
  for nonlinear Schrödinger equation, 240

Hirota's method (*contd.*)
  for Sine–Gordon equation, 240
  for three-wave interaction equation, 240
hydrodynamics, 32–41
  equation of, 32
hyperbolic equation, 20, 39, 41–55, 76
  nonlinear, and characteristic curve, 41–55
hyperbolic system of conservation laws, 62

incident wave for direct problem, 167
incoming wave, 63
increase of entropy, 18, 56, 61
increment operator for Bäcklund transformation, 219
interior derivative, 45
invariant far field, 96, 103
  shock as, 96
  soliton as, 103
inverse problem in inverse scattering method, 162*ff*
inverse scattering method, 159–180, 200–211, 223–230
  for AKNS equation, 225–230
  Bäcklund transformation and, 223–231
  conservation laws and, 180–188
  direct problem in, 162*ff*
  eigenvalue equation for, 160–164, 201, 225–230
  inverse problem in, 162*ff*
  for KdV equation, 159–180
  for nonlinear Schrödinger equation, 200–211
  scattering data for, 162*ff*, 202*ff*
  for Sine–Gordon equation, 224–229
  time evolution of scattering data in, 162–164, 203, 225–228
ion-sound wave, 110

jump condition, 62
  see also shock condition; Rankine–Hugoniot relation

Kadomtsev–Petviashvili equation (weakly two-dimensional KdV equation), 238
  Hirota's method for, 238
kink soliton, 157
kink solution, 124
  of Sine–Gordon equation, 157, 220–222
Klein–Gordon equation, 51, 211*ff*, 214*ff*
  nonlinear Klein–Gordon equation, 128
Korteweg–de Vries (KdV) equation, 13, 100*ff*, 159*ff*
  AKNS formalism for, 226
  canonical form of, 180–188
  conservation laws of, 180–188
  dispersion relation of, 100
  eigenvalue equation for, 160*ff*
  Hirota's method for, 232–237
  inverse scattering method for, 159–180
  Miura transformation for, 159
  N-soliton solution of, 174–177
  periodic solution of, 101, 188–197
  soliton solution of, 101, 165, 167, 174
  two-soliton solution of, 165–167

Lagrange density, 184
Lagrange equation, 184
Lamb diagram, 216
lattice dynamics, 188–197
Lax pair, 199, 201
  and eigenvalue equation for inverse scattering method, 197*ff*
  theory of, 197–200
light cone, 80
linear dispersion relation, 117, 125, 139
linear wave modulation, 114–116
Lipschitz continuous, 46

Mach wave, 87
magnetoacoustic (magnetohydrodynamic) shock, 71
   fast, 55
magnetoacoustic wave, 55, 56
magnetohydrodynamic (magnetoacoustic) shock, 71
magnetohydrodynamics, 54–55
Manley–Rowe relation, 144
matching (resonance) condition for three-wave interaction, 133, 137
Maxwell's equation, 80, 137, 153
method of characteristics, 23–32
microscopic causality, 49
Minkowski light cone, 83
Miura transformation, 159, 224
   as Bäcklund transformation, 224
mode coupling, equations for, 141ff
modified KdV equation, 106, 159, 224, 226, 230, 231, 239
   AKNS equation for, 226
   Hirota's method for, 231, 239
   Miura transormation for, 159, 224
modified Volterra equation, 240
   Hirota's method for, 240
modulation of plane wave
   linear, 114–116
   nonlinear, 116–132
modulational instability, 122–124
momentum density of wave, 142
multi-soliton, *see* N-soliton

N-kink solution of Sine–Gordon equation, 222
N-soliton solution
   by Bäcklund transformation, 220
   of Boussinesq equation, 238
   by Hirota's method, 231–241
   of Kadomtsev–Petviashvili equation, 238
   of KdV equation, 174–177, 233–237
   of model equation for shallow water, 238

   of nonlinear Schrödinger equation, 208–210, 248
nonlinear circuit, 195
nonlinear dispersion relation, 125
nonlinear exceptional case, 53, 98, 110
nonlinear hyperbolic equations and characteristic curves, 41–55
nonlinear Klein–Gordon equation, 128
nonlinear modulation of wave, 116–132
nonlinear optics, 130, 133, 151
   see also modulation of plane wave; self-focusing of wave; self-induced transparency; three-wave interaction
nonlinear polarization, 137–140
nonlinear refractive index, 130, 133
nonlinear Schrödinger equation, 116–127, 130–132, 200–211, 225, 226, 228, 240
   AKNS formalism for, 226–227
   canonical form for, 211
   conservation laws of, 210–211
   eigenvalue equation for, 201ff
   generalized, 121
   Hirota's method for, 240
   inverse scattering method for, 200–211
   N-soliton solution of, 208–210
   soliton solution of, 122, 207–208
nonlinear string, vibration of, 39, 110
normal velocity, 76, 79
   surface of, 79

one-soliton, *see* soliton
operator equation, 199
outgoing wave, 63
overtaking of wave, 1–8, 12, 52

parabolic equation, 39, 51
parametric excitation (amplification), 133–151
   threshold value for, 137
   see also three-wave interaction

pendulums connected by torsional spring, 155
periodic boundary condition, 188
periodic solution, 188–197, 231
 of KdV equation, 101*ff*, 188–197
 and recurrence, 188–197
permanent (progressive) wave, 2, 101
 persistent, 101
phase-difference of soliton by collision, 176, 210, 222
phase-mixing, 9, 22
plane wave, 79, 114–132, 138
 linear modulation of, 114–116
 nonlinear modulation of, 116–132
 quasi- (monochromatic), 115, 138
plasma, 23
 two-fluid model of, 242–244
 see also electron plasma wave; ion-sound wave; magnetohydrodynamics; modulation of plane wave; self-focusing of wave; three-wave interaction
polarization vector, 137–140
 of medium, 137
 nonlinear, 137–140
Prandtl number, 68
principal part of partial differential equation, 52
progress wave, 2, 101
progressive wave solution of KdV equation, 104

quasilinear equation, 52
quasi- (monochromatic) wave, 115, 138

range of influence, 49
Rankine–Hugoniot relation, 59*ff*
 generalized, 63*ff*
ray velocity, 76
recurrence (of solution), 188, 191, 195
 and periodic boundary, 188–197

recurrence distance, 191
recurrence time, 189
reductive perturbation method (RPM), 93, 127, 149
 for dissipative system, 107
 for long waves, 107–114
 for strongly dispersive system, 116–130
 for three-wave interaction, 149–151
 for weakly dispersive system, 107
reflected wave for direct problem, 167, 168
reflection coefficient for direct problem, 162, 163
reflectionless potential, 165
refractive index, nonlinear, 130, 133
resonance condition for three-wave interaction, 133, 137
restricted Bäcklund transformation, 215
Reynolds number, 55, 68
Ricatti equation, 160, 181, 210
 relation with Bäcklund transformation and AKNS equation, 224–230
Riemann invariant, 24, 27, 28, 30, 31, 35, 38, 43
 generalized, 44
Riemann problem, 71

scattering data for inverse scattering method, 162
 continuous spectrum as, 162*ff*, 204
 eigenstate as, 162*ff*
 of KdV equation, 162–178
 of nonlinear Schrödinger equation, 202–210
 time evolution of, 162–164, 203, 225–228
Schrödinger equation, 11
 for inverse scattering method of KdV equation, 160, 197, 226

INDEX

Schrödinger field for linear wave modulation, 114–116
Schrödinger operator, 198
  for inverse scattering method of KdV equation, 160, 198–200
Schrödinger-type eigenvalue equation, 160
secular term, 91
self-adjointness of Schrödinger operator, 160, 198
self-focusing of wave, 130–132
  condition for, 131
self-induced transparency, 151
  and Sine–Gordon equation, 151–158
self-modulation, *see* modulation of plane wave
semilinear equation, 52
shallow water wave, 40, 238
  dispersive, 113
shock, *see* shock wave
shock condition, 19, 57*ff*, 62
  see also Rankine–Hugoniot relation
shock solution of Burgers equation, 14, 68, 96
shock tube, 71
shock wave, 13, 14–22, 55–72
  in gas dynamics, 55–72
  generalized shock, 63
  as invariant far field, 96
  magnetohydrodynamic, 71, 72
  Rankine–Hugoniot relation of, 59*ff*
  similarity law of, 95, 96
  strong shock, 68
  structure of, 67, 68
  weak dissipation and, 14–22
  weak shock, 68, 69
similarity law, 94
  of KdV equation, 103, 178
  of shock, 95–96
simple wave, 29*ff*, 35, 39, 44
Sine–Gordon equation, 151–158, 211–229, 240
  AKNS equation for, 225, 229

anti-kink solution of, 157, 220–222
and Bäcklund transformation, 211–229
eigenvalue equation for, 225
Hirota's method for, 240
inverse scattering method for, 225
kink solution (soliton) of, 157, 220–222
N-kink solution of, 222
Ricatti equation for, 225, 229
self-induced transparency and, 151–158
three-soliton solution of, 240
two-kink solution of, 220–222
slow magnetoacoustic wave, 55, 66
smooth solution, 49
solitary wave, 14, 101, 149
soliton, 101–104
  bright soliton, 222
  compressive soliton, 102, 222
  dark soliton, 222
  envelope soliton, 122, 123, 201
  expansion soliton, 102, 123, 222
  Hirota's method for multi-soliton, 231–241
  as invariant far field, 103
  of KdV equation, 101–104, 162–178
  of nonlinear Schrödinger equation, 122, 201–211
  phase difference of soliton by collision, 176, 210, 222
  see also N-soliton solution: two-soliton solution
  recurrence of, 188–191
  of three-wave interaction, 149, 230, 240
similarity, law of, 103, 178
sound velocity, 32
speed of wave front, 47
string
  nonlinear elastic, 39
  dispersive, 110
strong dispersion, 100, 117

strong shock, 68
strong solution, 15, 17
strongly dispersive system, 116–130
  reductive perturbation method for, 116–130
supersonic/subsonic flow, 61
supersonic soliton, 102
surface of normal velocity, 79

Taylor solution, 68
thermal conductivity, 32
three-wave interaction, 133–151, 157, 230, 240
  reductive perturbation method for, 149–151
  resonance condition for, 133, 137
  soliton of, 149, 230, 240
  time-averaging method for, 133–143
threshold value
  for parametric excitation, 137
  for self-focusing, 131
time-averaging method for three-wave interaction, 133–143
time evolution of scattering data in inverse scattering method, 162–164, 203–204, 225
Toda lattice, 192, 194, 195, 197
  soliton solution of, 195
torsional motion of connected pendulums, 155
totally hyperbolic system, 73
transmission coefficient in direct problem, 162, 163
transmitted wave in direct problem, 165, 167
two-fluid model of plasma, 242–244
two-kink solution of Sine–Gordon equation, 220–222
two-level atom, system of, 152

two-soliton solution, 165–167, 210, 235, 239,
  by Hirota's method, 231–241
  of KdV equation, 165–167
  of nonlinear Schrödinger equation, 210
  see also N-soliton solution

unitarily equivalent systems, 198, 200
unitary transformation, 198

vibration of nonlinear string, 39, 110
viscosity, 32, 56
  coefficient of, 32
Volterra equation, modified, 240

wave
  energy density of, 142
  modulation of, 114–130, 132
  momentum density of, 142
  number density of wave quanta, 143
  self-focusing of, 130–132
wave equation, 23, 53
wave front, 46*ff*, 77
  speed of, 47
  surface of, 77
wave modulation, *see* modulation of plane wave
wave packet, 52, 116*ff*
wave quanta, 143
wave train, 104, 162, 184
wave train solution, 104
weak dispersion, 100
weak dissipation and shock wave, 14–22
weak shock, 68, 69
weak solution, 17, 63
weakly dispersive system, 13, 99–107
  reductive perturbation method for, 99–107